RELIABILITY AND RISK ANALYSIS

RELIABILITY AND RISK ANALYSIS

TERJE AVEN

Rogaland University Centre, Stavanger, Norway

ELSEVIER APPLIED SCIENCE
LONDON and NEW YORK

ELSEVIER SCIENCE PUBLISHERS LTD
Crown House, Linton Road, Barking, Essex IG11 8JU, England

WITH 17 TABLES AND 65 ILLUSTRATIONS

British Library Cataloguing in Publication Data

Aven, T.
Reliability and Risk Analysis
I. Title
620

ISBN 1-85166-896-9

Library of Congress CIP data applied for

Printed and bound in Great Britain at The Alden Press, Oxford

Preface

Analysis of reliability and risk is an important and integral part of planning, construction and operation of all technical systems. To be able to perform such analyses systematically and scientifically, there is usually a need for special methods and models. This book presents the most important of these. Particular emphasis has been placed on the ideas and the motivation for the use of the various methods and models. It has been an objective to compile a book which provides practising engineers and engineering graduates with the concepts and basic techniques for evaluating reliability and risk. It is hoped that the material presented will make them so familiar with the subject that they can carry out various types of analyses themselves and understand and make use of the more detailed applications and additional material which is available in the journals and publications associated with their own discipline. It has also been an objective to put reliability and risk analyses in context - how such analyses should be used in design and operation of components and systems.

The material presented is modern and a large part of the book is at research level. The book focuses on analysis of repairable systems, not only non-repairable systems which have traditionally been given most attention in textbooks on reliability theory. Since most real-life systems are repairable, methods for analysing repairable systems are an important area of research.

The book presents general methods, with most applications taken from offshore petroleum activities.

Reliability and risk analysis comprises a large number of fields and application areas. It has not been the purpose of this book to give a total overview of all these fields and application areas. This book does not, for example, cover structural reliability and software reliability.

What topics are covered in this book? Chapter 1 introduces some important concepts and gives a general description of a reliability analysis and a risk analysis. In Chapter 2 some basic reliability and risk analysis methods are presented. These include Failure mode and effect analysis, Fault tree analysis, Event tree analysis and Cause consequence analysis. Chapter 3 presents the most common techniques for calculating system reliability. Models for analysing periodically tested systems, and systems where the components are repaired or replaced at failure are studied. Methods for identification of critical components and events are also discussed. Chapter 4 studies multistate models where the components and the system are allowed to have an arbitrary number of states, not only two as in the traditional models described in Chapter 3. Analysis of flow network systems is emphasized. Such networks can be used to model e.g. gas and oil production and transportation systems. Chapter 5 reviews some common models for optimizing the limiting availability, or more generally the long run expected cost per time unit, when the system is undergoing preventive maintenance. The problem is to find the optimal replacement time of the system. Chapter 6 presents an overview of the most important elements of safety management. By safety management is understood all actions taken to obtain and maintain a safety level in accordance with goals and criteria. Emphasis is put on describing the role of reliability and risk analysis as a part of safety management. Chapter 7 presents some applications of reliability and risk analysis. The examples are taken from the petroleum activity in the North Sea.

Chapters 1, 2, 6 and part of Chapter 7 can be read without background in probability theory and statistics. Chapters 3, 4, 5 and part of Chapter 7 assume, however, a knowledge of elementary probability theory and some basic concepts such as failure rate, lifetime distribution, etc. Appendices A and B present a short overview of this material.

A reliability analysis and a risk analysis usually requires a statistical analysis of reliability data to estimate lifetime distributions, repair time distributions, mean lifetimes, etc. Appendix C describes how such analyses can be carried out. In Appendix D Markov models are studied. Markov models are used to establish some of the results obtained in Chapter 3. Appendix E presents some more advanced parts of the theory of stochastic processes, including renewal theory, renewal reward theory and counting processes. This theory is needed to prove some results in Chapters 3 and 5.

A number of examples and exercises have been included to illustrate, clarify, and supplement the theory and methods.

This book is based on a Norwegian version with the same title, issued in 1991 [8]. The present version is, however, more advanced and includes many parts that are not covered in the original [8].

The Norwegian version of the book originated from a course in Reliability and Risk Analysis at Rogaland University Centre, Stavanger. The Norwegian version is also used in safety education at the University of Oslo and at the University of Trondheim, Norwegian Institute of Technology.

Parts of the Norwegian book were written when I was working in Statoil, the Norwegian State Oil Company. I would like to thank Hans J. Grundt and Jo Hulbækdal in Statoil for many valuable suggestions and comments.

Stavanger

Terje Aven

Contents

Chapter 1

INTRODUCTION TO RELIABILITY AND RISK ANALYSIS

1.1 Historical perspective

After World War I the first reliability evaluations were performed in a technological context. Comparisons of single and multi-engined aircraft were then made, based on accident rates per hours of flying time.

But it was during World War II that the first formal reliability analysis was reported. This happened in Germany in connection with the development of the V-1 missile, work led by Werner von Braun. The first series of ten missiles were totally unreliable; they all blew up on the launching pads or fell into the English Channel. The reliability of the missiles was originally considered in the light of the old saw: "A chain is no stronger than its weakest link". After some time one realized that a large number of relatively strong "links" can be more unreliable than one weak "link". A mathematician, Robert Lusser, was called in as a consultant to analyse the system, and he produced the product law of reliability. This law is relevant to systems which function only if their components function, and it is valid under certain conditions. The law states that the reliability of such a system equals the product of the reliabilities of the components of the system. Thus, if the system comprises a large number of components, the system reliability can be quite low, even with high component reliabilities. Work based on this idea resulted in great improvement of the reliability of the V-1 missiles. Finally, a "success" rate of 60% was achieved.

In the United States, one tried to improve low system reliability by intensifying the demands on the individual components. Better design, stronger materials, harder and smoother wearing surfaces, more advanced inspection instruments etc. were required. In this way a higher system reliability was achieved, but a systematic analysis of the problem was not carried out at that time.

During the Korean War, the American Department of Defence found that unreliable equipment required a tremendous amount of maintenance. It found that the cost was 2 dollars per year to maintain every dollar's worth of electronic equipment: for an equipment life of ten years, it cost 20 million dollars to maintain every million dollars of purchase value of the equipment. It was thereby demonstrated that it is normally wiser to design for reliability than it is to wait and repair equipment after failure.

An enterprise that started in the 1950s and accelerated in the 1960s was the collection and systematization of component, system and human failure data.

In that period there was in the United States a growing interest in risk and safety, particularly as related to airspace and nuclear power. In the race against the Russians to put the first human being on the moon, it was very important that the launching of the manned spacecraft was successful. A considerable amount of money was therefore invested in educating personnel to analyse reliability and risk. A special committee was organized for engineers working with reliability. The first journal in the field, *IEEE Transactions on Reliability*, appeared in 1963, and several books were published during the 1960s.

In 1974 an extensive risk analysis of nuclear power plants was completed, "WASH-1400, The Reactor Safety Study" [135]. The project, which was led by Professor N. Rasmussen and cost millions of dollars, represents, despite its weak points, an epoch-making study. In the study a large number of undesirable events were analysed; these events were numerically ranked according to their probability of occurrence, and their potential consequences for the public assessed. The analysis techniques applied in the study have been widely used by other industries.

In most industrialized countries one is today intensively concerned with reliability and risk analysis, and there has been a rapid development of the field. In Norway we are also involved, mainly due to the petroleum activity. A large number of studies have been carried out since the 1970s within this industry. These studies have primarily focused on the traditional safety aspects: safety for human life, environment and material assets. In recent years there has also been much attention devoted to the safety of gas and oil production (delivery) for existing and planned production and transportation systems. Several Norwegian researchers related to the petroleum industry possess expertise on a high international level.

For more literature on reliability and risk analysis in a historical perspective, see e.g. the references [24,59,64,65,70,118,122].

1.2 Objectives of reliability and risk analyses

Some main objectives of reliability and risk analyses are:

- To provide a basis for prioritizing between alternative solutions and actions
- To provide a basis for deciding whether reliability and risk are acceptable
- To provide a basis for evaluating the profitability of a project

- To provide a basis for the development of safe and effective procedures for the operation or the monitoring of the process or the equipment
- To undertake a systematic description of undesirable events and their potential consequences
- To achieve improved system knowledge as a result of analysis of connection and interaction of the components in the system
- To develop competence and motivation for systematic safety follow-up

The primary objective of reliability and risk analyses is to provide a basis for decisions regarding choice of solutions and actions. The benefit of the analyses will, however, depend on how well they are planned and how they are followed up, i.e. how efficient the safety management process is. The purpose of safety management is to ensure that the work with and consideration of safety is taken care of systematically and coordinated with other activities, such that a desired level of safety is achieved and maintained.

1.3 Subject areas. Modelling

The execution of a reliability/risk analysis involves several subject areas:

- Knowledge about the technical and operational aspects of the system and the technical and operational factors that can lead to failures
- Knowledge about analysis methods and techniques, and basic mathematical and statistical theory
- In many cases there is need for knowledge about behaviour disciplines and organization, especially knowledge about how people act under pressure
- Economic calculations will often be included in the analysis to estimate the cost related to various undesirable events and proposed actions
- To estimate system reliability and risk, relevant reliability data (accident data) must be available
- When analysing a large, complicated system, the work often becomes so extensive that computers are required

In reliability and risk analysis system-models are used. These models might be graphical (networks of different types) or mathematical. The latter type of model is necessary in order later to incorporate experience data and use mathematical/statistical methods to estimate reliability and risk for specific systems. When developing such models it is important to strike a balance between the following two desired properties:

- *Simplicity.* The model must be sufficiently simple to be able to be used to study the system by mathematical/statistical methods

- *Accuracy.* The model must be a sufficiently accurate representation of the system

We must always be fully aware that we work with an idealized, simplified model of the system, that the results we obtain, strictly speaking, are only valid in the model, and that they therefore are "correct" only to the extent that the model is realistic.

The modelling task in a reliability or risk analysis can be divided into four main subtasks:

- To define the system to be analysed. To identify the factors that should be taken into account.

- To identify the system performance measures to be used. (How is reliability/risk to be measured?)

- To establish the approach (model) for evaluating these performance measures

- To perform the evaluation of the specified performance measures

These subtasks are common to most modelling tasks, not only within reliability and risk analysis.

1.4 Basic concepts

In this section we shall be looking more closely into the following concepts:

- Probability and frequency
- Accident and near misses
- Reliability
- Reliability analysis
- Reliability management
- Risk
- Risk analysis
- Safety
- Safety management
- Acceptance criteria

1.4.1 Probability and frequency

Probabilities are used when considering future events with more than one possible outcome. In a given situation only one of these outcomes will occur; in advance we cannot say which. Such situations are called stochastic, as opposed to deterministic situations where the outcome is determined in advance. The probability of an event is a measure of the chance that an event will occur. It is measured as a value in the interval [0,1]. Probabilities are usually assessed (estimated) by experience data, such as accident statistics and the operating statistics of components and systems.

A frequency expresses an average number of events per unit of time or per operation. The connection between frequency and probability is illustrated in the following example. Assume that we for a specific company have calculated a frequency of accidents leading to personnel injuries, at 7 per year, i.e. $7/8760 = 0.0008$ per hour. The probability that such an accident will occur during one hour can therefore be assessed at $0.0008 = 0.08\%$. Such a probability interpretation of the frequency value can be justified when this value is small; how small depends on the desired accuracy. As a rule of thumb one often use "less than 0.10".

Probabilities and frequencies are more thoroughly examined in Appendix A.

1.4.2 Accidents and near misses

The word accident here means an undesirable event which may lead to loss of human life, personal injuries, significant damage to the environment or significant economic loss. A near miss is an undesirable event without loss of life and personal injuries, and insignificant damage to the environment and insignificant economic loss, but which with small changes in the situation or in the state of the system might have resulted in an accident.

An unignited hydrocarbon leakage in a process plant will normally be categorized as a near miss. An ignited leakage will probably give rise to an accident.

1.4.3 Reliability

The word reliability is in everyday speech used relatively imprecisely. Here we use the word as a *characteristic of the ability of a component or a system to perform a specific function.* The level of this ability may be measured in different ways in different situations, for example as

- average lifetime
- average number of failures per unit of time (frequency of failures)
- probability that the component/system will function at a specific point in time
- average portion of time that the component/system will function

In Chapters 3 and 4 we define the various reliability measures in a precise, mathematical way.

For specific examples of application of the reliability concept, see Examples 1.5 and 1.6.

1.4.4 Reliability analysis

A reliability analysis is a systematic analysis of reliability. Planning, execution and use of reliability analyses are described in detail in Section 1.5.

1.4.5 Reliability management

With reliability management we understand all systematic measures undertaken in order to attain and maintain a reliability level that conforms with defined policies, goals and acceptance criteria. Reliability analyses constitute an important part of reliability management.

Reliability management is part of safety management.

1.4.6 Risk

The word risk as applied in everyday speech is also relatively imprecise. In this book risk is used to express *the danger that undesirable events represents to human beings, the environment and economic values.*

Risk can be expressed quantitatively in different ways, but usually by means of the frequency (probability) and consequence of undesirable events. In order to get a more thorough understanding of the risk concept, we shall look at a general risk model.

We consider an activity related to a system, for example a company producing machines of a certain type, or an offshore oil-producing platform. Undesirable events can occur as a result of component and subsystems failures and lead to loss of human life, personal injury, damage to the environment and/or loss of economic values. We would like to convert the consequences to a numerical loss using the same scale of measurement. This conversion is, however, difficult to carry out in practice and is usually omitted. Instead the different categories of consequences (loss) are specified separately, for example loss of life, loss of material assets, etc. The model is illustrated in Figure 1.1.

Risk may be expressed by the consequence spectrum $(K_1,F_1),(K_2,F_2),\ldots$, where F_i designates the frequency of undesirable events leading to the consequence K_i, or possibly the probability that an undesirable event shall occur which gives the consequence K_i.

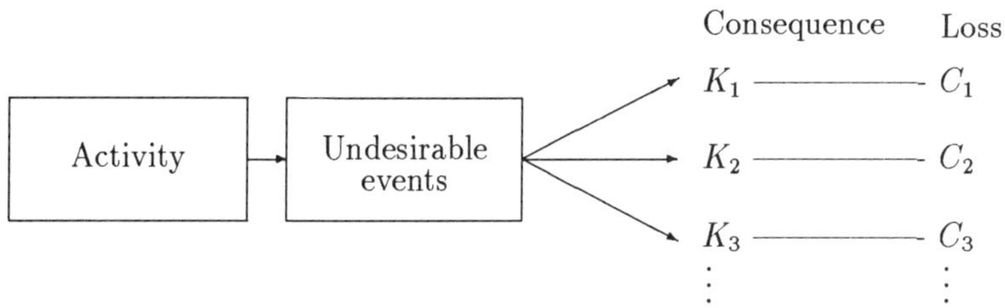

Figure 1.1: General risk model

If the losses C_1,C_2,..., are determined, we often use *statistically expected (mean) loss* as a measure of risk. This value is calculated by multiplying the losses by the corresponding frequencies (probabilities) and summing over all the relevant consequences, i.e.

$$\text{Statistically expected loss} = C_1 \times F_1 + C_2 \times F_2 + \ldots$$

This way of expressing risk has the advantage that there is only one value, so that evaluations and comparisons of risk can easily be performed.

Risk expressed by statistically expected loss should normally be reported in addition to the consequence spectrum. The expected value can only replace a full risk description in the form of a consequence spectrum if the variation in loss values (C values) is small or the activity is repeated a large number of times, cf. Example 1.1.

In situations where the risk is related to loss of life, the so-called FAR value (FAR = Fatal Accident Rate) is often used to measure the risk level. The FAR value is defined as the statistically expected number of accidental deaths per 100 million (10^8) exposed hours. At the time the FAR value was introduced, 10^8 hours corresponded to the time 1000 persons spent at their place of work for a lifetime. Today it takes 1400 persons to reach 100 million working hours.

See Appendix A, Problem A.4.5, for a possible generalization of the risk measure "Statistically expected loss".

In the following we give some concrete examples of the use of the risk concept.

Example 1.1
A casino offers the following game:

One ball is selected at random from among a total of 1000 balls, of which 999 are white and 1 red. If the red ball is chosen, the player wins 5010 dollars, otherwise nothing. Each game costs 10 dollars.

If we look at this activity from the casino's point of view, we will specify the consequence spectrum:

$$(5000, 1/1000), (-10, 999/1000)$$

On the other hand, if we look at the activity from the player's point of view, the consequence spectrum will be:

$$(10, 999/1000), (-5000, 1/1000)$$

For the player this consequence spectrum represents a complete description of the risk associated with the activity; there is a probability of 1/1000 of winning 5000 dollars with a stake of 10 dollars. For the casino the statistically expected loss (profit) will be a more informative measure, as this value equals approximately the average loss (profit) per game after a large number of games. We find that the statistically expected profit for the casino is about 5 dollars per game, which means that the casino in the long run will make a good profit on this activity.

Example 1.2

Consider two activities which are to be compared with respect to risk. Assume that the following consequence spectra have been calculated for the two activities:

Activity 1: $(2,10^{-2}),(100,10^{-4})$
Activity 2: $(10,10^{-3}),(20,10^{-3})$

The consequence spectrum for activity 1 says that there is a "high" probability (1%) of an event occurring resulting in a "small" loss (2), and there is a "small" probability (0.01%) for an event occurring resulting in a "large" loss (100). The consequence spectrum for activity 2 says that there is a "moderately high" probability (0.2%) of an event occurring resulting in a "moderately large" loss (10-20). Thus the consequence spectra for the two activities are very different. The statistically expected loss for the two activities is, however, identical, namely 0.03.

This simple example shows that by limiting the attention to statistically expected loss, the risk picture may be incomplete.

Example 1.3

In recent years there have been about 400 accidental deaths per year in road traffic in Norway. This quantity is a measure of the risk associated with this particular activity. The number 400 will be a good estimate for the statistically expected number of next year's accidental road traffic deaths in Norway. It will probably also be a good estimate for the next year's actual *number of accidental road traffic deaths*, since the variation in the number of accidental deaths is relatively small from year to year. If we look at the average number of accidental deaths per 24 hours instead of per year, we get a frequency of approximately 1. This frequency will be a good estimate for the statistically expected number of accidental road traffic deaths in this period, but it will be a poor estimate of the actual number of deaths in this period since the number of accidental deaths varies considerably from day to day.

In a consequence spectrum description for this activity the consequence K_i represents a certain number of fatalities, for example 1 fatality, 2 fatalities, 3-4 fatalities, more than 4 fatalities, and F_i represents the frequency of accidents resulting in this number of fatalities. In most cases it will, however, be sufficient to report the number of accidental deaths per year (or a derived quantity, e.g. FAR) as a measure of risk for this activity.

To compare this activity with corresponding activities in other countries, and with other activities, for example other transportation methods, there is a need for a common value of measurement. One way of doing this is to use the FAR value. Refer to Section 2.1 for a discussion of this.

Example 1.4

In Chapter 7 an example of a risk analysis of an offshore oil and gas producing platform is presented. The analysis focuses on risk related to loss of human life. The risk is expressed by means of accident frequencies for various loss of life categories: 1-2 fatalities, 3-5 fatalities, 6-20 fatalities, 21-100 fatalities and more than 100 fatalities. For example, a frequency of approximately 2×10^{-2} per year is calculated for accidents occurring with 3-5 fatalities; or expressed in a different way, the probability that an accident resulting in 3-5 fatalities will occur during one year is approximately 2%. FAR values are also calculated.

For this example the consequence spectrum description gives valuable information which the FAR value does not express in isolation, for example the probability of a major accident with more than 100 fatalities.

Example 1.5

Example 7.2 in Chapter 7 presents an analysis of reliability and deliverability of a gas production and transportation system in the North Sea. The risk associated with the activity is measured on the basis of the amount of gas lost due to failures occurring in the system. The consequence spectrum is here described such that K_i expresses lost volume of a certain quantity during a certain period of time, while F_i equals the probability that K_i will occur. In addition, the statistically expected lost volume of gas is calculated. The lost volumes are converted to an economic loss.

For this example it is reasonable to say that the reliability of the system is described by the risk associated with the activity.

Example 1.6

Chapter 7 also offers an example of a reliability analysis of process instrumentation on an offshore oil producing platform. Two design alternatives are compared with respect to reliability and cost. As a measure of reliability (risk) the statistically expected lost production during one year as a result of system failures is used.

1.4.7 Risk analysis

Risk analysis is systematic analysis of risk. Planning, execution and use of risk analysis are described in detail in Section 1.6.

1.4.8 Safety

The word safety can be interpreted in many different ways, depending on the situation considered. In this book we use the word as a *characteristic of the ability to avoid damage and loss as a consequence of undesirable events*, whether these are due to randomly occurring events or to deliberate actions. The damage and loss can be related to:

- The life and health of human beings
- The biological and physical environment
- Economic values

High or low safety is therefore closely related to the risk level; high risk means low safety, and vice versa.

1.4.9 Safety management

With safety management we understand all systematic measures undertaken in order to attain and maintain a safety level that conforms with defined objectives, goals and acceptance criteria. Safety management is discussed in detail in Chapter 6, see also Section 7.4.2.

1.4.10 Acceptance criteria

Acceptance criteria are verbal or numerical statements/quantities which express the level of risk or reliability deemed acceptable or desired. The acceptance criteria can be expressed through for example governmental or company requirements. When designing and constructing for example a large process plant the criteria might be:

- Statistical average for the industry
- That the risk level is to be on the same level as for similar plants/enterprises
- A specific probability for accidental deaths
- A certain number of accidents per year
- A certain number of injuries related to operations which from experience have a high risk level
- A certain number of fires

- Economic optimization with respect to damage to material assets and production loss
- A certain amount of discharge of a poisonous material
- Quality criteria (limitations for acceptable working situations, social conditions of work)
- Related to organization, training, etc.

Figure 1.2 shows an example of how acceptance criteria can be formulated based on accident frequencies and number of fatalities for a given activity. The two curves are called F-N (Frequency - Number of fatalities)-curves. The figure shows for example that one accident on average per 10 years which results in at least one fatality is not acceptable, while one accident on average per 1000 years which results in at least one fatality is acceptable.

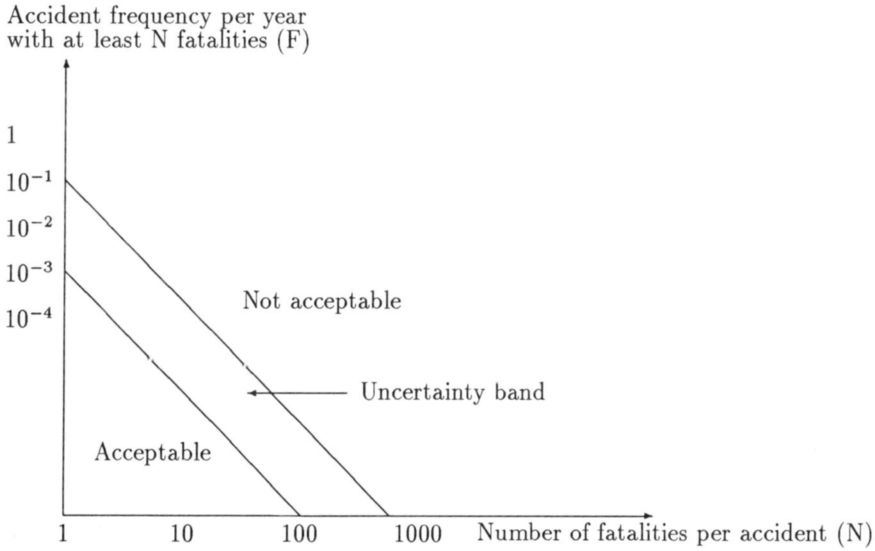

Figure 1.2: Example of an acceptance criterion

The level of risk considered acceptable in the "uncertainty band" depends to a certain extent on the difficulty of making further improvements. This is reflected in the often quoted ALARP principle; any risk must be reduced as far as reasonably practicable or to a level which is "as low as reasonably practicable" (ALARP).

Usually probability criteria (frequency criteria) are based on experience from other related activities. The comparison between established reliability/risk and the acceptance criteria requires normally that the reliability/risk is estimated by carrying out a reliability/risk analysis. This is always the case in the design and manufacturing/construction phase. During operations the criterion may be compared with real, observed values.

The question of what is acceptable safety is simple in many situations. In our daily lifes we take many decisions, for example as car drivers, as pedestrians, in a work situation, at home, etc. This is simple because we take decisions without thinking about the situation in its full depth, it happens more and less automatically. When we make the problems more concrete, for example through a risk analysis, the decision becomes more difficult. Safety and the question of what is safe enough are judged subjectively. As long as we also judge the risk subjectively, there is no conflict. But when we have carried out a risk analysis, we face the dilemma that the subjective decision basis must be formulated in the same form as the results from the risk analysis. This can be problematic since there is a break with the principles which are the basis of our tradition in this field.

The acceptance criteria will depend on our understanding of and attitude to the risk level of the activity. Important factors here are the benefit associated with the activity, the likelihood of a large accident, whether technology is old or new, or whether the risk is voluntary or not.

Different figures have been suggested for what is considered an acceptable risk level, but these figures have not obtained common endorsement. The problem is that such risk levels cannot be determined in general, but have to be specified separately for the individual activities and divided into various personnel groups. The demands for safety vary; for example, a person who lives close to a hazardous factory will require higher safety against hazards associated with the factory activity than persons who work in the factory.

The mechanisms that represent the basis of our understanding of what is safe enough are complex. It has to do with our experience and knowledge, it has to do with our imperfect senses and intellect, and it has to do with our physical and psychological environment.

Laws, regulations and rules for the activity/enterprise put important restrictions on what is considered an acceptable risk level. The risk level of the trade will also play an important role. Usually it will be unacceptable for a company to have a significantly higher risk level than the rest of the companies in the trade, even though it may be lower than the risk level in other trades.

Acceptance criteria are further discussed in Section 6.2.1 and 7.4.3.

1.5 Planning, execution and use of reliability analysis

This section covers planning, execution and use of reliability analysis. Each of these stages are divided into substages as shown in Figure 1.3.

Several activities under the two first stages will elapse in parallel and are highly integrated, for example the cause analysis and the data analysis. A total reliability analysis will include all the stages in Figure 1.3. Depending on the application, parts of the following description will therefore not be relevant.

A reliability analysis with a subsequent evaluation of the results, is an iterative process. The results from an analysis may lead to design changes in the system, and a new analysis is required, and so forth.

PLANNING	Objectives System definition Time planning Organization of the work
EXECUTION	System description Definition of system failure Assumptions Cause analysis Data collection and analysis Presentation of results
USE	Reliability evaluation Decisions

Figure 1.3: Planning, execution and use of reliability analyses

1.5.1 Planning of reliability analysis

Objectives

Before an analysis starts, the objectives of the analysis must be formulated. Usually the objectives will be to

- predict/estimate the reliability of a system

- compare alternative solutions with respect to reliability
- identify critical events/components

The reliability measures to be reported should be identified. In many cases there will be several ways of expressing the reliability.

When the objectives are formulated, any limitations to the scope of the analysis must be taken into consideration, such as lack of available resources, time limits and lack of data and information. This may be necessary in order to balance the complexity and size of the problem on one hand with the scope, ambitions and accuracy of the analysis on the other. The results from a reliability analysis can be used as

- decision basis for a choice between alternative solutions and actions in order to optimize reliability and costs
- documentation of achieved reliability
- basis for determining requirements on reliability of equipment and systems
- basis for risk analysis

The decisions that have to be made and possible criteria that may affect the decisions, for example reliability acceptance criteria, economic framework conditions or available time, should be identified. Such criteria can be formulated quantitatively or qualitatively.

System definition

The system to be analysed must be defined and limitations be made. Limitations are understood to be the limits or points of contact with surrounding areas.

The operating conditions that are to be included in the analysis must be determined. Examples of different operating conditions are start up, normal operation, running down, testing, maintenance, deviations and emergency situations.

The time period to be analysed must also be determined.

Time planning

Good time planning is essential for the quality and usefulness of reliability analyses. Reliability analyses provide a basis for decision making, and the analysis must therefore start so early that the results of the analysis are available when the actual decision is to be taken.

Organization of work

A working group must be established. This group must have knowledge about reliability analysis and about the system. Other specialized competence, for example experts in mathematical statistics, will be required in some cases.

A plan for the reliability analysis should be drawn up. The plan should cover activities, delegation of duties, scientific progress, time limits and milestones, reports and budget. It is important that the responsibility for control and approval of the work is determined.

1.5.2 Execution of reliability analysis

System description

A description of the system to be analysed should be available. This description of the system should include all technical, environmental, organizational and human circumstances that are relevant for the analysis. The scope and detail required for the description of the system depends on the objectives of the analysis, the phase for which the analysis is being done, the documentary data and the restrictions imposed by the methods used.

Definition of system failure

From the objectives of the analysis and the system definition, it will usually be clear what is meant by system failure. Notice that often there will be more than one type of system failure or level of system failure. For example, if the system is a fire detector system, the following two types of failure are relevant:

- The system does not give alarm on demand
- Spurious alarm

If the system is a gas/oil production system, the possible production capacities represent different levels of system failure.

Assumptions and suppositions

It is important that all assumptions, suppositions and simplifications made in connection with the analysis are documented. Incomplete documentation may result in misuse of the results of the analysis, and any updating of the analysis will be more difficult.

Cause analysis

The cause analysis identifies the events (combination of component failures) leading to system failures. The cause analysis is completed with the amount of detail that is appropriate, considering for example:

- The goal of the analysis and the decisions to be made
- The limitations made at an earlier stage in the analysis
- The availability of relevant and accurate data
- Probability of failure (certain failure events can be excluded since they have very low probability)

The system definition will often reveal simple series of events that may lead to system failures, but when the series of causes are more complex, special methods must be used. The most important of these are:

- Use of statistics
- Fault tree analysis (FTA)
- Failure mode and effect analysis (FMEA)
- Cause consequence analysis (CCA) / Event tree analysis (ETA)

These techniques are described in Chapter 2. The choice of technique depend on the actual system and the objectives of the analysis, cf. the discussion part in Sections 2.2-2.4.

If a quantitative analysis of reliability is to be carried out, a probabilistic model of the system must be established. The system will usually be represented by fault trees, reliability block diagrams, cause consequence diagram, event trees or flow networks. Some common models are described in Chapters 2, 3, 4 and 5. When the model with the data is established, the calculations can begin to estimate system reliability and identify critical components and events. The calculations can be carried out by means of analytical methods or Monte Carlo simulation. How the calculations are done within the various models, are explained in Chapters 3, 4 and 5.

If the purpose of the reliability analysis is simply to estimate system reliability and sufficient experience data for the actual system or similar systems are available, the cause analysis can be very simple. The system reliability is then estimated directly from relevant statistics.

Data collection and analysis

To perform a quantitative analysis, failure and repair data for the components in the model must be collected and analysed. Possible sources of data are data banks, handbooks, suppliers' specifications, information one has registered oneself, data from similar items, and expert evaluations. Often there will be a need for a statistical analysis of data to determine lifetime distribution, identify trends, etc.

Data collection and analysis is the topic of Appendix C, see also Section 2.1.

Presentation of results

The results of the analysis should be presented in a way which is appropriate for the decision makers, and such that they can be compared with possible acceptance criteria and other criteria that might be of importance for the decisions to be made.

All the essential assumptions, suppositions and simplifications that have been made should be summarized so that the area of validity and limitations of the analysis are made clear.

1.5.3 Use of reliability analysis

Reliability evaluation

If the results of a reliability analysis are to provide a suitable basis for decisions concerning choice of design alternatives, system modifications, etc., there is a need for a *reliability evaluation*. Such an evaluation comprises a comparison of the result of the analysis with acceptance criteria and other criteria that can affect the decisions to be made, or possibly a comparison of alternative solutions in an economic optimization process. The reliability evaluation may for example include an analysis of the profitability of technical or operational modifications where the system reliability represents a part of the data input in this analysis, and discussion of positive and negative affects of proposed actions.

Decisions

To what extent decisions with respect to choice of system design, system modifications, etc. is a direct consequence of the reliability evaluation, will depend on how the decision process functions. Effective reliability management will ensure that the results of the reliability analysis will be followed up and the right decisions taken.

1.6 Planning, execution and use of risk analysis

Figure 1.4 shows the different stages in planning, execution and use of risk analysis. To avoid repetition of the description of reliability analysis in Section 1.5, we will be content with describing in detail the two "new" stages "Identification of undesirable events" and "Consequence analysis". For the other stages refer to Section 1.5 with "reliability" replaced by "risk"/"safety", and "system" replaced by "activity".

The following two cause analysis methods should be mentioned in addition to those mentioned in Section 1.5:

- Management Oversight and Risk Tree (MORT)
- Safety Management and Organization Review Technique (SMORT)

These methods are presented in Chapter 2.

Identification of undesirable events

A risk analysis often begins by identifying which types of undesirable events are to be included in the analysis. An example of an undesirable event is a gas leakage in a process plant. Next, there may be mapping of where these events can occur within the subject under analysis. This part of the risk analysis provides a basis for selecting which undesirable events to include in the remainder of the analysis, and ensure that all relevant events are included in the analysis.

PLANNING	Objectives Definition of activity (subject of analysis) Time planning Organization of work
EXECUTION	Description of activity (subject of analysis) Identification of undesirable events Assumptions and suppositions Cause analysis Consequence analysis Data collection and analysis Presentation of results
USE	Risk evaluation/safety evaluation Decisions

Figure 1.4: Planning, execution and use of risk analysis

Examples of methods for identification of undesirable events are:

- Accident statistics
- Coarse/preliminary risk analysis
- Job safety analysis
- Failure mode and effect analysis (FMEA)
- Hazard and operability studies (HAZOP)
- Management Oversight and Risk Tree (MORT)
- Safety Management and Organization Review Technique (SMORT)

These methods are studied in Chapter 2. The cause and consequence analyses follow the identification of undesirable events.

Consequence analysis
In the consequence analysis possible series of consequences resulting from the undesirable

events are described. The analysis is completed with an appropriate amount of detail considering, for example:

- The objective of the risk analysis and the decisions to be made
- The limitations made at an earlier stage in the analysis
- The availability of relevant and accurate data
- The consequences of the undesirable events (some events can be omitted from the analysis because they have insignificant consequences or negligible probabilities)

In a consequence analysis we look at the effects the undesirable events may have on the system, other systems, human beings and environment. Possible effects are:

- Increased load on other components in a technical system
- Blast waves, flames, heat
- Flow from liquids and gases

The immediate effects of an event can be determined in a coarse risk analysis or a FMEA. But often failures in a system activate a more complex series of events, for example as a result of activation of alarm systems and other safety systems. Methods developed to analyse such series of events are:

- Event tree analysis
- Cause consequence analysis

These methods are studied in detail in Chapter 2.

To determine the physical effect of an event, for example related to gas dispersion, explosion pressure, etc., special techniques and computer codes are used. Some useful references to such techniques are [2,89,126].

The number of stages in an identified series of events depend on the complexity of the safety and control-systems (barriers). In a process plant with a high level of monitoring and control/alarm systems, the number of stages in the series of events can be relatively high, before the effects "leave the system".

If a quantitative analysis of risk is to be carried out, a probabilistic model of the activity must be established. Usually models based on cause consequence diagrams and event trees are applied. When the model with the data is established, the calculations can begin to estimate the risk and identify critical events.

If the purpose of the risk analysis is simply to estimate risk and there are available sufficient experience data for the actual activity or similar activities, the consequence analysis can be very simple. The risk is than estimated directly from relevant statistics.

1.7 Discussion

Reliability and risk analyses are systematic methods that can give a valuable decision basis for choosing one out of a number of possible alternatives. But there are also some problems associated with the use of such analyses.

One of the most serious problems is to obtain good experience data, for example for component failure rates and repair distributions. The data sources one has registered oneself are seldom sufficient. To carry out an analysis it is therefore necessary to use other sources.

Some data are available from so-called data bank systems. The System Reliability Services (SRS) in England is perhaps the most well-known such system. For other data bank systems, see Appendix C. In addition to the data available from such data bank systems, it will often be necessary to include expert opinions in the analysis. For example, to estimate the repair time distribution for a certain type of component, expert opinions will usually give better estimates than estimates obtained from a data bank. This is due to the fact that the data from a data bank system is very general and not necessarily relevant for the special situation to be modelled.

It is occasionally claimed that reliability and risk analyses are of little value since the uncertainty of the input data, for example the reliability data of the components, is so large. Against this one can argue that reliability and risk are factors that have to evaluated when decisions concerning for example choice of design are to be taken, whether the uncertainties are large or small. Therefore, the aim must be to provide the best input data that are available, and possibly perform sensitivity analyses with respect to those data that are most uncertain.

When for example comparing various design alternatives, it is the differences in system reliability/risk that are of interest. The uncertainties related to these differences are in general much lower than for the absolute values.

A reliability/risk analysis can be performed objectively within specified assumptions and suppositions. It is impossible to determine assumptions and suppositions on an objective basis, and they will consequently be a subjective element of the analysis. Therefore it is important to consider the results of the analysis together with the assumptions and suppositions.

Reliability and risk analyses have traditionally been used to establish whether the level of reliability or risk is acceptable. A potential disadvantage using reliability and risk analysis for verification purposes lies in the somewhat passive nature of this approach: the risk study is performed separately from the design process, often at a stage when any changes recommended would be very costly and sometimes even impossible to incorporate. Furthermore, unless the study results turn out to be "acceptable", the report is soon forgotten. Its purpose was to verify compliance. Once that has been done, there is no need use it any more. This is of course not really true, since most quality risk analyses provide a lot of useful information and follow-up items, regardless of whether the results at one particular point in time turned out to be acceptable.

The problem described above can be particularly bad if reliability/risk analysis is adopted in an organisation which does not have a reasonable safety management system, which ensures that the decision-making and follow-up associated with the analysis happens in a systematic way. Risk analysis cannot exist in a vacuum: it needs the framework which provides inputs and handles outputs in a constructive way. Safety management provides that framework.

Sometimes when risk analysis is used to verify acceptable risk, the study report is subject to public scrutiny. The risk profile will describe the consequences of potential accidents and the associated probabilities of occurrence. It is a general experience that the public will tend to put the emphasis on the "worst case" consequences, without paying too much attention to the fact that the probability of such an accident ever taking place is very low. This must be realised right from the start, and considerable thought should go into developing an unbiased presentation of the results with relevant reference to an understandable basis for comparison of the results.

There are many factors that can lead to reliability and risk estimates which deviate significantly from the true reliability and risk values. The most important of these factors can be divided into three main groups:

1. **Faulty or insufficient system or activity definition.** In the earliest phases of a project there will be limited information about technical solutions, operation and maintenance philosophies, logistic conditions, etc. The analyses must therefore be based on a number of assumptions and suppositions. As the project proceeds, more information will be available and the contribution to this type of analysis error can be reduced to a minimum.

2. **Inaccuracy of the model** can be due to omission of some relevant events that can affect reliability or risk, or due to inaccurate modelling of the corresponding consequences. It can also be a result of limitations and inaccuracy of the analysis-tool (for example the computer code).

3. **A poor data basis** can be the result of using data that are not representative for the actual equipment/event, short time of observation as a basis for applied data, or data collected from non-representative operating and environmental conditions.

The above factors are information sensitive. More information about the system, more representative data and more detailed analyses will reduce the analysis error. It is important to identify the main causes of this error, and determine if it is cost-effective to get more information to improve the decision basis. The analysis error will be reduced in the later project phases, but even for a system/plant in operation, it will not be completely eliminated; modelling inaccuracy and data uncertainty will always be present.

To reduce the analysis error to a minimum one must:

- Use sufficient detailed and correct information about the actual system

- Use representative reliability and accident data

- Reduce modelling inaccuracy by extensive quality assurance of data tools, double-controlling the analysis model, and critical evaluations in connection with the interpretations of the analysis results.

The definitions of the key concepts discussed in this chapter are in accordance with the definitions given in e.g. [125]. The description of the planning, execution and the use of risk analysis follows to a large extent the lines of [125].

1.8 Problems

1. Discuss briefly various ways of reporting risk.
   Give examples of situations where it is natural to use the different measures.

2. In a company there have been 6 accidental deaths in the last 10 years. In this period of time altogether 10000 man-years (1 man-year = 1600 hours) have been registered. Calculate observed FAR value for this period of time.

3. In trades A and B the following accident statistics have been registered:

Trade	Number of fatalities	Number of man-year
A	16	50000
B	5	10000

   a) Define an appropriate risk measure and calculate this based on the above statistics for each trade and for the two trades together.
   Do you see any problems in making a comparison of the risk of loss of human life based only on this measure?

   b) In a process plant the fatality risk associated with the operation is estimated to 10 FAR. Five hundred persons work at the plant. Each person is exposed to risk for 1600 hours per year. This risk can be reduced by 10% by introducing some risk reducing measures. These measures will cost 2 million dollars per year.

   How many operating years correspond to 100 million hours? By introducing these risk reducing measures, what is the cost per saved expected life?

4. Within a trade the following accident statistics have been collected for a period of ten years:

Accident category	1 fatality	2-4 fatalities	5-9 fatalities	10-20 fatalities
Number of accidents	140	10	6	3

In this trade an average of 2000 plants per year were in operation in this ten-year period.

a) Draw up a table or a figure showing the relation between accident category and frequency of accident per plant-year. Based on the information presented in the text, estimate the probability that in a specific plant an accident with 10-20 fatalities occurs during one year.

b) Calculate an approximate FAR value based on the accident statistics above. The average number of persons per plant is 100.

5. In a company there are 100 workers. Each worker is assumed to be exposed to risk for 1600 hours in a year. Let Y be the number of accidental deaths in one year.

   a) Show that
$$FAR \times (1600 \times 10^{-8}) = \frac{EY}{100}$$

   b) Use this formula and the result from problem 4 in Appendix A to give an alternative interpretation of the FAR value.

   c) Generalize the result to an arbitrary number of persons and to an arbitrary exposure time.

6. Discuss the advantages and disadvantages of using quantitative risk acceptance criteria.

7. To evaluate the quality of a reliability/risk analysis, what will you emphasis the most?

Chapter 2

BASIC RELIABILITY AND RISK ANALYSIS METHODS

This chapter gives an overview of some of the most common reliability and risk analysis methods, including:

- Use of statistics
- Failure mode and effect analysis
- Fault tree analysis
- Cause consequence analysis/event tree analysis
- Coarse risk analysis/Preliminary risk analysis
- Hazard and Operability Studies

We also briefly discuss methods for analysing human reliability.

2.1 Use of statistics

Reliability and accident statistics are the basis for all quantitative analysis of reliability and risk. This section describes some important aspects relating to the use of such statistics. Refer to text books in statistics and Appendix C for a more comprehensive description of statistical analysis of reliability and accident data.

The presentation in this section is divided into three parts:

1. Accident data
2. Near misses
3. Reliability data

The references [47,56,92,107,124] have been used as a basis for the content of this section.

2.1.1 Accident data

Accident statistics can be used to:

- Monitor the safety level
- Identify hazardous working situations and equipment
- Analyse accident causes
- Evaluate the effect of risk reducing measures
- Compare alternative area of efforts and measures

Statistics over the total number of accidents associated with the activity within different consequence categories (loss of life, personal injuries, material losses, etc.), provide a measure of the extent of the accident problem. If the statistics are related to time periods, they can also identify time trends. More detailed information can be obtained by dividing the accident data into subgroups, related to e.g. occupation, sex, age, operations, type of injury, etc. The possibility of making a meaningful division into subgroups will depend on the amount of data available. The more data available, the more subgroups can be defined, cf. the discussion at the end of this subsection.

The practice with respect to registration and analysis of accident statistics varies considerably between different enterprises and companies. Some companies have realized that an internal company accident data system is a useful tool to ensure effective safety management. An important element of such a system is systematic and qualified investigation of accidents.

Correct and relevant accident data are necessary to ensure high quality of the risk analyses. Continuous collection and registration of accidents is therefore important.

It is possible to present a number of interesting examples of accident statistics. Below we show two examples based on statistics from [47,92] and [107], respectively.

Example 2.1: General accident statistics for Norway
Accidents present undoubtedly a large health problem. Accidents are the most frequent cause of death of children, young people and adults up to 40-50 years. Each year about 2000 persons are killed as a result of accidents in Norway, which is about 5 fatalities per 10000 persons. The accidents cost society billions of dollars per year. Table 2.1 shows accidental deaths in Norway, Finland, Sweden and Denmark in 1985, distributed over the most important accident types. We see that Norway has the highest total accidental death rate, for occupational accidents, and home and leisure accidents. Sweden shows the best results in these statistics.

Table 2.1: Accidental deaths per million inhabitants in the Nordic countries in 1985

Accident types	Number of deaths per million inhabitants in:			
	Norway	Finland	Denmark	Sweden
Motor vehicle accidents	100	106	147	98
Occupational accidents	24	16	12	11
Home and leisure accidents	370	352	293	223
Total	494	474	452	332

The accidental death rate increased in the 1950s and 60s until the end of 1970, when the rate started to decrease. The trend is, however, different for males and females, see Table 2.2. The number of men killed in accidents in the period 1951-55 was more than the double of the number of women killed in the same period, whereas there were about one and a half times more men in the period 1981-85. The accident rate for men has decreased after a peak in 1971-75. For women the accident rate has been increasing during the whole period.

Table 2.2: Accidental deaths in Norway per million inhabitants distributed by sex and period of time

Period of time	Number of deaths per million inhabitants - average per year		
	Men	Female	Total
1951-1955	628	276	451
1956-1960	623	296	460
1961-1965	646	333	489
1966-1970	645	341	492
1971-1975	675	349	511
1976-1980	607	357	481
1981-1985	573	372	472

The accident numbers presented in Tables 2.1 and 2.2 are normalised in the sense that the number of deaths is divided by the number of persons in the population. This is done in order to ensure real comparisons. Such a normalisation is also carried out for the data in Table 2.3, where the accident statistics for various occupational and travelling activities are presented. How the number of person-hours related to the different activities have been calculated, is explained in [47], cf. also the discussion at the end of this subsection.

Table 2.3: Number of fatalities per 100 million person-hours (observed FAR) for different activities, 1981-1986.
Persons in the age range 15-69 years.

Activity	Observed FAR
ALL OCCUPATIONAL ACTIVITIES	2.5
Agriculture and forestry	2.3
Fishing	63.0
Oil activities	19.0
Industry	1.1
Service	1.5
Railway transport	6.2
Shipping	11.0
Aviation	50.0
ALL TRAVELLING	27.0
Travelling on roads	27.0
Pedestrians	14
Cyclists	28.0
Motorcyclists	280.0
Car drivers	25.0
Car passengers	29.0
Travelling by train	4.7
Travelling by ship	9.1
Domestic air services	56.0
TIME SPENT IN THE HOME	1.0
ACTIVITIES ELSEWHERE	8.2
ALL DISEASES	44.0
SUICIDE	2.0

We see from Table 2.3 that the observed FAR value for travelling activities is about 10 times higher than the observed FAR value for all occupational activities. Occupations within the transport sector have a high observed FAR value, the same for fishing, and the oil activity. For industry, and service activities outside the transport sector, the death rate is not significantly higher than for the time spent in the home.

Example 2.2: Accident statistics for the petroleum activity
The Norwegian Petroleum Directorate presents in its annual reports comprehensive statistics on personal injuries. Table 2.4 gives an example of such data. From the table we see for example that the injury frequency is highest for the drilling and building/maintenance functions. For drilling the statistics show a clear, positive trend.

Table 2.4: Number of accidental injuries per 1000 man years, distributed according to functions (1980-1988). Production installations

FUNCTION	1980	1981	1982	1983	1984	1985	1986	1987	1988
Administration/ production	19.6	19.2	16.1	24.5	15.5	13.9	22.6	23.1	29.7
Drilling	135.2	105.6	106.3	80.0	111.8	93.9	89.0	65.7	58.4
Catering	26.1	17.0	40.1	34.3	32.3	46.7	52.6	42.8	30.0
Building/ maintenance	62.3	43.1	65.7	51.7	62.5	85.2	67.0	71.6	59.5
Total	64.6	46.6	62.5	51.0	58.7	69.8	62.0	60.4	50.5

The data are further grouped according to e.g.:

- Operator employees/contractor employees
- Injury event/occupation
- Injury event/injured part of the body

Similar statistics are collected by the various operators on the Norwegian Continental Shelf, for the total activity and for each production installation.

Discussion

Statistics as described in the Examples 2.1 and 2.2 can be very useful in accident prevention. To improve safety efficiently, a picture of the risk related to the activity should be available: what type of accidents that might occur, the frequency of these, possible time trends, where they might occur, where the probability of an accident is particularly high and the consequences particularly severe.

We must, however, be very careful when drawing conclusions from statistics. Too often it is seen that statistics are misused. Some important aspects in this context are discussed in the following:

- The future situation can be totally different from the one the statistics are based on, such that direct use of the statistics to say something about the risk level today or in the future may give wrong conclusions.

- The statistics presented in Tables 2.1-2.4 say something about the risk level related to the activity. What risk the individual person, say Ole Olsen, is exposed to, the statistics do not say anything about. For example, Ole Olsen might not be in the traffic at all, so that the probability that he will be killed in the traffic the next year is zero.

- To be able to perform real comparisons of the risk level between different groups, a common measurement scale must be established. This can for example be done by using the FAR value. It is, however, often difficult to assess the number of exposed hours in a "fair way". For example: in Table 2.3 an accidental death frequency is calculated based on 100 million *working hours*. But personnel that are working on an offshore installation are also exposed to many types of accident in their leisure time on the installation, e.g. risk related to accidental events such as fire and explosion, ship collisions, and structural collapse. For some accident types it would therefore be more correct to count the number of hours that the personnel stays on the installation, cf. Section 7.1. However, the use of this method of measuring the exposed hours, means that the FAR value for personnel in oil activities is related to a larger portion of a year than the FAR value for e.g. onshore industry. In order to avoid this problem and to be able to make a real comparison of the risk level for different occupational activities, we must look at the totality of life and work and divide this into different activities (work, travelling to work, leisure, staying at home, etc.).

- It can be very difficult to make comparisons of the injury frequency between different activities (for example companies) that are not using the same routines for accident reporting. Different criteria for reporting, i.e. what type of injuries are to be reported, can have a considerable effect on the total number of injuries. Furthermore, the statistics will seldom include all events. Some events that should have been reported, may have been left out in some way or other. The extent of this problem depends on the quality of the reporting system. One way of improving the reliability of the data is to restrict attention to accidents with severe injuries.

- In situations with little data, random variations may give widely differing results. Statistical methods can be used to test whether for example an observed difference in the injury frequency for two groups is *statistically significant* or is due to random variations. Such methods are described in general text books on statistics.

  The major accidents may give particularly large variations in the risk numbers. An accident resulting in a large number of fatalities may have a dramatic effect on the statistics. If for example we had included 1980 in the statistics in Table 2.3, then as a result of the Alexander Kielland accident, where 123 persons were killed, the observed FAR value for the oil activity would have increased by a factor of about 4. If we had considered a longer period, the contribution from the Alexander Kielland accident would of course be reduced. By making the data basis larger, we will obtain a more correct weighting of the major accidents. However, in this way the data may be of less relevance for the estimation of the risk level today and in the future.

  Similarly, care must be taken when drawing conclusions from the data in Table 2.1, which is based on one year only. If statistics were available for a longer period of time, and the data showed the same tendency as Table 2.1, we would be able to draw more reliable conclusions.

- Accident statistics can in some cases be used to analyse causes of specific types of accident. Statistics as such do not prove any causal connection. If for example a statistical correlation between injury frequency and age has been identified, it would not be correct to say that age is the cause of the difference in the injury frequency. The statistics shows that factors related to age, e.g. type of occupation, experience and training, may be of importance. The results provide a basis for asking new and more precise questions.

  Analysis of the accident frequency over time may also give indications of casual mechanisms. By radical changes in the working situation, for example by introduction of shift work, new equipment or other changes, possible trends in the accident level can be studied. It is, however, difficult to draw reliable conclusions about the effects of such changes, because it is impossible to avoid influences from other factors that may have changed in the same period of time.

- Statistical data that do not state sources and assumptions are usually of little interest. It must, for example, be clear what accidents have been included in each accident category.

Insight into the causes of accidents is important in order to choose the correct preventive measures, and thereby improve safety. The statistical analysis of accident data can be a good way to achieve this insight. Even with little data investigation of causes of accidents, effective risk-reducing measures may be generated.

Accident data may to some extent be used to analyse the effect of accident-preventing measures. If we had enough data, it would be possible to estimate the effect of a measure by disregarding specific accidental events. This might, however, be difficult to carry out in practice if the measure were not directly related to one or more of the factors that the statistical data covers.

2.1.2 Near misses

Statistics on near misses can give valuable information about hazardous conditions and ideas for preventing measures. The number of near misses is normally much higher than the number of accidents, therefore the near misses can give a considerable extension of the information basis for hazardous conditions.

Near misses have implications for the risk related to the activity. Investigations of near misses in traffic [127] seem to indicate that such events can give a relatively good picture of where accidents might occur, but that they do not give a good indication of the frequency of accidents at those places.

When analysing near misses the following questions should be addressed:

- Why did the event not result in an accident?
- How extensive could the damage have been in the worst case?

It is particularly important to consider cases where an accident did not occur as a result of pure chance, and where the damage could have been considerable under other conditions.

2.1.3 Reliability data

There are many points of resemblance between accident data and reliability data, and much of the material in Section 2.1.1 is also relevant for reliability data. To avoid repetition, we will in this section limit ourselves to comment on some selected topics.

Reliability statistics can be used to:

- Monitor the reliability level
- Identify critical components, equipment and systems
- Analyse causes of failure
- Evaluate the effect of measures and prioritize between different measures

Reliability data originate from individual failures and test reports on components, equipment and systems. The data are registered in connection with repair and maintenance activities, and with testing of equipment.

The failure and test reports are analysed in order to estimate for example failure frequency, statistically expected portion of time the unit is not functioning, statistically expected time to failure, probability that the unit is functioning on demand, and probability distribution for the time to failure. In Figure 2.1 the observed failure frequency for a compressor unit is illustrated graphically by means of a histogram, as a function of time. We see from the figure that the failure frequency increases. In addition to the data of Figure

2.1, statistics on the portion of time the compressor has not been been functioning would have been informative.

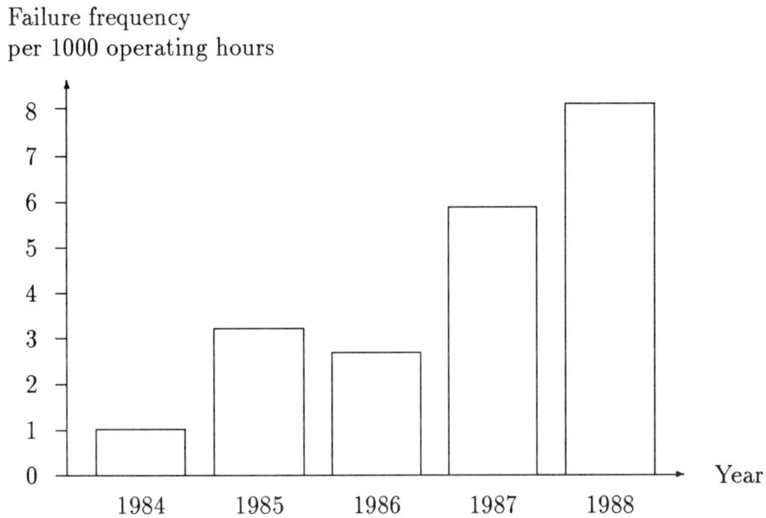

Figure 2.1: Failure frequency for a compressor unit

In Figure 2.2 the lifetime of a large number of components of a specific type has been illustrated graphically using a histogram. From the figure we can for example see that approximately 50% of the components have a lifetime of 500 to 1000 hours.

Ideally, reliability data to be used in an analysis should satisfy the following requirements:

1. The data originate from relatively similar units with respect to type and model

2. The data originate from equipment that has been exposed to comparable operating and maintenance conditions

3. The data basis is large

Unfortunately, these requirements are conflicting, a consequence of requirements 1 and 2 is that the data become more limited.

For a presentation of some key methods of analysing reliability data and a presentation of some relevant data base sources, see Appendix C.

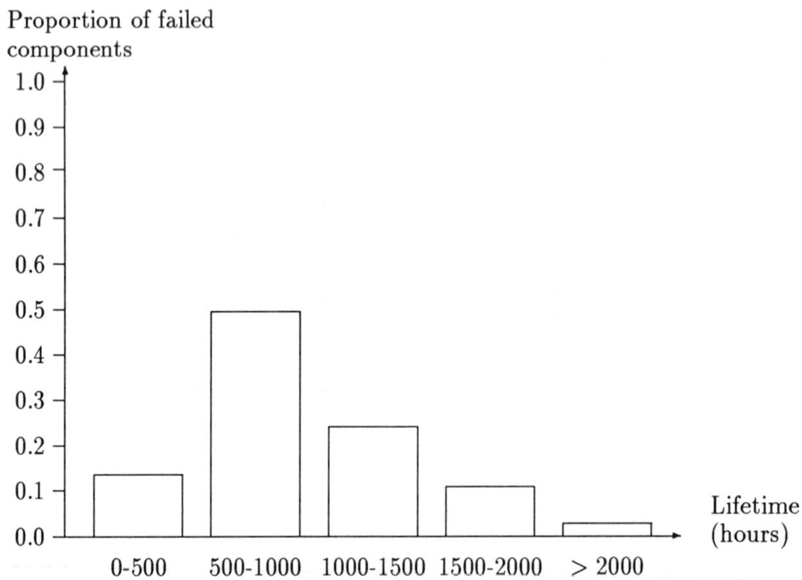

Figure 2.2: Distribution of the time to failure for a specific component type

2.2 Failure mode and effect analysis

2.2.1 Introduction

Failure mode and effect analysis (FMEA) was developed in the 1950s and was one of the first systematic methods used to analyse failures in technical systems. The method has appeared under different names and with somewhat different content. The presentation in this section is basically identical with the one presented in e.g. [61,126]. Other references to the method are [36,96]. If we describe or rank the criticality of the various failures in the failure mode and effect analysis, the analysis is often referred to as an FMECA (Failure Modes, Effects and Criticality Analysis). The criticality is a function of the failure effect and the frequency/probability, cf. Section 2.2.3. The difference between an FMEA and an FMECA is not distinct, and we find it inappropriate to distinguish between these two. In the following we will also use the term FMEA when the analysis includes a description/ranking of criticality.

In several enterprises, it is nowadays a requirement that an FMEA be included as part of the design process, and that the results from the analysis be part of the system documentation.

2.2.2 Basic principles for FMEA

FMEA is a simple analysis method to reveal possible failures and to predict the failure effects on the system as a whole. The method is inductive; for each component of the system we investigate what will happen if this component fails. The method represents a systematic analysis of the components of the system to identify all significant failure modes and to see how important they are for the system performance. Only one component is considered at a time, the other components are then assumed to function perfectly. FMEA is therefore not suitable for revealing critical combinations of component failures.

2.2.3 Execution of an FMEA

To ensure a systematic and full study of the system, a specific FMEA form is used. The content of this form may be suited to each application. The FMEA form may for example include the columns shown in Figure 2.3. The execution of an FMEA is illustrated by going through the content of the form. The starting point of the analysis would be the information available from the system definition (functional diagrams, descriptions of components, etc.), cf. Section 1.5.2.

Identification (column 1). Here the specific component is identified by a description and/or number. It is also possible to refer to a system drawing or a functional diagram.

Function, operational state (column 2). The function of the component, i.e. its working tasks in the system, is briefly described.

The state of the component when the system is in normal operation is described, e.g. whether it is in continuous operation mode or in stand-by mode.

Failure modes (column 3). All the possible ways the component can fail to perform its function are listed under this column. Only the failure modes that can be observed from "outside" are included. The internal failure modes are to be considered as causes of failure. These causes can possibly be listed under a separate column. In some cases it will also be of interest to look at the basic physical and chemical processes that can lead to failure (failure mechanisms), such as corrosion. Often we also state how the different failure modes of the component are detected, and by whom.

Example: In a chemical process plant a specific valve is considered as a component in the system. The function of the valve is to open and close at demand. "The valve does not open at demand" and "the valve does not close at a demand" are relevant failure modes, as well as "the valve opens when not intended" and "the valve closes when not intended". However, "washer bursts" is an example of the cause of a specific failure mode.

Effect on other units in the system (column 4). In those cases where the specific failure mode affects other components in the system it is stated in this column. Emphasis should be given to identification of failure propagation which does not follow the functional chains of the functional diagrams. For example: increased load on the remaining pillars that are supporting a common load when a pillar collapses; vibration in a pumping house may induce failure of the driving unit of the pump, etc.

Effect on system (column 5). In this column we describe how the system is influenced by the specific failure mode. The operational state of the system as a result of failure, is to be stated, for example, whether the system is in the operational state, changed to another operational mode, or not in an operational state.

Corrective measures (column 6). Here we describe what has been done or what can be done to correct the failure, or possibly to reduce the consequences of the failure. We may also list measures that are aimed at reducing the probability that the failure will occur.

Failure frequency (column 7). Under this column we state the estimated frequency (probability) for the specific failure mode and consequence. Instead of presenting frequencies for all the different failure modes, we may give a total frequency and relative frequencies (in percentages) for the different failure modes.

SYSTEM/EQUIPMENT: EXECUTED BY:
REF. DIAGRAM/DRAWING NO.: DATE: PAGE: OF:

Identification	Function/ operational state	Failure mode	Effect on other units in the system	Effect on the system	Corrective measures	Failure frequency	Failure effect ranking	Remarks
1	2	3	4	5	6	7	8	9

Figure 2.3

Failure effect ranking (column 8). The failure is ranked according to its effect, with respect to reliability and safety, the possibilities of mitigating the failure, the length of repair time, production loss, etc. We might for example use the following grouping of failure effects:

Small: A failure that does not reduce the functional ability of the system more than normally is accepted.

Significant: A failure that reduces the functional ability of the system beyond the acceptable level, but the consequences can be corrected and controlled.

Critical: A failure that reduces the functional ability of the system beyond the acceptable level and which creates an unacceptable condition, either operational or with respect to safety.

Usually it will be necessary to define the ranking in more detail in a specific situation.

Remarks (column 9). Here we state for example assumptions and suppositions.

By comparing failure frequency (probability) and failure effect (consequence) the criticality of the specific failure mode can be determined. This is performed by using a diagram as shown in Figure 2.4. The most critical failure modes will be found at the bottom right-hand corner of this diagram.

Probability/frequency	Consequence category Small	Significant	Critical
Very unlikely, Once per 1000 year or more infrequently			
Unlikely, Once per 100 year			
Quite likely, Once per 10 year			
Likely, Once per year			
Frequently, Once per month or more frequently			

Figure 2.4: Example of a probability-consequence diagram

2.2.4 Resources

As already mentioned the FMEA is quite simple to carry out, but it can be extremely demanding, even for systems of moderate size. As a basis for the analysis, the analyst must have a thorough knowledge of the mode of operation of the system and the environmental and operational conditions. It is usually necessary to have detailed drawings of the system and the individual components. For process systems P&I (Process and Instrumentation) diagrams will normally be required. To fill in column 8 "Failure frequency" relevant data sources must be available.

2.2.5 Example 2.3: Storage tank

Figure 2.5 shows an open container for preliminary storage of fluid for use in the production process. The consumption of fluid in the process is not constant, and the liquid level in the tank will therefore vary. Filling the tank is automatically controlled and can be described as follows: when the liquid level reaches a certain height - "normal level", then the Level Switch High (LSH) will be activated and send a closure signal to the valve V1. The fluid supply to the tank then stops. If this mechanism does not function and the liquid level continues to increase to "abnormal level", then the Level Switch High High (LSHH) will be activated and send a closure signal to valve V2. The fluid supply to the tank then stops. At the same time the LSHH sends an opening signal to valve V3 so that the fluid is drained. The draining pipe has a higher capacity than the supply pipe.

A simple FMEA has been carried out for this system. Figures 2.6 and 2.7 show the completed forms for the components LSH, LSHH, V1, V2 and V3. The following ranking has been used:

1. No fluid supply
2. The fluid in the tank is drained
3. The liquid level may increase to an abnormal height
4. The tank is over-filled if not the valve V1 closes

The consequence categories are very rough. For example there is no indication of the length of a break in the fluid supply. Only failure modes related to *normal* operational state have been included, for example the failure mode "does not open" is not included for the valves V1 and V2.

Now, what are the result of the analysis? Firstly, the analysis of the relevant components have given a good understanding of the type of component failures that might occur, and their effect. The analysis demonstrates that the system has a high reliability. The probability that the buffer storage is over-filled is small. The component LSHH seems to be the most critical component in the system, see Section 2.3 and Section 3.5. This example is also discussed in the following sections.

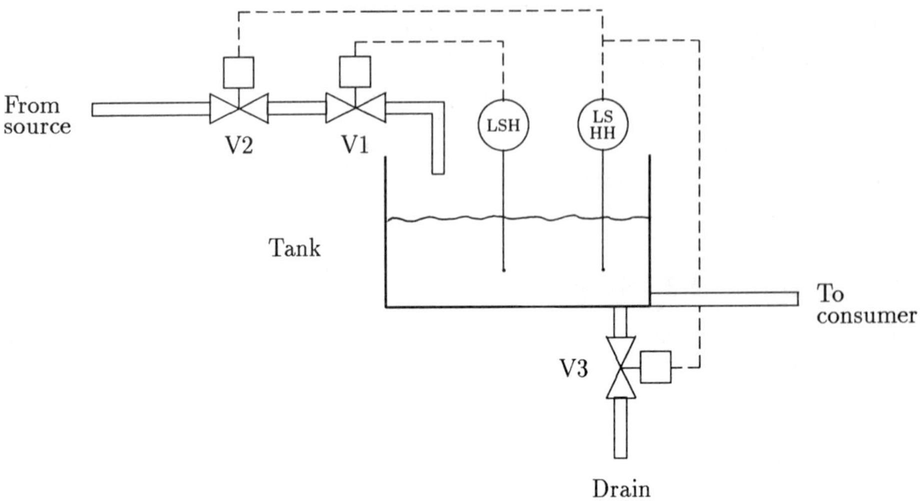

Figure 2.5

2.2.6 Discussion

An FMEA gives no guarantee that all critical component failures have been revealed. There are many reasons for this, for example lack of information and knowledge about the design and operation of components and system. The analyst's imagination and ability to identify possible problems may of course also be a limiting factor. By means of a systematic review as represented by FMEA, most weaknesses of the system as a result of individual component failures will be revealed.

In an FMEA attention is in many cases focused too much on technical failures, whereas the contributions of human failure are often overlooked. This may to some extent be compensated by including human functions as components in the system.

The strong point of the FMEA is the fact that it gives a systematic overview of the important failures in the system, and that it forces the designer to evaluate the reliability of his system. In addition it represents a good basis for more comprehensive quantitative analyses, such as fault tree analyses and cause consequence analyses.

SYSTEM/EQUIPMENT: Storage tank EXECUTED BY: TAV
REF. DIAGRAM/DRAWING.NO.: DATE: 08.08.90 PAGE: 1 OF: 2

Identification	Function/ operational state	Failure mode	Effect on other units in the system	Effect on the system	Corrective measures	Failure frequency	Failure effect ranking	Remarks
1	2	3	4	5	6	7	8	9
LSH	Switch that sends stop signal to V1 if the liquid level is high	Does not send signal when the liq. level is high	V1 does not close	The liquid level may increase abnormally		1% of total number of demands	3	
		Sends signal when the liquid level is not high	V1 closes when not intended	The fluid supply stops		Once per year on average	1	
LSHH	Switch that sends stop signal to V2 and open signal to V3 if the liquid level is abnormally high	Does not send signal when the liquid level is abnormally high	V2 does not close. V3 does not open	The tank is overfilled if V1 does not close		1% of total number of demands	4	
		Sends signal when the liquid level is not abnormally high	V2 closes when not intended. V3 opens when not intended	The tank is drained		Once every 2nd year on average	2	
V1	Stops the fluid supply when the liquid level is high. The valve is normally open	Does not close at a signal		The liquid level may increase abnormally		2% of total number of demands	3	
		Closes when not intended		The fluid supply stops		Once in 10 years on average	1	
		Significant leakage		The fluid supply stops		Once in 10 years on average	1	

Figure 2.6

SYSTEM/EQUIPMENT: Storage tank EXECUTED BY: TAV

REF. DIAGRAM/DRAWING.NO.: DATE: 08.08.90 PAGE: 2 OF: 2

Identification	Function/ operational state	Failure mode	Effect on other units in the system	Effect on the system	Corrective measures	Failure frequency	Failure effect ranking	Remarks
1	2	3	4	5	6	7	8	9
V2	Stop the supply when the liquid level is abnormally high. The valve is normally open	Does not close at a signal		Undesired supply to the tank. The fluid is drained if V3 opens		2% of total number of demands	2	
		Closes when not intended		The fluid supply stops		Once in 10 years on average	1	
		Significant leakage		The fluid supply stops		Once in 10 years on average	1	
V3	Drain the fluid when the liquid level is abnormally high. The valve is normally closed	Does not open at a signal		Undesired supply to the storage		2% of total number of demands	2	
		Opens when not intended		The fluid is drained		Once in 10 years on average	2	
		Significant leakage		The fluid supply stops. The fluid is drained		Once in 10 years on average	1,2	

Figure 2.7

An FMEA can be unsuitable for analysing systems with much redundancy (several components that can perform the same function such that failure of one unit does not result in system failure). In such systems it will not be so interesting to analyse individual component failures, since these cannot directly affect the function of the system. The interest then is focused on a combination of two or more events that together can cause system failure. Example 2.3 shows, however, that an FMEA can also give valuable information about possible failures and their effect for a system with some redundancy. The system in Example 2.3 is redundant in that to avoid over-filling the tank it is sufficient that valve 1 closes, or that valve 2 closes or valve 3 opens. The analysis is a good starting point for a fault tree analysis or a cause consequence analysis (event tree analysis).

Perhaps the main disadvantage of using the FMEA method is that all components are analysed and documented, including the failures that are of little or no consequence. An FMEA can therefore be very demanding. The amount of documentation can be extensive. This problem can be reduced by proper component definitions. For the system in Example 2.3 we could have defined the system components by the different parts of the valves V1, V2 and V3 and the level switches LSH and LSHH. This would, however, increase the extent of the analysis considerably, without obtaining more insight into possible undesirable events at the system level.

For larger systems it may be an advantage to define sub-systems (system functions). An initial FMEA may than be related to failures of these sub-systems. Detailed FMEA studies can then be carried out for specific sub-systems.

Today there exist many computer based tools for execution of an FMEA. In Norway the most well-known system is the programme CARA [123].

2.3 Fault tree analysis

2.3.1 Introduction

The fault tree analysis (FTA) method was developed by Bell Telephone Laboratories in 1962 when they performed a safety evaluation of the Minuteman Launch Control System. The Boeing company further developed the technique and made use of computer programs for both quantitative and qualitative fault tree analysis. Since the 1970s fault tree analysis has become very widespread and is today the most widely used reliability and risk analysis method. The applications of the method include most industries. The space industry and the nuclear power industry have perhaps been the two that have used fault tree analysis the most.

Fault tree analysis is presented in many textbooks. The presentation in this book is to a large extent identical with the one given in [53,64,65,126].

A fault tree is a logical diagram which shows the relation between system failure, i.e. a specific undesirable event in the system, and failures of the components of the system. The undesirable event constitutes the *top event* of the tree and the different component failures constitute the *basic event* of the tree. For example, for a production process the top event might be that the process stops, and one basic event that one particular motor fails. A basic event does not necessarily represent a pure component failure; it may also represent human errors or failures that are due to external loads, such as extreme environmental conditions.

The results from the fault tree analysis include:

- A list of possible combinations of component failures/basic events which will ensure that the top event occurs
- Identification of critical components/events
- The unreliability of the system, i.e. the probability that the top event will occur

A fault tree analysis often comprises several stages:

1. Definition of the top event and framework conditions
2. Construction of the fault tree
3. Identification of the minimal cut sets
4. Qualitative analysis of the fault tree
5. Quantitative analysis of the fault tree

Before we look closer into these stages, we will describe the basic elements of a fault tree.

2.3.2 Description of a fault tree

A fault tree comprises symbols which show the basic events of the system, and the relation between these events and the state of the system. The graphic symbols that show the relation, are called logical gates. The output from a logical gate is determined by the input states. The graphic symbols vary somewhat depending on the standard that is used. In the following we will apply the American standard. Table 2.5 shows the most important symbols in a fault tree, with the interpretations of the symbols.

A *primary* failure is often defined as a failure occurring under normal operation, a failure which the component itself is responsible for, whereas a *secondary* failure is defined as a failure that is not caused by the component itself, a failure occurring as a result of extreme environmental conditions, insufficient maintenance, etc., where the failure causes are not studied further. The secondary failure is due to loads that exceed the design specifications or because the maintenance is not as prescribed.

To distinguish between primary and secondary failure based on these definitions is often problematic and inappropriate. We will therefore in the following not distinguish between primary failure and secondary failure. A circle will be used to denote both failure types. Thus a circle refers to a fault event that is sufficiently basic for no further development to be necessary.

The inhibit gate describes a causal relation between failure events. For the output event to occur, the input event must occur and a condition must be satisfied. An inhibit gate can be replaced by an And gate.

A fault tree that comprises only And and Or gates can alternatively be represented by a reliability block diagram. This is a logical diagram which shows the functional ability of a system. Each component in the system is illustrated by a rectangle as shown in Figure 2.8.

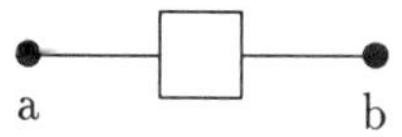

Figure 2.8: Functional element in a reliability block diagram

If there is a connection from a to b in Figure 2.8, this means that the component is functioning based on the criteria that apply for the particular analysis. Usually "functioning" means absence from one or more failure modes.

A presentation of some equivalent fault trees and reliability block diagrams is shown in Figure 2.9.

Table 2.5: Fault tree symbols

	Symbol	Interpretation
Logical symbols	"Or" gate 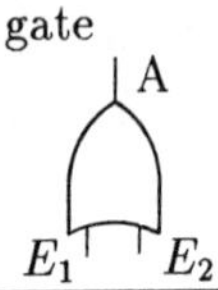	The output event A occurs if at least one of the input events E_i occurs. The number of input events is arbitrary.
	"And" gate 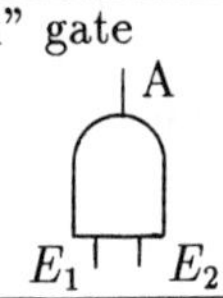	The output event A occurs if all input events E_i occur. The number of input events is arbitrary.
	"Inhibit" gate 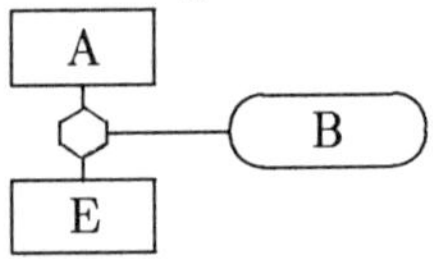	The output event A occurs if the input event E occurs and the condition B is present.
Basic (input) events	"Normal" input 	Symbol for primary failure state.
	"Secondary" input 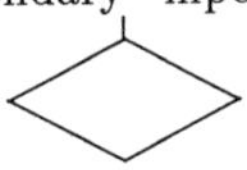	Symbol for secondary failure state.
Description of event	"Comment" rectangle 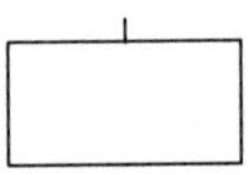	Events (states) are described in the rectangle. The rectangles are usually placed above all logical gates and input events.
Transfer symbols	"Transfer" in  "Transfer" out 	Transfer symbols for further development of a cause sequence; used when the same branch occurs at several places in the tree, and when the tree must be drawn on several pages.

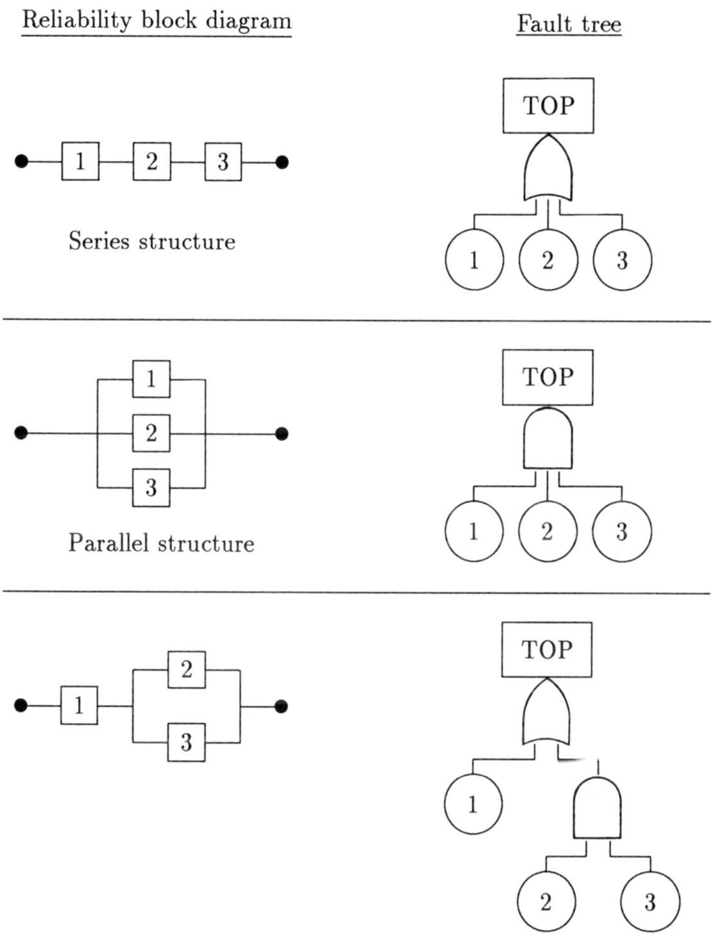

Figure 2.9: Connection between reliability block diagram and fault tree

2.3.3 Definition of undesirable events and framework conditions

It is important that the top event is defined in a clear and unique way. As an example, the top event "Fire" is usually too general and imprecise. The description of the top event should give answer to the following questions:

What : Describe which type of undesirable event that occurs, for example fire

Where: Describe where the undesirable event occurs, for example in the gas compression area

When : Describe when the undesirable event occurs, for example under normal operation

Thus a more precise definition of the top event would be:

"Fire in the gas compression area under normal operation"

To secure a successful fault tree analysis, the framework conditions of the analysis must be identified. With framework conditions we mean for example:

- Definition of the physical boundaries of the system: what parts of the system shall be included in the analysis and what parts shall not be considered?
- Definition of initial conditions: for example, which valves are open/closed and which pumps are functioning?
- Limitations with respect to external loads: which external loads shall be considered? By external loads we mean earthquake, lightning, war, etc.
- The level of detail: how detailed shall the analysis be? For example, would it be sufficient to identify the cause of failure "valve failure", or should we go further and divide into failures caused by valve house, actuator, etc.? Normally the level of detail will be determined from the level the necessary data are available, cf. Section 1.5.2.

2.3.4 Fault tree construction

The top event is the starting point when constructing the fault tree. Next we must identify the possible failures (events) that are the direct causes of the top event. These events are linked to the top event by a logical gate. Then we work successively down to the basic events on the component level. The analysis is deductive, and is carried out by repeatedly asking: "How can this happen?" or "What are the causes of this event?" The development of the causal sequence stops when we have reached the desired level of detail.

It is very important to "think locally", and to develop the fault tree carefully in a step-by-step approach. This will not only provide the best opportunity for a complete fault model, but also ensure that the tree is drawn up in a systematic way which a reader of the study can appreciate. Avoid gate-to-gate connections, i.e. connecting one gate directly to the next without providing an intermediate event in between. If transfer symbols are used to transfer between different pages of the fault tree, then the event description should be repeated on the following page. This will all help the reader to better understand (and verify) that the fault tree is correct.

Some common mistakes in fault tree construction are:

- Over-rapid development of one branch of a tree without proceeding down level by level systematically (tendency to want to reach basic events too rapidly and not to use broad sub-event descriptions)
- Omission of an important failure mechanism from lack of skill or a false assumption of negligible contribution
- Inappropriate balance between component failures and human errors
- Failure to recognize dependence of events

Example 2.4
We consider the system described in Example 2.3. As a first task we would like to construct a fault tree for the system with the top event "Over-filling of the tank" and basic events corresponding to failures of V1, V2, V3, LSH and LSHH.

Figure 2.10 shows a fault tree for this top event, with associated reliability block diagram. Note that we disregard the possibility of failures of the transfer of signals from LSH to V1 and from LSHH to V2 and V3.

Figure 2.11 shows a fault tree for the top event "Undesired stoppage of the fluid supply" with its associated reliability block diagram.

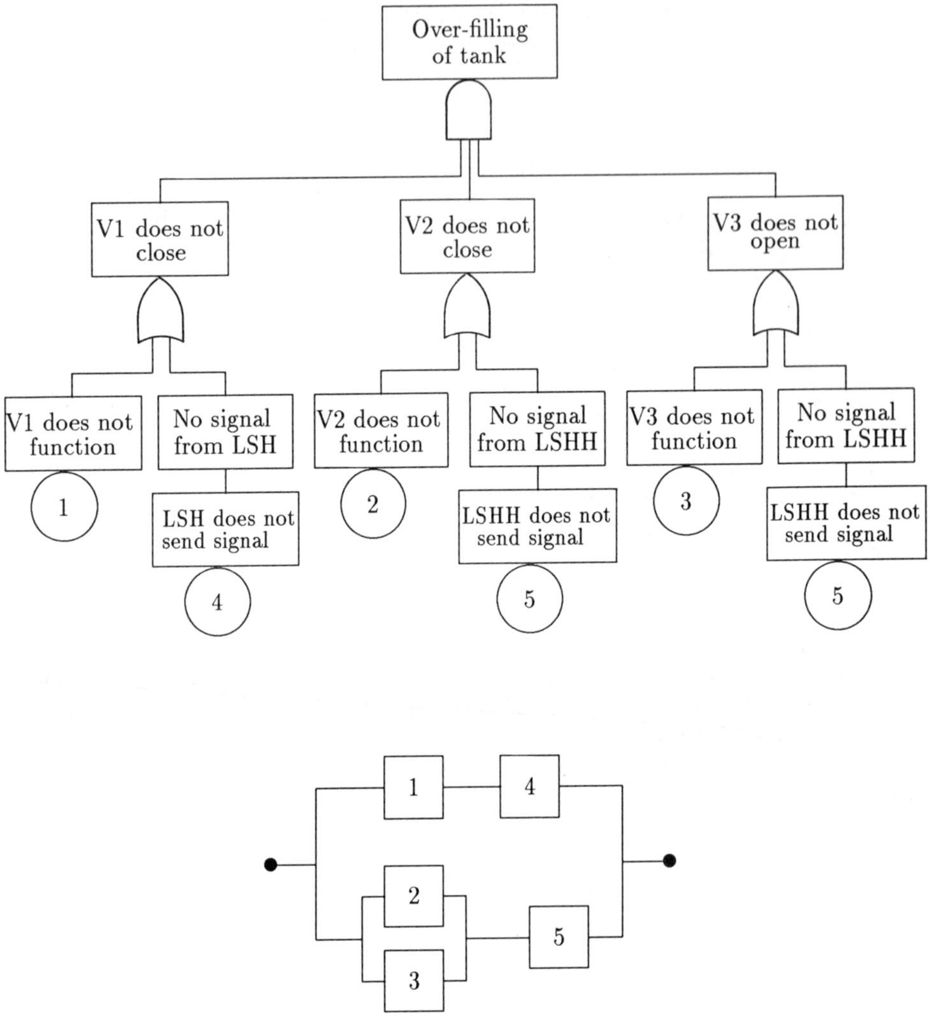

Figure 2.10: Fault tree for the top event "Over-filling of the tank", with its associated reliability diagram

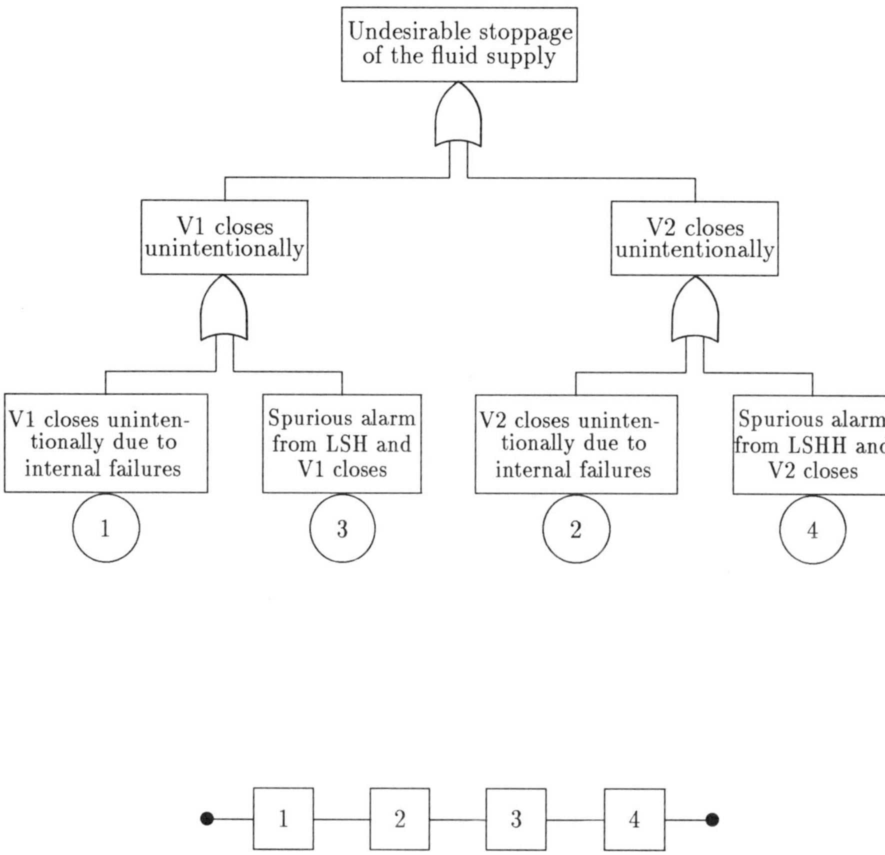

Figure 2.11: Fault tree for the top event "Undesired stoppage of the fluid supply", with its associated reliability block diagram

2.3.5 Minimal cut sets

A fault tree gives valuable information about which failure combinations can result in an undesirable event. Such a failure combination is called a cut set:

> A cut set in a fault tree is a set of basic events the occurrence of which ensures that the top event occurs. A cut set is minimal if it cannot be reduced and still ensures the occurrence of the top event.

For the corresponding reliability block diagram this definition is equivalent to saying:

> A cut set is a combination of components the failures of which ensure that the system fails. A cut set is minimal if it cannot be reduced without losing its status as a cut set.

For simple fault trees the minimal cut sets can be determined directly from the fault tree or from the associated reliability block diagram. In most cases it would be most convenient to use the reliability block diagram. For more complex fault trees there is a need for an algorithm. The most well-known computer based algorithm is MOCUS [64]. For other algorithms refer to [64,88,123].

Example 2.4 continued
The minimal cut sets for the tree in Figure 2.10 are determined directly from the fault tree or its associated reliability diagram:

$$\{1,5\},\{4,5\},\{1,2,3\},\{2,3,4\}$$

For the tree in Figure 2.11 we see that the minimal cut sets are:

$$\{1\},\{2\},\{3\},\{4\}$$

2.3.6 Qualitative analysis of the fault tree

Once the fault tree is constructed, the structure of the tree can be examined qualitatively to understand the failure mechanisms. This information is valuable as it provides a powerful insight into the identified possible modes of failure (i.e., all the combinations of events that lead to the top event).

For simple trees consisting of only a few gates, qualitative examination is possible by inspection. It is frequently found that simply going through the exercise of constructing the fault tree and reviewing its structure provides a tremendous benefit in terms of realising the failure potential of a system.

However, for larger fault trees, a more formal approach is needed. The principal means for fault tree evaluation is the complete set of minimal cut sets. System failure occurs when all the events in at least one minimal cut set occurs. The system can therefore be viewed as a series structure of the minimal cut parallel structures, as shown in Figures 2.12 and

2.13. This representation is particularly important when a quantitative analysis is to be performed, see the following section.

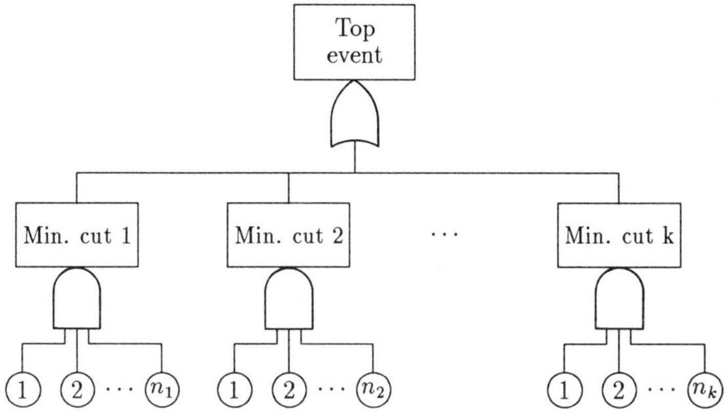

Figure 2.12: The minimal cut representation of the fault tree

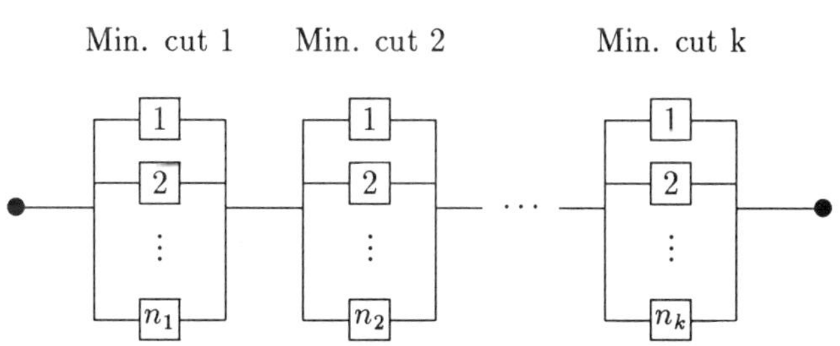

Figure 2.13: The minimal cut representation of the reliability block diagram

The number of events in a cut set is called the order of the cut set. The minimal cut sets are ranked according to their order. It is argued that single event cut sets (single jeopardy), are highly undesirable as only one failure can lead to the top event, two event cut sets (double jeopardy) are better, etc. Further ranking based on human error, and active and passive equipment failure is also common. The qualitative approach is potentially misleading. It may be that larger cut sets have a higher failure probability than smaller ones; this requires quantitative evaluation.

Common-cause failures are due to a single event affecting multiple events in the fault tree. This might be a power failure error miscalibrating all sensors. At an obvious level, if power failure appears in two arms of a fault tree joined by an And gate, and a "gate-by-gate" method is followed, then the final result may be evaluated incorrectly. Less obviously, elements not included in the FTA may lead to common-cause failure: common manufacturer, common location, etc., cf. Section 3.6.

2.3.7 Quantitative analysis of the fault tree

If we can estimate the probabilities of the basic events of the fault tree, then we can perform a quantitative analysis. Usually we would like to calculate:

- The probability that the top event will occur
- The reliability importance (criticality) of the basic events (components) of the tree.

Chapter 3 presents theory for the execution of the quantitative analysis. In this chapter we shall limit ourselves to indicate a useful approximation method for calculating the probability of the top event, and to briefly discuss how we can identify the most critical components.

Approximate method for calculating the probability of the top event

If the probability of the top event is small and the basic events are stochastically independent, then the probability of the top event can be approximated by summing the probabilities of the minimal cut set failures; for a specific minimal cut set this probability is found by multiplying together the probabilities of the basic events that constitute the cut set. The basic events are stochastically independent if the probability that a basic event will occur is not dependent on whether one or more of the other basic events have occurred.

Using this method we disregard the possibility that two or more minimal cut sets will be in the failure state at the same time.

Example 2.4 continued.

Again we look at the fault tree in Figure 2.10. Assume that the probabilities of the basic events 1,2,...,5 occurring, are as given in the FMEA, i.e. 2%, 2%, 2%, 1% and 1%. Under the assumption of independent basic events, we find that the probability that the top event "Over-filling of the tank" will occur is approximately equal to:

$$0.02 \times 0.01 + 0.01 \times 0.01 + 0.02 \times 0.02 \times 0.02 + 0.01 \times 0.02 \times 0.02$$
$$= 0.03 \times 10^{-2} = 0.03\%$$

This means that if the liquid level increases and reaches a high level about 25 times a year, then the probability of over-filling the tank during a one year period would be approximately 0.75% (= 25 × 0.03%). The percentage 0.75 can be viewed as a risk measure related to the activity. We are allowed to sum the unreliabilities for the 25 cases because the probability that the top event will occur two or more times during one year is negligible compared to 0.75%, cf. Appendix A, Formulae (A.3) and (A.4).

In Chapter 3 we will show that the error introduced by using this approximation method for calculating the probability of the top event, is negligible in this case.

We see from the above calculations of the probability of the top event that component 5 (LSHH) is the most important component from a reliability point of view, in the sense that the probability of the top event would be most reduced by an improvement in the reliability of this component.

2.3.8 Resources

A fault tree analysis can be performed by one analyst. Normally, it would, however, be better if the analysis were undertaken by a group of 2-4 persons. The group must comprise at least one experienced fault tree analyst. The other can be technical personnel familiar with the system and its operating conditions. The FTA work will depend on the complexity of the system and the top events to be analysed. An FTA of a minor process plant may take 4-10 days to carry out. For larger systems, the analysis will take weeks. The analysis work requires:

- Thorough knowledge of the system
- Knowledge of the failure modes of the equipment of the system, and their effect on the system. Such information may for example be available from an FMEA.
- If a quantitative analysis is to be performed, then relevant reliability data must be available.
- An efficient computer program for construction and analysis of the fault tree will be useful. There exists a number of such programs, see e.g. [64,88]. In Norway the most common fault tree program is CARA [123].

2.3.9 Discussion

A fault tree or the associated reliability block diagram provides a clear and well arranged picture of the combinations of equipment failures and other events that can lead to a specific undesirable event in the system. The fault tree is easily understood by persons with no knowledge of the technique in advance. The fault tree analysis is well documented and simple to use.

One of the advantages of using the FTA is that the persons undertaking the analysis are forced to understand the system. Many weak points in the system are revealed and already corrected at the construction phase of the tree.

A fault tree gives a static "picture" of the failure combinations that can cause the top event to occur. The fault tree analysis method is not suitable for analysing systems with dynamic properties. For example, it is difficult to analyse a stand-by system using fault tree analysis. Periodical testing and maintenance are also difficult to take into account in a fault tree analysis. Another problem is treatment of common-mode failures.

2.4 Cause consequence analysis/Event tree analysis

2.4.1 Introduction

Cause Consequence Analysis (CCA) was invented by RISØ Laboratories in Denmark to be used in risk analysis of nuclear power stations. CCA is, however, a general method that can be used for widely different technological systems. The method is particularly suitable for analysing systems and situations where the time and sequence of events is important.

CCA includes Event Tree Analysis (ETA) as a special case. Event trees are discussed in Sections 2.4.6 and 2.4.8.

There exist different versions of CCA. The presentation in this section is basically the same as the one we find in e.g. [53,64,126].

CCA combines cause analysis (described by fault trees) and consequence analysis (described by event trees), and is therefore both a deductive and an inductive analysis method. The starting point of the analysis is an initiating event (undesirable event), that may lead to a hazard. Examples of initiating events are a pipe leakage, a vessel rupture or an explosion. The "consequence tracing" part of the CCA involves taking the initiating event, and following the resulting chains of events through the system. Consideration is given to the various barriers and safety functions that have been constructed to prevent or limit the consequences of the initiating event.

The purpose of CCA is to identify chains of events that can result in undesirable consequences. If we know the probabilities of the various events that are included in the diagram, we can calculate the probability for the various chains of events. Then system reliability and risk can be calculated.

A CCA includes several stages:

1. Definition of initiating event and framework conditions

2. Construction of the CCA diagram

3. Qualitative analysis of the CCA diagram

4. Quantitative analysis of the CCA diagram

These stages will be discussed in the following, after a brief presentation of a CCA diagram.

2.4.2 CCA diagram

As already mentioned, CCA is a marriage between fault tree analysis and consequence (event tree) analysis. The symbols that are used in FTA is described in Section 2.3. The symbols that are special for CCA are presented in Figure 2.14. The most important symbol is the *branching operator*. It describes a point where the chain of events may branch, taking two paths. The outcome is "yes" if a specific condition is satisfied, and "no" otherwise. The condition is analysed using a fault tree. It is therefore suitable to define the condition "negatively", for example "the fire alarm does not function", such that the rectangle that includes the condition also represents the top event of the associated fault tree.

The event descriptions are small "flags" in the diagrams to describe an event or a state that exists at a particular point in time. *The delay operator* indicates the time delay necessary to ensure that the input event will result in the output event. *The consequence operator* represents the end points of the CCA diagrams. To which level the consequences should be developed and in what detail they are to be described, depends on the purpose of the analysis. Often it is found suitable to modularize the problem, such that a consequence description in one diagram constitutes the initiating event in a subsequent diagram. Such an approach may be appropriate when analysing large systems, or if the sequence of events are very comprehensive. An example of a CCA diagram is shown in Figure 2.15. The event descriptions are omitted in the figure.

2.4.3 Definition of initiating events and framework conditions

The definition of initiating events is an important part of the analysis. An initiating event should give rise to several chains of events. If the chosen initiating event results in only one undesirable consequence, it would normally be better to perform a fault tree analysis to determine the causes of the undesirable consequence.

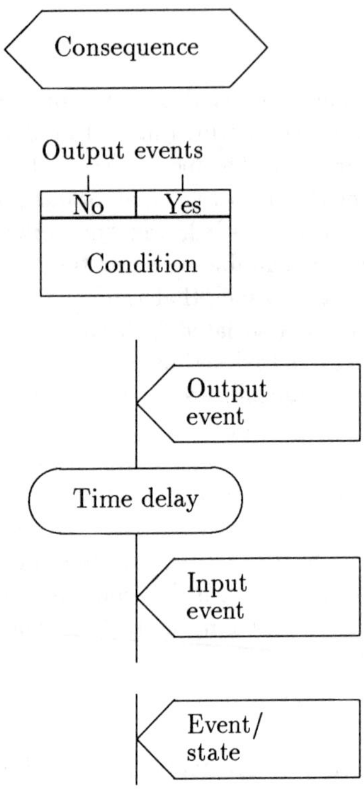

CONSEQUENCE DESCRIPTION

Description of a consequence, end point in the diagram.

BRANCHING OPERATOR

The output is "yes" if the condition is satisfied, "no" otherwise.

THE DELAY OPERATOR

Indicates the time delay that is necessary to ensure that the input event will result in the output event.

EVENT DESCRIPTION

The event/state that exists at a specific point in time is described.

Figure 2.14: CCA symbols

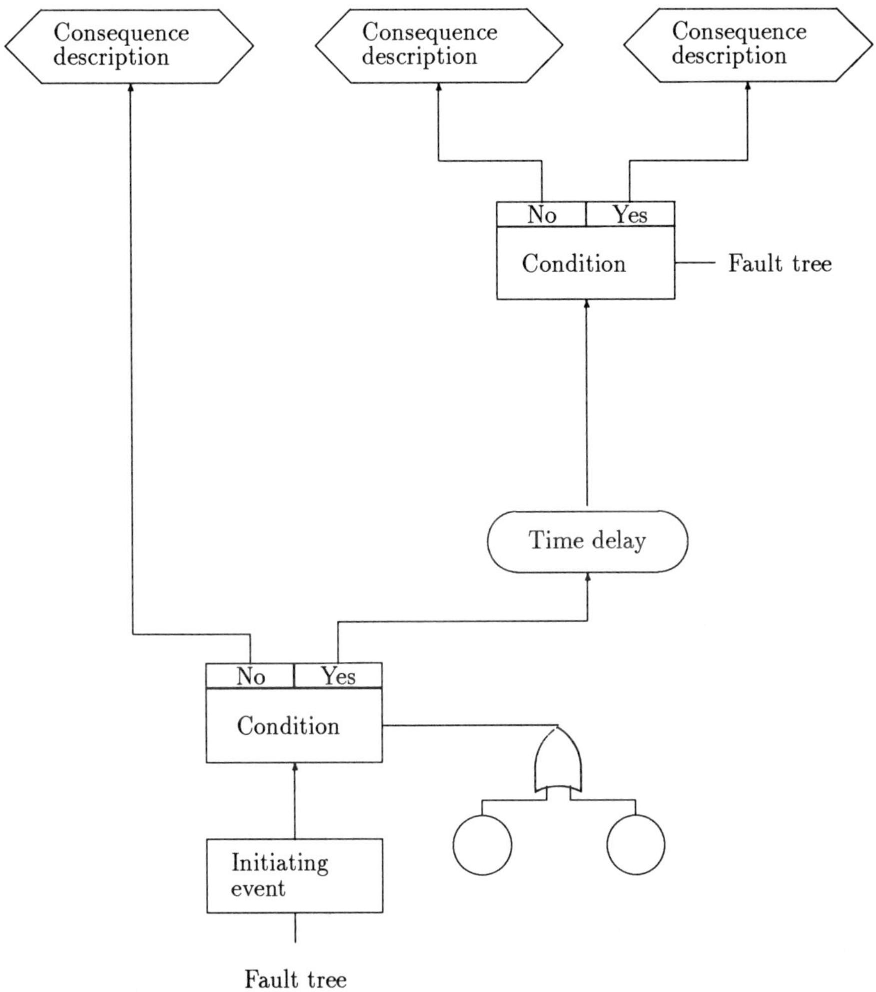

Figure 2.15: Example of a CCA diagram

The initiating event is usually identified and considered to be a possible undesirable event already in the planning phase of the system. The system therefore comprises barriers, safety systems or procedures which are to prevent or limit the consequences of the initiating event.

In general the initiating event is identified using experience, historical records, and hazard identification methods such as HAZOP.

For the system in Example 2.3, a natural choice of the initiating event would be "the liquid level increases".

For the framework conditions refer to Section 2.3.3.

2.4.4 Construction of the CCA diagram

The CCA diagram shows the development of the chains of events in chronological order. The diagram starts with the initiating event and proceeds via the various safety functions and the hazard promoting factors. Safety functions may be of many types:

- Automatic safety systems, for example automatic fire fighting systems
- Alarms to alert operators
- Operator action to mitigate incident
- Barriers or containment to limit effect of initiating event

Hazard promoting factors are more varied and would include for example:

- Ignition or no ignition of release
- Explosion or flash fire
- Liquid spill contained in dike or not
- Daytime or night-time
- Meteorological condition

To be able to perform the quantitative analysis of a CCA diagram in a simple way, common basic events in the various fault trees of the diagram should be avoided. A natural choice of branching operator would, however, in many cases lead to such a diagram, cf. Example 2.5 below. When performing the quantitative analysis, it would then be best to modify the diagram so that none of the fault trees have common basic events. To ensure this we must define each of the common basic events as branching conditions.

The construction of the diagram is shown by an example.

Example 2.5
We consider the system described in Example 2.3.

We would like to perform a CCA for this system, and define "the liquid level increases" as the initiating event.

There are many ways of constructing the CCA diagram for this system. The Figures 2.16 and 2.17 present two solutions. The relevant fault trees are shown in Figure 2.10. For the diagram in Figure 2.16 we have defined two barriers, valve 1 and the valves 2 and 3. Note that for the first diagram there is only one undesirable consequence, namely "over-filling of the tank".

It would also be natural to choose failures of the three "safety barriers" valve 1, valve 2, and valve 3 as branching conditions. This would, however, have led to a common basic event for the two last mentioned barriers, "no signal from LSHH", and as a result of this, the quantitative analysis would have been more complicated.

The CCA for this example is relatively rough. For example we could have analysed further how much liquid could be released by over-filling and the possible consequences of this.

2.4.5 Qualitative analysis of the CCA diagram

The CCA diagram shows the different chains of events that can result from an initiating event. Each chain of events ends in a consequence description, and the whole spectrum, from the most severe consequence, to the least severe consequence where the situation has become stable again, can be described in the diagram. The diagram provides a good basis for evaluating the safety functions of the system.

A CCA diagram can be used to study whether there exist failure sources that affect more than one safety function at the same time, or cause the initiating event to occur and at the same time induce failure in one or more of the safety functions. This is done by comparing the minimal cut sets of the fault trees in the diagram. A minimal cut set that causes failure in several safety functions (possibly also the initiating event) may be much more critical with respect to reliability and safety than a minimal cut set that only affects one top event.

2.4.6 Quantitative analysis of the CCA diagram

If we have available estimates for the probabilities of the various events in the diagram, then we can perform a quantitative analysis. We will use an example to demonstrate how this is done.

Example 2.5 continued
First we look at the diagram in Figure 2.16. The diagram can be simplified as shown in Figure 2.18, so that the structure of the diagram becomes more clear. The diagram in Figure 2.18 is called an *event tree*. The interpretation of the tree is the same as for the CCA diagram.

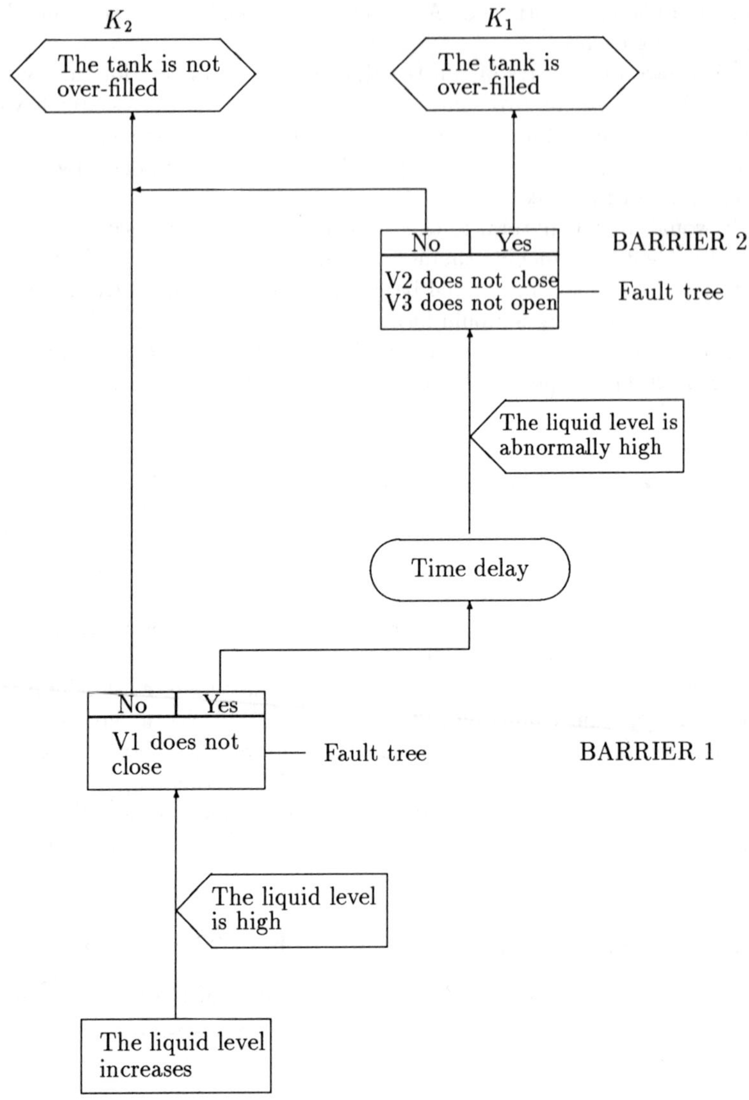

Figure 2.16: CCA diagram for the system described in Example 2.3

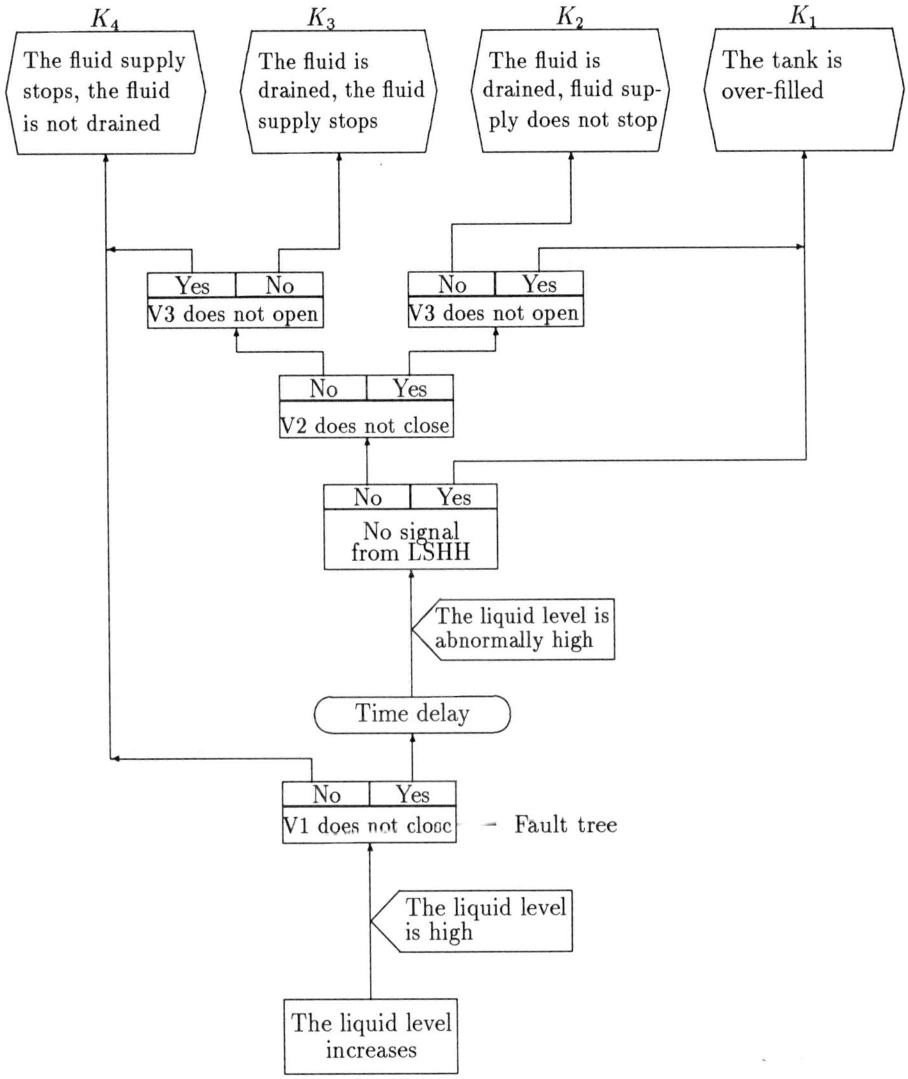

Figure 2.17: Alternative CCA diagram for the system described in Example 2.3

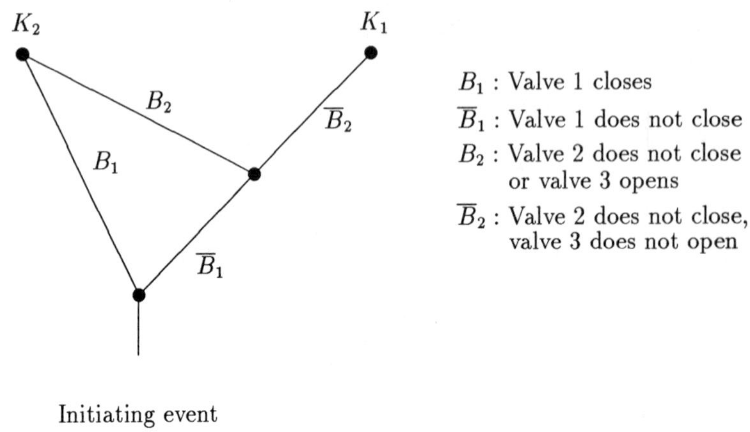

Figure 2.18: CCA diagram in Figure 2.16 transformed to an event tree

We would like to calculate the probability of the various consequences in the diagram, given the initiating event "the liquid level increases", based on the same reliability data and assumptions made earlier, see Sections 2.2 and 2.3. To be able to carry out these calculations, we must know the probabilities for the various events in the tree. These can be found by calculating the probability of the top event in the two trees which are included in the CCA diagram. We find:

1. The probability that valve 1 does not close, is approximately equal to $2\% + 1\% = 3\%$ (minimal cut sets {1} and {4}),

2. The probability that valve 2 does not close and valve 3 does not open, is approximately equal to $0.01 + 0.02 \times 0.02 = 1\%$ (minimal cut sets {5} and {2,3})

In Figure 2.19 the calculated probabilities have been inserted in the event tree.

Now we can calculate the probability that a specific chain of events will occur by multiplying the probabilities for the events in the chain. The probability of a specific consequence can then be calculated by summing the calculated probabilities for those chains of events that lead to the specific consequence. These rules for event trees follow directly from basic probability rules, see Appendix A for details.

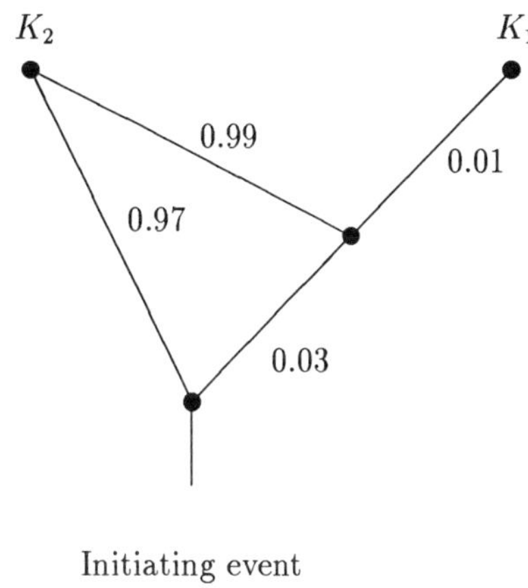

Figure 2.19: The event tree in Figure 2.18 with inserted probabilities

Simple calculations give:

1. The probability of $K_1 = 0.03 \times 0.01 = 0.03\%$
2. The probability of $K_2 = 0.97 + 0.03 \times 0.99 = 99.97\%$ ($= 100\% - 0.03\%$)

The calculations agree with the results established in the fault tree analysis. The results means that if the frequency of the initiating event is for example 25 per year, as assumed under the FTA, the frequency of "over-filling of the tank" equals 0.75×10^{-2} per year, i.e. once per 133 years on average.

Similarly, we calculate the probabilities for the various consequences for the diagram in Figure 2.17. The associated event tree is shown in Figure 2.20, with the probabilities of the various branches of the tree inserted. Note for example that the probability that V2 will not close when it is given that a signal is received from LSHH, equals 2%. This probability corresponds to the probability of the basic event 2 in the fault tree in Figure 2.10. We find:

$$
\begin{aligned}
K_1: &\quad 0.03 \times (0.01 + 0.99 \times 0.02 \times 0.02) &= &\quad 0.03\% \\
K_2: &\quad 0.03 \times 0.99 \times 0.02 \times 0.98 &= &\quad 0.06\% \\
K_3: &\quad 0.03 \times 0.99 \times 0.98 \times 0.98 &= &\quad 2.85\% \\
K_4: &\quad 0.97 + 0.03 \times 0.99 \times 0.98 \times 0.02 &= &\quad 97.06\%
\end{aligned}
$$

2.4.7 Resources

Refer to the corresponding section for fault tree analysis, Section 2.3. With obvious changes, the resource section for the fault tree analysis also applies to the CCA. The computer program CARA [123] also includes a CCA module.

2.4.8 Discussion

A CCA diagram (or an event tree) is a graphical representation of a logic model which identifies and quantifies possible outcomes following an initiating event. The event tree provides systematic coverage of the time sequence of event propagation, either through a series of protective system actions, normal plant functions and operator interventions, or where loss of containment has occurred, through the range of possible consequences. Consequences can be direct (fires, explosions, etc.) or indirect (domino incidents on adjacent plants).

CCA diagram (event tree) structure is the same as that used in decision tree analysis. Each event following the initiating event is conditional on the occurrence of its precursor event. Outcomes of each precursor event are most often binary (SUCCESS or FAILURE, YES or NO), but can also include multiple outcomes (for example, 100%, 20% or 0% in the operation of a control valve).

CCA (ETA) have found widespread applications in risk assessments for both the nuclear and process industries. Two distinct applications can be identified. The "pre-incident" application examines the protective or other systems in place which would prevent incident precursors developing into an actual incident. The CCA (ETA) of such systems is often sufficient in itself for the purposes of estimating the safety of the system. The "post-incident" application is used to work out the many possible outcomes for flammable and toxic releases. The CCA (ETA) is rarely sufficient of itself in this application; it contributes to the assessment of the frequency of incidents used in the risk calculation. Nuclear studies, such as the Reactor Safety Study [135], have highlighted the pre-incident use of event trees to demonstrate the effectiveness of successive levels of protective systems. Chemical industry risk assessments have highlighted the post-incident uses, although protective systems are also investigated this way. The technique of human reliability analysis makes extensive use of CCA (ETA).

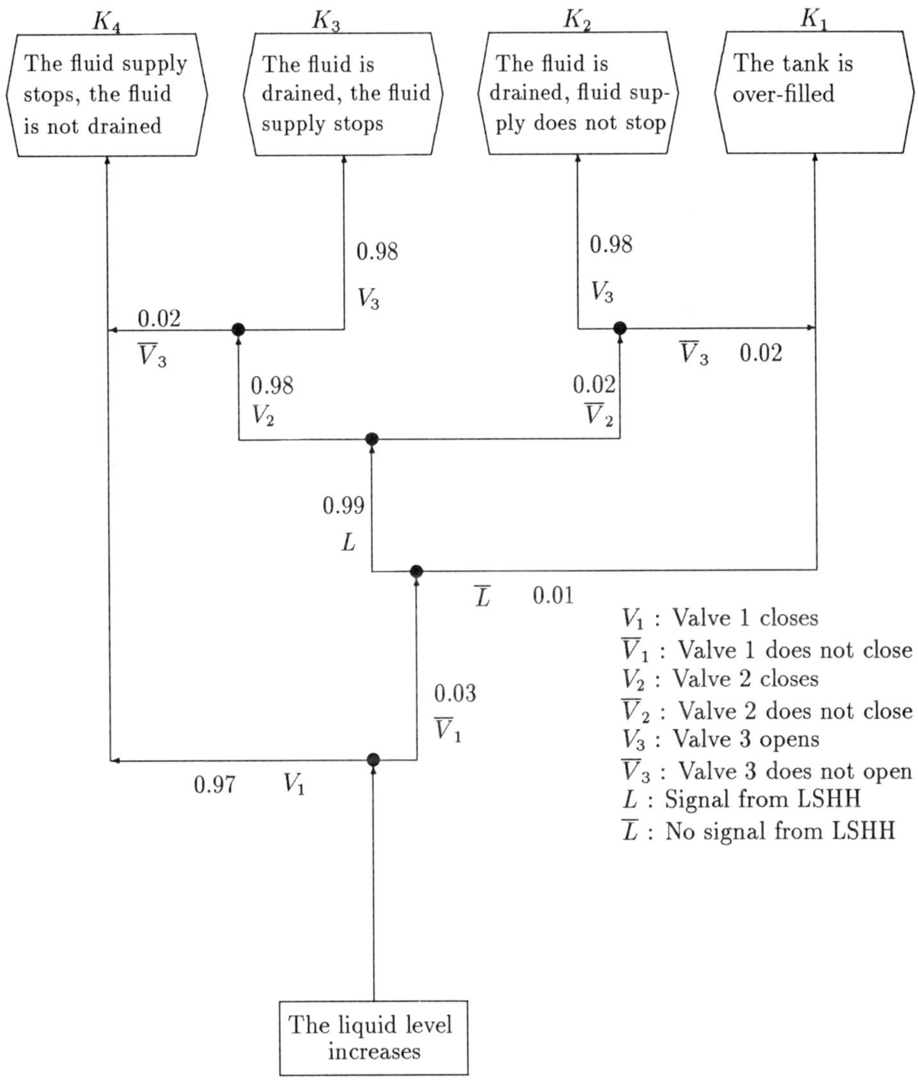

Figure 2.20: The CCA diagram in Figure 2.17 transformed to an event tree with inserted branching probabilities

2.5 Other reliability and risk analysis methods

2.5.1 Coarse risk analysis/Preliminary risk analysis

Every risk analysis starts with a hazard identification. To ensure that the identification provides a good basis for the further work with the risk analysis, it must be systematic. The Coarse risk analysis or the Preliminary risk analysis is a common method for performing such a systematic identification.

A coarse risk analysis gives an overview of hazardous conditions related to an activity. The results from a coarse risk analysis can provide the basis for determining which categories of hazard should be looked into more closely and which analysis methods are most suitable. The coarse risk analysis requires relatively small resources, but it may still give valuable information about the risk related to the activity. The coarse risk analysis is preferably undertaken at an early phase of a project. The aim is to reveal potential hazards early in the project development so that they can be eliminated, reduced or controlled early in the project development. But a coarse risk analysis can also be used in other phases. The method has shown to be very useful in relation to working environment problems, cf. Example 2.7 below. In the following we give a short description of the method, based on [2,61,64,126,138].

The work sequence in a coarse risk analysis is to first identify the possible undesirable events, and then analysis each of these separately. For each identified undesirable event we study possible causes and consequences. We discuss possible improvements, or preventive measures. In the analysis work the following factors should be considered if they are relevant:

- Hazardous equipment and hazardous materials, for example explosive materials, poisonous materials, high pressure systems
- Safety problems related to connections between plant equipment and materials, for example reaction between different chemical substances
- External factors which can affect the equipment or materials, for example earthquake, vibrations, lightning, moisture
- Operation, testing, maintenance, shutdown procedures, related to, for example, human errors, placing of equipment, access to equipment, personal safety equipment
- Safety equipment, such as redundant equipment, fire fighting equipment, protection equipment

Tools that can be used to identify the undesirable events are for example:

- Relevant statistics
- Information about machines and equipment

- Experience from similar analyses

In some cases we would like to evaluate the identified hazards with respect to probability and effect (consequence). To do this we can make use of various types of probability consequence spectra, for example the one presented in Figure 2.4 or in Example 2.6 below. By evaluating probability and consequence together, the identified hazards can be ranked according to risk.

Example 2.6 [2,126]
We consider a process plant where H_2S is to be used in the process. The analyst has otherwise little information about the process plant. He/she identifies poisonous release of H_2S as a possible hazard. The release can be due to:

- Rupture or leakage of a vessel
- Excess of H_2S in the process
- Rupture/leakage in the H_2S supply
- Leakage when filling H_2S

The effects of these causes are then to be determined, and preventive measures are to be proposed. Figure 2.21 shows a possible result of this analysis. The form is one of many that is used for coarse risk analysis.

Example 2.7 [138]
The working environment committee at the Packing factory ltd. has found that in the bag department, which has about 90 employees, the number of injuries is too high. The committee therefore decides to implement some injury preventive measures to reduce the injury rate in the department. It is, however, not clear where such measures should be directed and which measures would most effectively prevent injuries. Opinions in the working environment committee differ widely.

In the bag department production is a series production of large, multiple-ply paper bags. Each production line comprises several machines. The raw material is paper rolls. The main operator task in the department is to monitor and adjust the machines. Some operators deal with manual handling of products. It is the operators who are operating the machines that are most exposed to injuries.

The working environment committee instructs the safety delegate of the company to work out a basis for decisions on preventing measures. The safety delegate is familiar with risk analyses and he uses such an analysis to establish the desired decision basis.

Hazard	Cause	Main effects	Preventing/improving action
Poisonous release	H_2S causes cylinder leakage	Large releases may cause severe accidents	-Alert system -Minimize storage at the site -Develop procedures for cylinder inspection
	H_2S not consumed in the process	As above	-Construct collection system for excess of H_2S -Control system to identify excess of H_2S, and stop the process

Figure 2.21: Simple coarse risk analysis (example of a completed form).

By means of the risk analysis the safety delegate will identify possible injuries that might occur in the department, where they might occur, possible causes, and the severity of the injuries. Figure 2.22 presents a summary of the analysed hazards. With respect to probability/frequency the classification is the same as the one used in Figure 2.4; with respect to consequence categories the following categories have been used:

I Does not result in injuries

II Minor injuries

III Major injuries

IV Death or total disability

V Death or total disability for several persons

The starting point for the hazard identification has been injury reports for the whole department for the last 9 years. The collection of near misses has also been carried out, and in addition the system description has been studied to identify additional hazards. For those hazards which are based on the injury reports, the classification is based on this statistics. For example, two injuries caused by crushing in the cutting machinery are registered. This gives a classification "likely". Judgement has been used for the hazards that are not based on the injury statistics.

In Figure 2.23 the hazards have been placed in a consequence probability diagram. The consequence categories are marked along the horizontal axis with consequence increasing to the right. Similarly, the frequency/probability is marked along the vertical axis, with frequency increasing downwards. The following conclusions are drawn:

- Highest risk: Hazard number 14, which equals "paper roll falls from tackle"
- Other contributors to high risk are: events that include crushing and catching in the machinery

The events with the lowest risk are numbers 9 and 15. Thus in general for this type of consequence probability diagram, the events with the highest risk are those in the bottom right corner, whereas the events with the lowest risk are placed in the upper left corner. We should, however, be very careful when drawing conclusions from the matrix since it is based on a very rough classification.

Note that hazard 15 has low risk in relation to personnel. If we focus on material assets or economic values, fires would contribute much more to risk.

Based on the risk analysis the safety delegate can now prioritize measures to prevent accidents.

2.5.2 Hazard and operability studies

One of the most important tasks within risk analysis is to identify weaknesses and hazards in a plant. Such an identification is, however, not always simple to carry out in practice, particularly nowadays with advanced technology and complicated plants. Often it is not sufficient to have engineering competence in one field. In complicated systems, events that alone can be harmless, can together give effects that is far above the design specifications. In addition it will be difficult to foresee all the possible hazards that might occur.

One of the most effective methods to identify potential hazards in a process plant is Hazard and operability (HAZOP) studies. The method is based on principles developed in [46], but has been further developed for applications by Imperial Chemical Industries [85,86]. For a good introduction to HAZOP, see [42,89]. In this section we shall give a brief presentation of the method.

A HAZOP study is a systematical analysis of how deviation from the design specifications in a system can arise, and an analysis of the risk potential of these deviations. Based on a set of *guidewords*, scenarios that may result in a hazard or an operational problem, is identified. The following guidewords are commonly used: NO/NOT, MORE OF/LESS OF, AS WELL AS, PART OF, REVERSE, AND OTHER THAN. The guidewords are related to process conditions, activities, materials, time and place. For example, when analysing a pipe from one unit to another in a process plant, we define the deviation "no throughput" based on the guideword NO/NOT, and the deviation "higher pressure than the design pressure" based on the guideword MORE OF.

No	Event	Cause	Consequence	Probability	Consequence categ.
1	Crushing in cutting machinery	Hand in running machine, e.g. due to inattention	Finger/hand injury	4	III
2	Crushing in pulling machinery	As 1	Finger/hand injury	4	III
3	Crushing at intermediate station	As 1	Finger/hand injury	3	III
4	Damage at guillotine	As 1	Finger/hand injury	3	III
5	Being caught at the glue station	Inattention	Hand/arm injury	3	III
6	Being caught in folding station	Missing cover + inattention	Significant body injuries	3	IV
7	Crushing between rollers	As 1	Finger/hand injury	4	III
8	Damage from machinery splinter	Rupture during operation	Major wounds	4	II
9	Knocks from edge, machine part, etc.	Inattention	Wounds, cuts	4	I
10	Hair or clothes being caught between rollers	Inattention	Significant body injuries	2	IV
11	Bodily damage from unobserved machinery start-up	Technical failure, noise, inattention	Significant body injuries	2	IV
12	Crushing when lifting roll	Inattention	Finger/hand injuries	4	III
13	Damage due to roll coming loose	Rupture of spindle, carelessness	Severe injuries	2	V
14	Damage due to dropping roll	Failure of tackle, inadequate fastening	Severe injuries	4	V
15	Paper fire	Ignition of paper dust + oil, weld sparks, smoking	Loss of casings/ sacks, destruction of machines	3	I

Figure 2.22: Summary of identified hazards - Example 2.7

	Consequence category				
Frequency-category	I	II	III	IV	V
1					
2				10,11	13
3	15		3,4,5	6	
4	9	8	1,2,7,12		14
5					

Figure 2.23: Probability consequence diagram - Example 2.7

Then causes and consequences of the deviations are studied. This is done by asking questions, for example for the first mentioned deviation in the pipe example above:

- What must happen to give rise to the deviation "no throughput" (cause)?
- Is such an event possible (relevance/probability)?
- What are the consequences of no throughput (consequence)?

To assist in formulating meaningful questions based on the guidewords, special sheets have been developed.

The principle that is used in a HAZOP study can be illustrated in the following way:

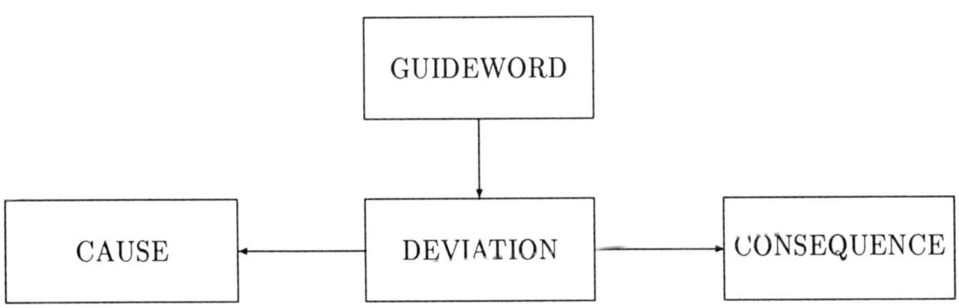

In a HAZOP study separate worksheets are used to document deviations, cause, consequence, and recommendations/decisions. These worksheets are a type of FMEA forms.

A HAZOP study is undertaken by a group of personnel led by a HAZOP leader. The leader should be experienced in using the technique, but does not necessarily need to have thorough knowledge of the actual process. The group comprises persons with detailed knowledge of the system to be analysed. Typically the group will consist of 5-6 persons in addition to the HAZOP leader.

Through a HAZOP study critical aspects of the design can be identified, which requires further analysis. Detailed, quantitative reliability and risk analyses will often be generated in this way.

A HAZOP study of a planned plant will as an FMEA normally be most useful if the analysis is undertaken after the Process and Instrumentation diagrams have been worked out. It is at this point that sufficient information about the way the plant is to be operated is available.

The HAZOP technique was originally developed for chemical process plants, but the method can also be applied to other installations and systems. In [43] is described how the HAZOP technique can be used for drilling activities.

A HAZOP study is demanding of time and resources. Nevertheless, it has been widely used in connection with the review of the design of process plants to achieve a safer, more effective and reliable plant.

2.5.3 Other analysis methods

In this section we give a brief presentation of some other methods for safety analysis:

- Job safety analysis
- Management Oversight and Risk Tree (MORT)
- Safety Management and Organization Review Technique (SMORT)

For a more thorough discussion of these methods refer to [4,76,82,124,126]. In these references other similar analysis methods are also presented.

Job safety analysis

The purpose of the job safety analysis is to identify hazards related to the execution of a job, evaluate these hazards and possibly propose measures to eliminate or control them. The idea of the method is to divide the job into a number of sub-jobs or tasks, and then perform an analysis for each task. The division into tasks is illustrated by the following example:

Change of a car wheel

1. Turn on the hand brake
2. Take out the spare wheel from the luggage compartment

3. Test the spare wheel (air pressure)
4. Remove the wheel cover
5. Ensure that the jack fits and is stable
6. Jack up the car - but not so much that the wheels leave the ground
7. Loosen wheel nuts
8. Jack up the car further - but not more than necessary
9. Take off the wheel, and so on.

The identification of hazards includes a check of:

- What type of damage may occur, for example crushing.
- Are special problems or deviations likely to occur?
- Is the task difficult or uncomfortable to carry out?
- Are there alternative ways of performing the task?

The identified hazards can be summarized in a rough measure of risk, for example in the following way:

0 Insignificant risk

1 Acceptable risk, actions unnecessary

2 The risk should be reduced

3 The risk must be reduced, there is need for immediate action

When evaluating the risk and the need for actions/risk reducing measures, consideration should be given to for example:

- Violation of statutory requirements
- Violation of requirements set by the company
- High risk documented by means of accident statistics
- High energy concentration
- Unreasonable requirements with respect to attention and vigilance for the operator
- Low tolerance of human errors in the technical system
- The solution of the problem is known and available

Special sheets have been developed for job safety analysis. Such sheets will typically include the following main points:

- Description of the job
- Accident experience (statistics)
- Accident potential
- Requirements
- Job sequence (tasks)
- Risk evaluation
- Actions/measures

Often the sheets include a list of possible actions, which are to be considered. The actions may for example be related to:

- Improved equipment and tools
- Improved and safer working routines and methods; can the work be carried out in a different way?
- Elimination of the need for specific tasks
- Improved work instructions
- Improved education and training
- Planning of the handling of difficult deviation situations
- Improved safety systems
- Suitable protection equipment

MORT

The purpose of the MORT (Management Oversight Risk Tree) technique is to:

- Analyse/investigate accidents and events
- Evaluate safety programmes (safety programmes are presented in Chapter 6)

MORT focuses on the many factors that contribute to the occurrence of an undesirable event or accident. This is done by a thorough identification of the undesirable energy sources that exist in the system, together with an evaluation of the efficiency of the barriers that exist in the system. MORT is based on experience from a wide spectrum of fields: nuclear power, space industry and development of advanced military systems. MORT has its origins in the USA in the early 1970s.

The MORT analysis is carried out by means of a fault tree, where the top event is "Damage, destruction, other costs, lost production or reduced credibility of the enterprise in the eyes of the society". The tree gives an overview of the factors or basic events (causes) that may lead to this top event. The MORT tree includes more than 1500 possible basic events. These basic events have in turn input to about 100 generic events, which have been increasingly identified in the fields of accident prevention, administration and management. Furthermore there are many evaluation criteria/questions in assessing whether a basic event in the MORT tree is acceptable (satisfactory).

As an example we will look closer into these three levels for some events related to the heading "barriers"; "a","b" and "c" refer to "generic problems", "basic events" and "criteria", respectively:

a) Were there barriers to persons or objects?

  b1) Were no barriers present?

  b2) The barrier fails/failed. Did the barrier prevent the energy transfer as it was designed to do?

  b3) Not in use. Were barriers used?

    c1) Were barriers not provided?

    c2) Were the provided barriers correctly used?

The MORT tree comprises three main branches:

Assumed and accepted risk (R-factors). The R-factors are the risks that have been analysed and accepted.

Specific control factors (S-factors). This branch of the tree assumes that an undesirable event or an accident has occurred. Failures (omissions or faults) that were supposed to prevent the accident are analysed. A key question is: What happened? Examples of factors that are investigated are: Barriers, maintenance, inspections and working management. The above example is taken from this part of the tree.

Management factors (M-factors,). This part of the tree is developed to evaluate a safety programme. Also in this case omissions and faults are investigated. But now the key question is: Why did it happen? Examples of M-factors are: Strategy, information flow, goals and acceptance criteria, activities and plans, documentation, data collection and analysis, and communication.

SMORT

The purpose of the SMORT (Safety Management Organization Review Technique) method is to investigate the safety of industrial activities. SMORT can be used to perform detailed investigation of accidents and near misses. The method is also suitable for safety audits and planning of safety measures. SMORT is based on MORT. MORT is based on a comprehensive tree structure, whereas SMORT is structured by means of analysis levels with associated checklists. SMORT has been developed and tested in Norwegian and Swedish industry and thereby adapted to include Scandinavian traditions in company organization. SMORT is undertaken by a group of persons that are involved in the problem to be studied, and have thorough knowledge about the different parts of the organization.

SMORT is based on three concepts:

Deviation: When the production is not functioning as planned and when the work deviates from the usual practice. In other words, deviation means disagreement between real conditions and specified requirements.

Risk condition: Conditions at a place of work or within an activity which may lead to an accident. Risk conditions are not limited to deviations. They may be conditions related to the design and construction of the technical equipment, the way the work is executed, maintenance routines, wage systems, management, etc.

Accepted risk: Risk conditions in a decision process that have been identified, evaluated and deemed acceptable.

A SMORT analysis may consist of up to four levels:

1. **The sequence of events. Accident situation.** Introductory investigation and description of what happened, deviation and problems related to the working environment, technical equipment, procedures, etc., i.e. an extension and complement to the ordinary accident reporting.
   Action: Correct and eliminate deviations and problems at the place of work where the accident occurred.

2. **Daily operation.** Identification of risk conditions in the daily operation. The analysis is based on the results from level 1.
   Action: Improvement of the operation organization and production system.

3. **Planning and building.** Identification of how many risk conditions in the production (daily operation) are a result of weaknesses in the organization and the routines in the planning, building and start-up of the production system.
   Action: Improvement of organization and routines for the planning and building work.

4. **Line management and working environment.** Identification of risk conditions that can be related to the line management and the company's top management, including administration and management of the working environment activity in the enterprise.
Action: Improve the enterprise's overall organization, management functions and routines, with emphasis on safety management.

How many levels are included in the analysis depends on how complicated the system is, and how much time and resources are allocated to the investigation. For each level, a collection of check lists has been developed. As an example, we present below an extract from such a list for level 2. The check list for this level is divided into the following main points:

1. The personnel
2. Technical equipment
3. Maintenance/repair of technical equipment
4. Materials/products
5. Activities
6. Place of work
7. Social environment
8. Working management
9. Safety

The extract is taken from the check list for the third main point "Maintenance/repair of technical equipment".

3.1 Resources

- Personnel resources
- Access to and development of technical tools for the maintenance work

3.2 Inspection, condition-based control

- Plan, execution

3.3 Routines/procedures for maintenance and repairs

- Plan, execution
- Strategy

- Work distribution between the operating personnel and the maintenance personnel
- Priority
- Inspection of performed maintenance work
- Criteria for the repair and maintenance activities to be planned

3.4 Safety routines for maintenance

- Plan, execution
- Job safety analysis
- Inspections and preparation for maintenance
- Instructions to the maintenance personnel

3.5 Co-ordination between operation and maintenance

- Distribution and transfer of responsibility and authority

3.6 Documentation of maintenance

- Documentation and follow-up of maintenance requirements

Based on these checklists, relevant questions are worked out. Examples of such questions in relation to point 3.4 above are:

- Is safety evaluation of the maintenance activity performed, i.e. is hazard identification of the working operations carried out?
- Does the maintenance require special work permits. Are the routines for work permits satisfactory, in view of the hazards that might occur?

The SMORT analysis includes data collection based on the checklists and their associated questions, in addition to evaluation of the results. The data collection is performed using interviews, studies of documents, investigations, etc. The need for actions/measures must be considered.

2.5.4 Special methods for analysing human reliability

By human reliability is meant the ability that a person has to perform a function. This ability might for example be expressed by the probability of performing the function correctly and within the time that the system requires.

A complete analysis of reliability/risk must take into account human reliability. This is done for example in a fault tree analysis by letting basic events represent human errors. In addition, human errors can cause basic events to occur. This is reflected in the probabilities of the basic events. Experience shows that many accidents and undesirable events can be traced back to human errors. It can, however, be difficult to quantify human reliability. This is due to the fact that human behaviour can be on different levels:

Skill-based behaviour. This is automatic behavioural sequences which we perform without conscious control. Skill-based behaviour is established through training and experience. Human reliability can be reasonably well estimated if experience data from the same working operation, possibly simulated under the same conditions, are available. The performance level can, however, vary considerably under stress or other unfavourable conditions.

Rule-based behaviour. The behaviour is controlled by rules that have been established through experience, transferred from other persons as an instruction, or developed in advance for the actual situation by analysis and evaluation of the problem. Reliability of rule-based behaviour may in many cases be estimated by decomposition of a complex task into simple operations. A number of tables for operator failure rates for simple operations have been developed, see e.g [41,130]. In addition, special techniques have been developed to predict the operator failure rates by means of expert evaluations in cases where experience data are not available, see e.g. [87].

Knowledge-based behaviour. This behaviour means problem analysis and evaluation in situations where there are no available rules that can steer behaviour, or where it is unclear which rules apply. The individual is developing a goal and a plan to meet this goal. The plan is chosen by practical or mental trial of different alternatives.

It is difficult to estimate the probability of correct performance of a task that requires problem analysis and planning. In general the probability of failure in stress situations is extremely high. In [114] the use of a probability of 1 in such situations is recommended.

Human errors only rarely result in undesirable consequences. This is partly due to the fact that the relation between human errors and undesirable consequences often depends on a number of various other conditions. Many errors, or the consequences of these, are discovered and corrective measures undertaken in time to avoid undesirable consequences or reduce the severity of the consequences. As a result most human errors are not registered. Therefore little data are available about human reliability based on operational experience, and we often have to use other data sources such as:

- Data from simulator training
- Data from experimental studies
- Expert evaluations

A number of special techniques are available for estimation and analysis of human reliability. Perhaps the most well-known of these are THERP (Technique for Human Error Rate Prediction). THERP has been developed to estimate the human reliability (the probability of performing a specified function) and evaluate the effect of human errors in complicated man-machine systems. The starting point of the analysis is one or more system failures. The method is an application of the event tree analysis. Based on an identification of the different steps in the work task, an event tree is developed where the branches of the tree correspond to these steps. Such steps can for example be:

- Intervention before/after alarm
- Intervention within/after two minutes
- Correct/wrong switch

To estimate the failure probabilities for the various steps of the working operation, consideration must be given to factors that can influence behaviour (Performing Shaping Factors - PSF). Examples of such PSFs are:

- Complexity of task
- Information quality
- Feedback
- Breaking stereotypes
- Distractions
- Time pressure
- Hazard exposure
- Task-relevant experience
- Personnel qualifications

For working operations where the different steps are carried out by the same person or by several persons working as a team, there will usually be dependency between the various steps of the working task. The dependency may for example arise if an error results in stress, and this in its turn increases the probability of failure in the subsequent steps of the task.

A detailed description of THERP is given in [130]. For other special techniques for analysing human reliability refer to [5,48,62,87,111,126]. For an interesting discussion of analysis of human reliability, see [129].

2.6 Problems

1. Discuss the statistical data in Table 2.2. Discuss the reasons for the observed time trend.

2. In company A, six persons have been killed as a result of accidents in a certain period of time. In the same period, there has been a total of 10000 man-years. Ole Jensen works in company A. After having looked at the statistics for accidental deaths, he states: "The probability that I will be killed in an accident at work during the next year, is $6/10000 = 0.06\%$". Comment on this statement.

3. Ole Olsen, who lives in Stavanger, is to travel to Oslo. He considers two ways of travelling: go by car or by plane. Ole looks at Table 2.3 and concludes that it is approximately twice as hazardous to go by plane as by car. He therefore decides to go by car. Discuss this reasoning.

4. In a large company an observed FAR value has been registered equal to 5 for a one-year period. A FAR value for the personnel of the company equal to 16 has been estimated in theory. Discuss reasons for this large difference between the observed and calculated FAR values.

5. Figure 2.24 shows a system for emergency shut-down of a gas pipeline from A to B.

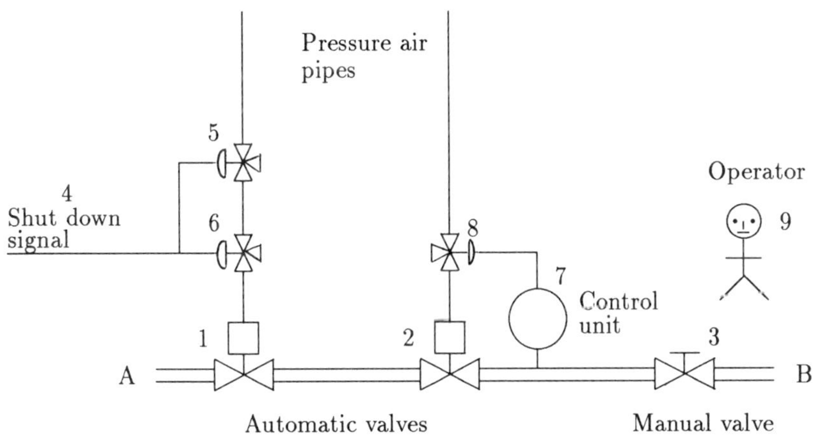

Figure 2.24

If a critical situation should occur, the gas process can be shut down by closing a minimum of one of the automatic valves 1 and 2 and/or the manual valve 3. Thus there are three barriers from A to B.

In a critical situation valve 2 will normally close first. It is done as a result of a local signal from the pilot valve 8 so that the air in the pressure air pipe is released. At the same time an emergency signal 4 is sent from a control room to the pilot valves

5 and 6 so that they open and release the air in the pressure air pipe. As a result valve 1 closes. The automatic valves 1 and 2 are kept open by means of air pressure. When the air pressure in the pipe disappears, the valves will close. If the pipeline has not been successfully shut down by means of valves 1 and 2, an operator 9 can manually close valve 3.

a) Perform an FMEA for this system.

b) Construct a fault tree for the system with the top event "The gas pipeline system is not shut down in a critical situation" and basic events that correspond to failures of the 9 "components" in Figure 2.24. Explain the meaning of the 9 basic events.

c) Assume the following probabilities for the basic events in the tree:

$$q_1 = 3 \times 10^{-2} \quad q_2 = 3 \times 10^{-2} \quad q_3 = 3 \times 10^{-2} \quad q_4 = 1 \times 10^{-2} \quad q_5 = 5 \times 10^{-2}$$
$$q_6 = 5 \times 10^{-2} \quad q_7 = 1 \times 10^{-2} \quad q_8 = 5 \times 10^{-2} \quad q_9 = 1 \times 10^{-2}$$

Here q_1 denotes the probability that basic event 1 will occur, q_2 denotes the probability that basic event 2 will occur, etc. Compute the probability of the top event. Assume stochastically independent events.

d) Construct a CCA diagram for this system. Choose as an initiating event: "A critical situation has occurred which necessitates shutdown of the pipeline system".

e) Calculate the probabilities for the various consequences based on the same assumptions as made under *c*).

6. Discuss whether we should include a full analysis of failure causes/failure mechanisms for all failure modes in an FMEA.

7. Consider a production system comprising four machines, each with three motors. The system does not function if one or more of the machines does not function. A machine does not function if a minimum of two of its motors do not function. Construct a fault tree for this system with the top event "The system does not function" and basic events corresponding to failures of the various motors. Construct also the associated reliability block diagram. Find the minimal cut sets.

8. Carry out the detailed calculations of the probabilities for the various consequences in Figure 2.20.

9. Construct a CCA diagram for the system in Example 2.5, with failure of the three "safety barriers" valve 1, valve 2 and valve 3 as branches in the tree.

10. In a risk analysis an event tree was used to analyse the risk related to loss of human life. The event tree terminates in 5 consequences: K_1, K_2, K_3, K_4 and K_5. Given the initiating event in the tree, the probability of these five consequences is

0.01, 0.02, 0.03, 0.06 and 0.08, respectively. The number of fatalities related to these consequences has been estimated at 10, 4, 2, 1 and 0, respectively.

Calculate the statistically expected number of fatalities per year related to the initiating event when the frequency of this event equals 0.1 per year. Give an interpretation of the result.

Chapter 3

BINARY MODELS FOR ANALYSING SYSTEM RELIABILITY

In this chapter we present some common models for analysis of system reliability. First we study the structural relations between a system and its components. Then methods for calculation of system reliability are studied, when the component reliabilities are known. Models for non-repairable and repairable systems are looked into. The components are assumed to be functioning/not functioning independently of each other, except in Section 3.6 where dependency is discussed.

The risk concept is not explicitly referred to in this chapter, but it is related to the material in this chapter in many ways. From the (un)reliability of a system one can often directly derive the risk associated with an activity, cf. Examples 1.5, 1.6 and 2.4. In other cases the system (un)reliability will be input-data in a risk calculation. A common situation is that system unreliability is entered in a branch probability in a CCA-diagram or an event tree, see for example Section 7.1.

This chapter covers the basic reliability theory as presented in most text books in the field, see e.g. [24,65]. But in addition we present some new results derived from various research papers. The presentation in this chapter is based on the appendices A, B and to some extent D and E.

3.1 Structural properties

We consider a system comprising n components which are numbered consecutively from 1 to n. In this chapter we will distinguish between only two states: a functioning state and a failure state. This dichotomy applies to the system as well as to each component. To indicate the state of the ith component, we assign a binary variable x_i to component i:

$$x_i = \begin{cases} 1 & \text{if component } i \text{ is in the functioning state} \\ 0 & \text{if component } i \text{ is in the failure state} \end{cases}$$

(The term *binary variable* refers to a variable taking on the values 0 and 1.) Similarly, the binary variable ϕ indicates the state of the system:

$$\phi = \begin{cases} 1 & \text{if the system is in the functioning state} \\ 0 & \text{if the system is in the failure state} \end{cases}$$

We assume that the state of the system is determined completely by the states of the components, so that we may write

$$\phi = \phi(\mathbf{x})$$

where $\mathbf{x} = (x_1, x_2, \ldots, x_n)$.

The function $\phi(\mathbf{x})$ is called the *structure function* of the system, or simply the structure. In the following we will often use the phrase structure in place of system.

Example 3.1
A system which is functioning if and only if each component is functioning, is called a *series* structure. The structure function for this system is given by:

$$\phi(\mathbf{x}) = x_1 \cdot x_2 \cdot \ldots \cdot x_n = \prod_{i=1}^{n} x_i$$

A series structure can be illustrated by the reliability block diagram in Figure 3.1. "Connection between a and b" means that the system functions, cf. Section 2.3.

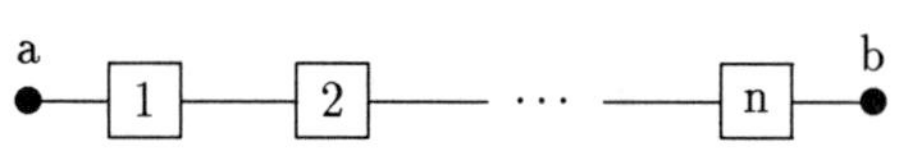

Figure 3.1: Series structure

Example 3.2
A system which is functioning if and only if at least one component is functioning, is called a parallel structure. The corresponding reliability block diagram is shown in Figure 3.2.

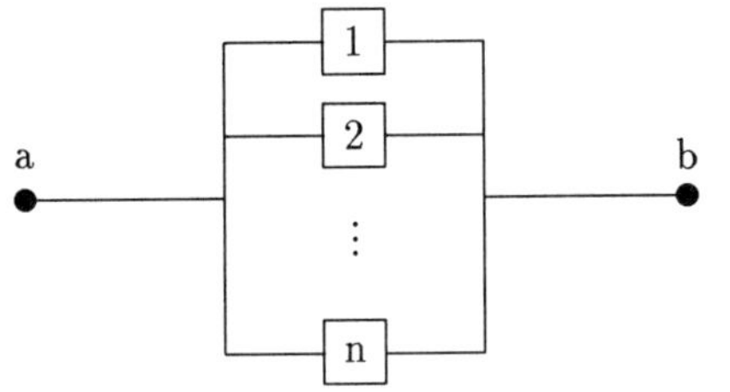

Figure 3.2: Parallel structure

The structure function is given by:

$$\phi(\mathbf{x}) = 1 - (1 - x_1)(1 - x_2) \cdots (1 - x_n) = 1 - \prod_{i=1}^{n}(1 - x_i) \tag{3.1}$$

The expression on the right hand side in (3.1) is often written $\coprod x_i$. Thus a parallel structure with two components has structure function

$$\phi(\mathbf{x}) = 1 - (1 - x_1)(1 - x_2) = \coprod_{i=1}^{2} x_i$$

which we also write

$$\phi(\mathbf{x}) = x_1 \amalg x_2$$

Example 3.3
A system which is functioning if and only if at least k out of n components are functioning, is called a *k*-out-of-*n* structure. A series structure is an *n*-out-of-*n* structure, and a parallel structure is a *1*-out-of-*n* structure.

The structure function for a *k*-out-of-*n* structure is given by:

$$\phi(\mathbf{x}) = \begin{cases} 1 & \text{if } \sum_{i=1}^{n} x_i \geq k \\ 0 & \text{if } \sum_{i=1}^{n} x_i < k \end{cases}$$

As an example, we will look at a *2*-out-of-*3* structure. This structure can be illustrated by the reliability block diagram shown in Figure 3.3. An aeroplane which is capable of functioning if and only if at least two of its three engines are functioning is an example of a *2*-out-of-*3* system.

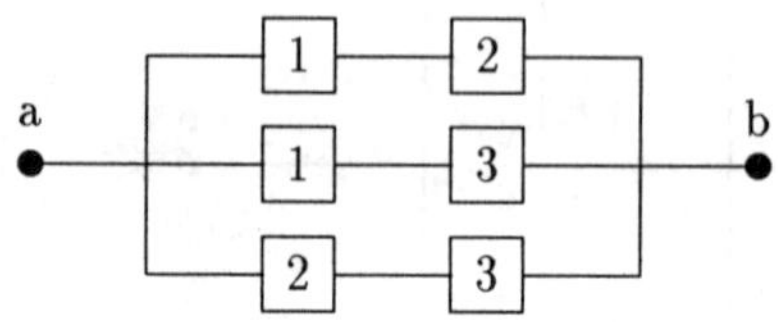

Figure 3.3: *2*-out-of-*3* structure

Monotone structure
A structure is monotone if

1. $\phi(\mathbf{0}) = 0$ and $\phi(\mathbf{1}) = 1$
2. The structure function is non-decreasing in each argument

Condition 1 says that if all the components are in the failure state, the system is in the failure state, and if all the components are in the functioning state, then the system is in the functioning state. Condition 2 says that the system cannot deteriorate (that is, change from the functioning state to the failed state) by improving the performance of a component (that is, replacing a failed component by a functioning component).

All the systems we have considered are monotone. In the following we shall limit ourselves to monotone systems.

Minimal cut sets and path sets
In accordance with the definition of cut sets and minimal cut sets in Section 2.3, we now define:

> A cut set K is a set of components whose failure causes the system to fail. A cut set is minimal if it cannot be reduced without losing its status as a cut set.

Analogously we define a path set and a minimal path set:

> A path set S is a set of components whose functioning ensures the functioning of the system. A path set is minimal if it cannot be reduced without losing its status as a path set.

Example 3.4
Consider the reliability block diagram for the buffer storage in Figure 2.5:

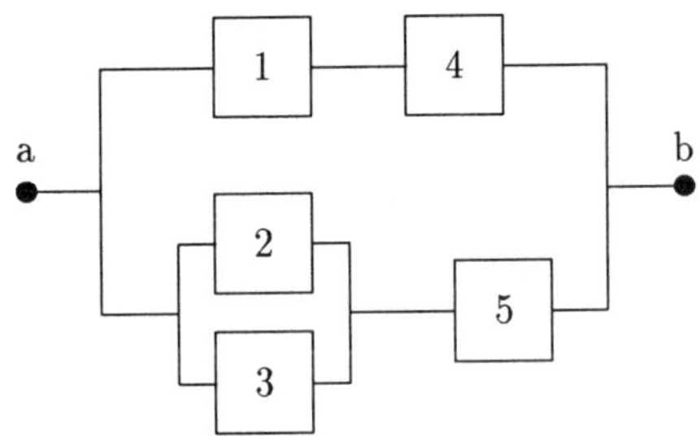

Figure 3.4: Reliability block diagram for Example 2.4

The minimal cut sets of the system are: {1,5}, {4,5}, {1,2,3} and {2,3,4}. Note that for example {1,4,5} is a cut set, but it is not minimal. The minimal path sets are {1,4}, {2,5} and {3,5}.

3.2 Computing system reliability

Assume that the state X_i of the ith component is a stochastic variable, $i = 1,2,\ldots,n$. Let

$p_i = P(X_i = 1) =$ Reliability of component i

$q_i = P(X_i = 0) =$ Unreliability of component i

$h = h(\mathbf{p}) = P(\phi(\mathbf{X}) = 1) =$ Reliability of the system

$g = g(\mathbf{q}) = P(\phi(\mathbf{X}) = 0) =$ Unreliability of the system

If we refer to a fixed point in time t, we write $X_i(t)$, $\mathbf{X}(t)$, and so on. The problem is to compute the system reliability h given the component reliabilities p_i. Often it will be more efficient to let the starting point of the calculation be the unreliabilities. Note that $h + g = 1$ and $p_i + q_i = 1$.

3.2.1 Computing exact system reliability

Before we present methods for computation of system reliability for a general structure, we will look closer into some special cases. We start with the series structure.

Reliability of a series structure

The reliability of a series structure of independent components equals the product of the component reliabilities, i.e.

$$h = \prod_{i=1}^{n} p_i \tag{3.2}$$

For a series structure the system functioning means that all the components function, hence

$$h = P(\phi = 1) = P(X_1 = 1, X_2 = 1, \ldots, X_n = 1) = \prod_{i=1}^{n} P(X_i = 1) = \prod_{i=1}^{n} p_i$$

The third equality holds since the X_i's are assumed to be independent. Thus if all the components have the same reliability p,

$$h = p^n$$

From this formula we see that the system reliability decreases rapidly with an increasing number of components, for example, for $n = 20$ and $p = 0.99$ we find

$$h = 0.99^{20} = 0.818$$

Reliability of a parallel structure

The unreliability of a parallel structure of independent components equals the product of the component unreliabilities, i.e.

$$g = \prod_{i=1}^{n} q_i \tag{3.3}$$

Thus the reliability of a parallel structure is given by

$$h = 1 - \prod_{i=1}^{n}(1 - p_i) = \coprod_{i=1}^{n} p_i$$

The proof of (3.3) is analogous with the proof of (3.2).

Reliability of a k-out-of-n structure

The reliability of a k-out-of-n structure of independent components which all have the same reliability p, equals

$$h = \sum_{i=k}^{n} \binom{n}{i} p^i (1-p)^{n-i}$$

This formula holds since $\sum_{i=1}^{n} X_i$ has a binomial distribution with parameters n and p under the given assumptions. The case that the component reliabilities are not equal is treated in the general setting below, and in the next section.

Reliability of an arbitrary series-parallel structure

By using the calculation formulae for a series structure and a parallel structure it is relatively straightforward to calculate the reliability of a series-parallel structure, i.e. a structure which can be divided into only series and parallel structures. Let us consider an example.

Example 3.4 continued

Consider again the reliability block diagram for the buffer storage in Figure 3.4. We make the same assumptions as earlier. The system can be viewed as a parallel structure of two independent components: the structure comprising the components 1 and 4, and the structure comprising the components 2, 3 and 5. The reliability of the former structure equals p_1p_4, whereas the reliability of the latter equals $(1 - q_2q_3)p_5$. Thus the system reliability is given by

$$h = 1 - \{1 - p_1p_4\}\{1 - (1 - q_2q_3)p_5\}$$

If for example a 2-out-of-3 structure of independent components with the same reliability p is in series with the above system, the total system reliability will be as above multiplied by the reliability of the 2-out-of-3 structure, which equals

$$\binom{3}{2} p^2(1-p) + \binom{3}{3} p^3(1-p)^0 = 3p^2(1-p) + p^3$$

Reliability of a general monotone structure

Computation of system reliability might be a formidable task (in fact, impracticable in some cases) unless an efficient method (algorithm) is used. Developing such methods is therefore an important area of research within reliability theory.

There exists a number of methods for reliability computation of a general structure. These methods are often divided into two main categories: methods based on the minimal cut (path) sets, and methods that are *not* based on the minimal cut (path) sets. For the former category, the reliability is found using a two-stage procedure:

1. Identification of the minimal cut (path) sets of a fault tree or block diagram

2. Computation of the reliability by using a method which is based on knowledge of the minimal cut (path) sets

As mentioned in Section 2.3.5, MOCUS [64] is the most common algorithm for identification of the minimal cut (path) sets for a fault tree. For a block diagram, some algorithms for identification of the minimal cut (path) sets are referred to in [75].

To compute exact system reliability based on the minimal cut sets or the minimal path sets, we can for example use one of the algorithms described in [1,15]. The algorithm [15] is used in the computer package CARA [123] and is presented in Appendix F. For smaller systems the so-called inclusion-exclusion method may also be applied, but this method is primarily a method for approximate calculations for systems that are either very reliable or unreliable, see Section 3.2.2.

The number of minimal cut sets or path sets for complex systems can be extremely large, but by using so-called modular decomposition it is often possible to significantly reduce the number and simplify the calculations. The strategy is then to find the minimal cut (path) sets for each module and compute the corresponding reliability. The reliability of the whole system is found by inserting the module reliabilities into the reliabilities for the overall module structure (the organizing structure).

If the number of minimal cut (path) sets is very large, some kind of cut-off procedure might be required, for example the neglect of minimal cut sets of a high order. An upper bound on the error in the results due to the cut-off should then be established, see e.g. [69].

It might be considered troublesome to go via the minimal cut (path) sets. One should, however, realize that the minimal cut (path) sets are often sought because they give valuable qualitative information, cf. Section 2.3.6. In addition, they are necessary to establish some bounds, cf. Section 3.6, and to calculate certain criticality measures, cf. Section 3.5 and [64].

If the problem is merely to compute system (un)reliability, then we can instead use a direct approach that is not based on the minimal cut or path sets. There are a number of such methods. The "state enumeration method" is conceptually the simplest of these. With this method reliability is calculated using

$$h = E\phi(\mathbf{X}) = \sum_{\mathbf{x}} \phi(\mathbf{x}) P(\mathbf{X} = \mathbf{x}) = \sum_{\mathbf{x}:\phi(\mathbf{x})=1} \prod_{i=1}^{n} p_i^{x_i}(1-p_i)^{1-x_i}$$

The method, however, is not suitable for larger systems since the number of terms in the sum can be extremely large, up to $2^n - 1$.

Of other methods we will confine ourselves to describing the so-called *factoring algorithm* (pivot-decomposition method). The basic idea of this method is to make a conditional probability argument using the relation

$$h(\mathbf{p}) = p_i h(1_i, \mathbf{p}) + (1 - p_i) h(0_i, \mathbf{p}) \tag{3.4}$$

where $h(x_i, \mathbf{p})$ equals the reliability of the system given that the state of component i is x_i. The formula (3.4) follows directly from a well-known probability rule, see Appendix A, formula (A.10). This process repeats until the system comprises only series-parallel structures. To illustrate the method we will give an example.

Example 3.5
Consider a bridge structure as given by the diagram shown in Figure 3.5 below.

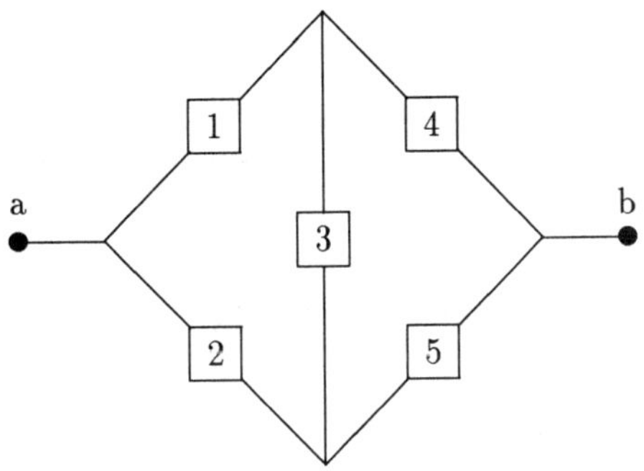

Figure 3.5: Bridge structure

If we first choose to pivot on component 3, the formula (3.4) holds with $i = 3$. But it is not difficult to see that given $x_3 = 1$, the system structure has the form

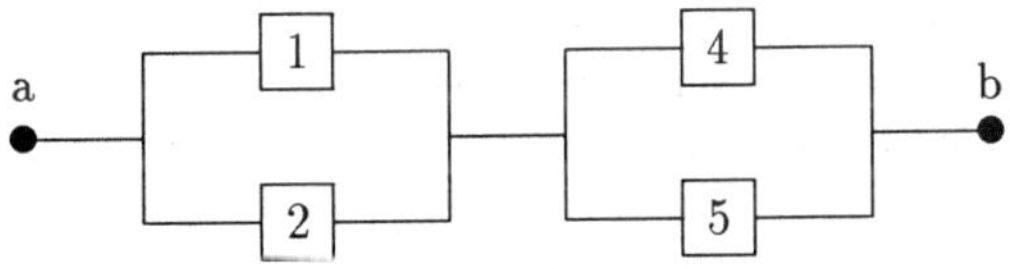

and that given $x_3 = 0$, the system structure has the form

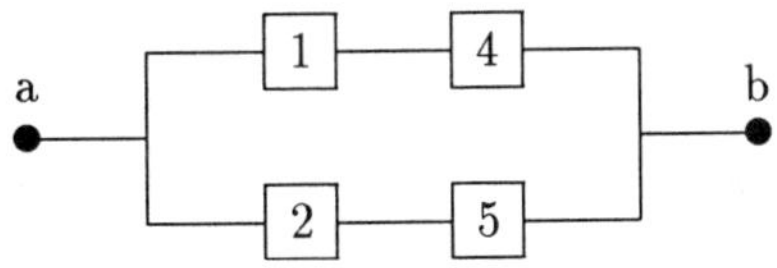

These two structure are both on series-parallel form, and it follows that

$$\begin{aligned} h(1_3, \mathbf{p}) &= (p_1 \amalg p_2)(p_4 \amalg p_5) \\ h(0_3, \mathbf{p}) &= p_1 p_4 \amalg p_2 p_5 \end{aligned}$$

Thus a formula for exact computation of $h(\mathbf{p})$ is established. Note that it was sufficient to perform only one pivotal decomposition in this case. If the structure given $x_3 = 1$, had not been in a series-parallel form, we would have had to perform another pivotal decomposition, and so on.

□

The above methods are analytical methods. The system reliability can also be studied by means of Monte Carlo simulation. Refer to Section 4.3 for a short discussion of the use of Monte Carlo simulation in reliability analyses.

3.2.2 Computing approximate system reliability

In most practical applications, the component reliabilities are high, in which case the computation method described in Section 2.3.7 is usually adopted, i.e.:

$$g \approx w_1$$

where

$$w_1 = \sum_{j=1}^{k} \prod_{i \in K_j} q_i$$

Here k is the number of minimal cut sets. Hence the system unreliability is found by summing the unreliabilities of the minimal cut sets; the unreliability of a minimal cut set equals the product of the unreliabilities of the components that are included in the minimal cut set.

The above method is based on the so-called *inclusion-exclusion method*, cf. Appendix A, Formulae (A.3) and (A.4). It follows from the inclusion-exclusion method that since

$$\begin{aligned} g &= \text{probability that there exists a failed minimal cut set} \\ &= P(\textstyle\bigcup_{j=1}^{k} \text{minimal cut set } j \text{ is not functioning}), \end{aligned}$$

we must have

$$w_1 - w_2 \leq g \leq w_1 \tag{3.5}$$

where

$$w_2 = \sum_{j<r} \prod_{i \in K_j \bigcup K_r} q_i$$

The term w_2 represents the sum of the unreliabilities of pairs of minimal cut sets; each of these unreliabilities equals the product of the unreliabilities of the components that are included in both minimal cut sets.

If the q_is are small, i.e. the p_is are large, the w_2 term will usually be negligible compared to w_1, such that

$$g \approx w_1 \quad \text{and} \quad h \approx 1 - w_1$$

Notice that w_1 is an upper limit for g. By using w_1 as an estimate for the system unreliability, we will overestimate the system unreliability. In most cases, such an underestimation of reliability is preferable compared to an overestimation of reliability.

If w_2 is not negligible compared to w_1, it is possible to develop new bounds based on the Formulae (A.3) and (A.4): $w_1 - w_2, w_1 - w_2 + w_3, \ldots$, which finally produce the exact value. With a large number of minimal cut sets, such calculations will, however, be extensive. The number of terms in the sum in w_j equals $\binom{k}{j}$. Thus the total number of terms can be

$$\sum_{j=1}^{k} \binom{k}{j} = (1+1)^k - 1 = 2^k - 1$$

There exist also other bounds and approximations for the system reliability. For example

$$1 - \prod_{j=1}^{k}(1 - \prod_{i \in K_j} q_i)$$

is an upper bound for g, and a good approximation for small values of the q_is [24]. This bound is always as good as or better than w_1.

For systems with low reliabilities, it is possible to establish similar results as given in this section, based on the minimal path sets.

Example 3.4 continued

The problem is to calculate the unreliability of the system of Figure 3.4 by means of the approximation method described above. We assume as earlier that $q_1 = q_2 = q_3 = 0.02$ and $q_4 = q_5 = 0.01$. In Section 2.3.7 we showed that the approximation w_1 gives $g \approx 0.03\%$. The error term w_2 becomes in this case

$$\begin{aligned} &q_1 q_4 q_5 + q_1 q_2 q_3 q_5 + q_1 q_2 q_3 q_4 q_5 + q_1 q_2 q_3 q_4 q_5 \\ &+ q_2 q_3 q_4 q_5 + q_1 q_2 q_3 q_4 = 2.2 \times 10^{-4} \end{aligned}$$

so that we can write $g = 0.03\%$.

3.3 Non-repairable systems

We consider a monotone system in a time interval $[0, t_0]$. The components of the system are "new" at time $t = 0$. A failed component stays in the failure state for the rest of the time interval. Thus the component is not repaired or replaced. The situation can for example describe a system with component failure states that can only be discovered by testing or inspection.

We assume that the lifetime of component i is determined by a lifetime distribution $F_i(t)$. If for example the distribution is exponential, then

$$F_i(t) = 1 - e^{-\lambda_i t}$$

where λ_i is the failure rate of component i. To calculate system reliability at a fixed point in time, i.e. the reliability function at this point, we can proceed as described in Section 3.2 with

$q_i = F_i(t)$ = Probability that component i is not functioning at time t

$p_i = R_i(t)$ = Probability that component i is functioning at time t

For a series system the reliability at time t thus take the form

$$h = \prod_{i=1}^{n} R_i(t) \tag{3.6}$$

But $R_i(t)$ can be expressed by means of the failure rate $z_i(t)$:

$$R_i(t) = e^{-\int_0^t z_i(u)\,du} \tag{3.7}$$

By putting (3.7) into formula (3.6) we obtain

$$h = e^{-\int_0^t [\sum_{i=1}^n z_i(u)]\,du} \tag{3.8}$$

From (3.8) we can conclude that *the failure rate of a series structure of independent components equals the sum of the failure rates of the components of the structure.* In particular this means that if component i has a constant failure rate λ_i, $i = 1, 2, \ldots, n$, then the series structure has constant failure rate $\sum \lambda_i$.

For a parallel structure we do not have a similar result. With constant failure rates of the components, the system will have a time dependent failure rate. The following simple example illustrates this.

Example 3.6
Consider a parallel system comprising two components with lifetimes which are independent and exponentially distributed with failure rates λ_1 and λ_2. The lifetime distribution for this system is given by

$$F(t) = (1 - e^{-\lambda_1 t})(1 - e^{-\lambda_2 t}) = 1 - e^{-\lambda_1 t} - e^{-\lambda_2 t} + e^{-(\lambda_1+\lambda_2)t}$$

From this expression we find that the failure rate of the system equals

$$z(t) = \frac{\lambda_1 e^{-\lambda_1 t} + \lambda_2 e^{-\lambda_2 t} - (\lambda_1 + \lambda_2)e^{-(\lambda_1+\lambda_2)t}}{e^{-\lambda_1 t} + e^{-\lambda_2 t} - e^{-(\lambda_1+\lambda_2)t}}$$

It can be shown that for $\lambda_1 \neq \lambda_2$, there exists a point in time t^* such that $z(t)$ is increasing in the interval $(0,t^*)$ and decreasing in the interval (t^*, ∞). The point in time t^* will depend on λ_1 and λ_2.

3.4 Repairable systems

For a repairable system it is common to speak about availability in place of reliability. In the following we will adopt this usage.

3.4.1 Replacement/repair at failures

We consider a monotone system comprising n independent components. A failed component is repaired or replaced. Let T_{ij}, $j = 1, 2, \ldots$, represent the length of the jth operation period, and let D_{ij}, $j = 1, 2, \ldots$, represent the length of the jth repair/replacement time for component i, see Figure 3.6.

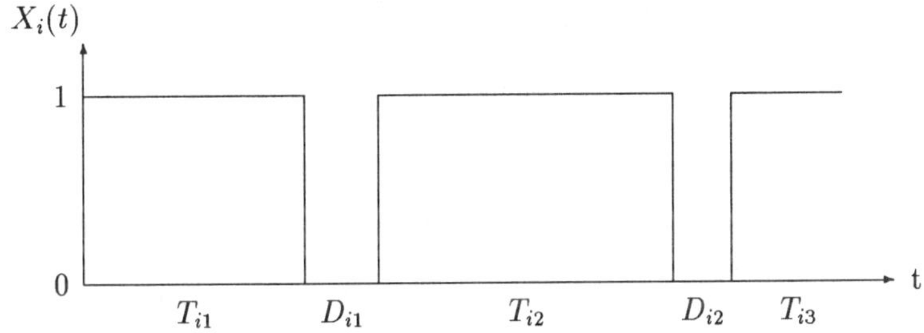

Figure 3.6: Failure and repair process for component i

We assume that the pairs (T_{ij}, D_{ij}), $j = 1, 2, \ldots$, are independent and identically distributed, with finite means. Note that we do not assume that T_{ij} and D_{ij} are independent. Let $MTTF_i$ (Mean Time To Failure) and $MTTR_i$ (Mean Time To Repair) denote the mean of the operation time T_{ij} and the mean of the repair time D_{ij}, respectively.

As a measure of component i's availability we usually use the so-called *limiting availability*, A_i, given by

$$A_i = \frac{MTTF_i}{MTTF_i + MTTR_i}$$

It follows from the strong law of large numbers that A_i equals the long run *portion of time that component i is functioning/operating*, see Appendix E.5.1. Hence the limiting availability can be interpreted as the portion of time the component is operating in the long run. If for example a component has a limiting availability of 98%, it means that the component will operate 98% of the time and undergo repair 2% of the time in the long run.

Furthermore, it can under a certain condition be shown that the availability at time t, $A_i(t)$, converges towards A_i when $t \to \infty$, see Appendix E.5.1. For most lifetime and repair time distributions the convergence will be fast, so that we can use A_i as an approximation to the availability at time t. The assumption that has to be fulfilled is that the distribution of $T_{ij} + D_{ij}$ has a density on some interval. This means that the distribution of $T_{ij} + D_{ij}$ must not be periodic, i.e. a discrete distribution with values $\{c, 2c, 3c, \ldots\}$ for some c.

If the availability at time t converges towards A_i, it can been shown that the average availability in [0,t], $(1/t)\int_0^t A_i(s)\,dt$, converges towards A_i when $t \to \infty$, see Appendix E, Result E.8. The average availability in [0,t] can be interpreted as the expected portion of time the component is operating in the time interval.

The limiting unavailability, $\overline{A}_i$, is defined by

$$\overline{A}_i = 1 - A_i = \frac{MTTR_i}{MTTF_i + MTTR_i}$$

Usually $MTTR$ is small compared to $MTTF$, so that

$$\overline{A}_i \approx \frac{MTTR_i}{MTTF_i} \tag{3.9}$$

If the underlying lifetime distribution is exponential, we can write (3.9) in the following form:

$$\overline{A}_i \approx \lambda_i\, MTTR_i \tag{3.10}$$

where λ_i represents the failure rate of component i. The formula (3.10) is used very often in practice when calculating unavailability.

Let $A(t) = h(A_1(t), A_2(t), \ldots, A_n(t))$. We see that $A(t)$ denotes the system availability at time t. Furthermore, let A denote the limiting system availability given by $A = h(A_1, A_2, \ldots, A_n)$. Then since the reliability function $h(\mathbf{p})$ is a linear function in each p_i, and therefore a continuous function, it follows that $A(t) \to A$ as $t \to \infty$, assuming that $A_i = \lim_{t\to\infty} A_i(t)$ exists for all i. Thus A equals approximately the probability that the system is functioning at a fixed point in time, and it can be calculated using the methods described in Section 3.2, with

$$p_i = A_i \text{ and } q_i = \overline{A}_i$$

The limiting system availability can also be interpreted as the expected portion of time the system is operating in the long run, or as the long run average availability, see Appendix E, Result E.8.

Example 3.7

In the design process of a system two alternatives are considered:

Alternative 1: Redundancy with two units of a certain type in parallel

Alternative 2: No redundancy, only one unit

For alternative 1 both units will normally run, even though the system functions with only one unit in operation. Operational experience for similar units has indicated a $MTTF =$ 1000 hours and a $MTTR = 50$ hours. The problem is to estimate the availability difference for the two alternatives, based on this experience.

We find it reasonable to use the above model and we do the following calculations. The limiting availability for one unit (alternative 2) is $50/1050 = 4.76\%$, whereas the limiting unavailability for the two units in parallel (alternative 1) is $0.048 \times 0.048 = 0.23\%$, giving a difference of about 4.5%.

□

Calculating the availability at a given point in time t, $A_i(t)$ is difficult except for a few special cases. If for example the lifetime and repair times are independent and they have exponential distributions, it can be shown that the availability of a unit at time t equals (see Appendix D, (D.3))

$$\frac{\mu}{\lambda+\mu} + \frac{\lambda}{\lambda+\mu} e^{-(\lambda+\mu)t} \tag{3.11}$$

where λ is the failure rate and μ is the repair rate of the unit. It is assumed that the unit is functioning at time zero. Note that as t increases, (3.11) is approaching the limiting availability. In practice the state of the components will usually be unknown at time zero, and it is then natural to estimate the availability at a certain point in time by the limiting availability.

In the above model, it is assumed that the components are operating independently of each other. This means for example that if a component does not function, i.e. is under repair or is replaced, the other components continue to operate. This assumption is in many cases not fulfilled, but the error is usually negligible for systems with high component availabilities. To give an indication of the error term, let us consider a series system, and let us assume that while a component is undergoing repair (replacement), all other components remain in "suspended animation". When repair of the failed component is completed, the remaining components resume operation. At any instant, they are not "as good as new", but rather as good as they were when the system stopped operating. We make the additional assumption that two or more components cannot fail at the same time. This will be true, for example, if all failure distributions are continuous.

Then it can be shown, see Appendix E.5, that the long run (expected) portion of time the system is functioning equals

$$\frac{1}{1+\sum_{i=1}^{n} \frac{MTTR_i}{MTTF_i}} \tag{3.12}$$

This expression is to be compared to

$$h(\mathbf{A}) = \prod_{i=1}^{n} \frac{MTTF_i}{MTTF_i + MTTR_i}$$

If the component availabilities are high, then

$$h(\mathbf{A}) \approx 1 - \sum_{i=1}^{n} \overline{A}_i$$

which is seen to be approximately equal to (3.12).

Refer to [24] for a thorough analysis of the limiting properties of a series system when the functioning components suspend operation during repair.

There exist several other informative system performance measures for this model than those presented above, see Chapter 4. In this section we will restrict attention to one such measure: the (expected) number of system failures in a time interval $(u, u+s]$.

Let $N(t)$ denote the number of system failures in $[0, t]$. It can be shown, see Appendix E.5.1 and Problem 3.15, that if the distribution of $T_{ij} + D_{ij}$ is not periodic and with probability one, two or more components cannot fail at the same time, then:

$$\lim_{t\to\infty} \frac{EN(t)}{t} = \sum_{i=1}^{n} \frac{h(1_i, \mathbf{A}) - h(0_i, \mathbf{A})}{MTTF_i + MTTR_i} \tag{3.13}$$

$$\lim_{u\to\infty} E[N(u+s) - N(u)] = s\sum_{i=1}^{n} \frac{h(1_i, \mathbf{A}) - h(0_i, \mathbf{A})}{MTTF_i + MTTR_i} \tag{3.14}$$

Furthermore, if the process ϕ is a regenerative process, then with probability one,

$$\lim_{t\to\infty} \frac{N(t)}{t} = \sum_{i=1}^{n} \frac{h(1_i, \mathbf{A}) - h(0_i, \mathbf{A})}{MTTF_i + MTTR_i}$$

From these formulae approximations for $N(t)$, $EN(t)$ and $E[N(u+s) - N(u)]$ can be established. If the distribution of T_{ij} has a constant failure rate λ_i, then ϕ is a regenerative process, and it follows from the proof in the appendix that the expected number of system failures in the period $(u, u+s]$ equals:

$$E[N(u+s) - N(u)] = \sum_{i=1}^{n} \int_{u}^{u+s} [h(1_i, \mathbf{A}(t)) - h(0_i, \mathbf{A}(t))]\, \lambda_i A_i(t) dt$$

3.4.2 Preventive maintenance policy

Various maintenance policies have been proposed to reduce the occurrence of system failure. Chapter 5 studies the problem of optimizing the maintenance policy in different situations. In this section we seek an expression for the system availability in a general minimal repair and replacement model. This model represents a modification of a model introduced in [34], and it includes several basic preventive maintenance models as special cases. The minimal repair concept is introduced in Appendix B.2. The model is described in the following.

Assume that component i is minimally repaired or replaced according to the following scheme. The component is "new" at time zero, and it is replaced whenever it reaches age T_i (planned replacement). The age is measured in operating time. The case $T_i = \infty$ corresponds to no planned replacement. The expected replacement time for a planned replacement equals u_i. If the component fails at age $t < T_i$, it is either replaced by a new component of the same type with probability $\alpha_i(t)$ (unplanned replacement) or it undergoes minimal repair with probability $\beta_i(t) = 1 - \alpha_i(t)$. Whether the component is replaced or minimally repaired at failure may depend on a number of factors, such as the extent and cost of the repair, but "on average" if the component fails at t it is replaced $100\alpha_i(t)\,\%$ of the times and minimally repaired $100\beta_i(t)\,\%$ of the times. The expected replacement time for an unplanned replacement is v_i. The expected length of a minimal repair at time t is $r_i(t)$. After a replacement, the procedure is repeated. We assume that all failures are instantly detected, and that the cycles generated by consecutive replacements are independent.

The problem is to calculate the system availability. To this end, let us first consider the problem of calculating the component availability. As in the previous section we are focusing on the limiting availability.

Let $N_i(t)$ represent the number of failures in [0,t], assuming that no replacement is performed in the time interval. The process $N_i(t)$ is assumed to be a non-homogeneous Poisson process with a failure intensity $\lambda_i(t)$. Furthermore, let L_i represent the operating time until replacement of the component, and let R_i and W_i represent the downtime associated with minimal repairs and downtime associated with the replacement, respectively.

It follows from Appendix E.5.2 that the limiting availability, A_i, is given by

$$\begin{aligned} A_i &= \lim_{t\to\infty} A_i(t) \\ &= \frac{E\,L_i}{E\,L_i + E\,R_i + E\,W_i} \qquad (3.15) \\ &= \frac{\int_0^{T_i} P(L_i > t)\,dt}{\int_0^{T_i} P(L_i > t)\,dt \,+\, \int_0^{T_i} r_i(t) P(L_i > t)\beta_i(t)\lambda_i(t)\,dt \,+\, v_i\,P(L_i < T_i) \,+\, u_i\,P(L_i = T_i)} \end{aligned}$$

where

$$\begin{aligned} P(L_i > t) &= e^{-\int_0^t \alpha_i(s)\lambda_i(s)\,ds}, \quad t < T_i \\ P(L_i = T_i) &= e^{-\int_0^{T_i} \alpha_i(s)\lambda_i(s)\,ds} \end{aligned}$$

To establish (3.15) it is necessary that the cycle length $L_i + R_i + W_i$ is non-periodic. All integrals are assumed to be finite.

Thus far we have obtained availability results for a component or a system treated as a single unit. To compute the limiting availability of a monotone system, we can proceed as in Section 3.2 with

$$p_i = A_i \text{ and } q_i = 1 - A_i$$

Some simple special cases

Case 1. If $\alpha_i(t) \equiv 0$, the model corresponds to the situation of minimal repairs with no unplanned replacement. In this case we have

$$A_i = \frac{T_i}{T_i + \int_0^{T_i} r_i(t)\lambda_i(t)\,dt + u_i}$$

Case 2. If $\alpha(t) \equiv 1$, then the model reduces to the classic age replacement situation considered by e.g. [25]. The component is replaced at failure or at time T_i, whichever comes first. There are no minimal repairs, so in this case we have $E\,R_i = 0$.

Example 3.8
Consider component i. Assume that $T_i = 30$, $u_i = 0.1$, $v_i = 0.2$ and that $\lambda_i(t)$, $\alpha_i(t)$ and $r_i(t)$ are given by

t	$\lambda_i(t)$	$\alpha_i(t)$	$r_i(t)$
$[0, 20)$	0.02	0.2	0.02
$[20, 25)$	0.05	0.4	0.04
$[25, 30)$	0.2	0.9	0.2

Then by straightforward calculation we find that $EL_i = 26.363$, $ER_i = 0.022$, $EW_i = 0.166$, which give $A_i = 99.3\%$.

3.4.3 A model for analysing standby systems

The model presented in this section is taken from [10]. For other models for analysing standby systems see [32,139] and Appendix D.

We consider a parallel system comprising n identical and independent components where one component serves as a standby. Hence, $n-1$ components are normally running. If a component fails, the standby component is immediately put into operation. A failed component is repaired. Furthermore, we assume:

- Switchover to the standby component is perfect, e.g. instantaneous and failure-free
- A standby component that has completed its repair is functioning at demand (it does not fail while waiting for its next service period)
- The components have a constant failure rate
- All failure and repair times are independent

We introduce the following notation:

J	Time period considered
$MTTF$	Mean time to failure of a component
λ	Failure rate of a component
$MTTR$	Mean time to repair a component
$\overline{a}$	$MTTR/MTTF = \lambda\, MTTR$
u_i	mean number of units to be undergoing repair at the same time when i components have failed, $i = 1, 2, \ldots, n$, $1 = u_1 \leq u_2 \leq \cdots u_n \leq n$. For example we might have a) $u_i \equiv i$, i.e., up to n simultaneous repairs can be carried out, or b) $u_i \equiv 1$, i.e., only one repair can be carried out at a time
U_i	$\prod_{j=1}^{i} u_j \quad (U_0 = 1)$

As a measure of system performance we introduce

$S_i =$ Expected portion of time in the long run that i out of the n components are not functioning

If the repair time distribution is exponential, it is possible to show, see appendix D, that

$$S_i = \frac{b_i}{1 + \sum_{j=1}^{n} b_j} \tag{3.16}$$

where

$$b_i = \begin{cases} \frac{(n-1)(n-1)!}{(n-i)!} \frac{1}{U_i} \overline{a}^i & i = 1, 2, \ldots, n \quad (0! = 1) \\ 1 & i = 0 \end{cases}$$

An exponential repair time distribution is seldom realistic, but the formula (3.16) is quite robust with respect to the repair time distribution, in the sense that the error introduced by using the above formula for other repair time distributions is usually very small. We will discuss the accuracy of the approximations later in this section.

Note that if $\overline{a}$ is small, then $S_i \approx b_i$ for $i \geq 1$.

Example 3.9

We consider alternative 1 in Example 3.7, but we now assume that one component serves as a standby and that the assumptions above hold with $u_i = i$ for all i. Then

$$\begin{aligned} S_0 &= 1/(1 + \overline{a} + \tfrac{1}{2}\overline{a}^2) &= 0.9512 &= 95.1\% \\ S_1 &= \overline{a}\, S_0 &= 0.0476 &= 4.8\% \\ S_2 &= \tfrac{1}{2}\overline{a}^2\, S_0 &= 0.0012 &= 0.1\% \end{aligned}$$

Thus we can expect that the system does not function, i.e. both components are under repair, in 0.1% of the time. This percentage is as expected somewhat lower than the limiting unavailability for alternative 1 in Example 3.7.

□

Next, we will generalize the above model by taking into account that components can be taken out of service due to preventive maintenance (PM), e.g. cleaning, adjusting, lubricating the parts, and replacement of parts. Such maintenance work is common for many types of equipment. We make the following additional assumptions:

- The components have a constant failure rate up to the time of the PM
- A standby component that has completed its PM is functioning at demand (it does not fail while waiting for its next service period)
- No more than one component is out for PM at the same time
- A component that is out for PM must complete its PM before it again can be put into operation; it is not immediately ready for operation on demand. (Note: If the component was immediately ready for operation on demand, we did not need to include the PM in the model.)

Let q denote the expected portion of time one of the components is out for PM in the time interval J. In view of the last assumption above, we must define the time the components are out for PM as the minimum time required to perform the PM.

Now, to compute S_i^{PM}, the expected portion of time that exactly i out of the n components are not functioning, i.e. is under repair or PM, the following approximation formula can be used:

$$S_i^{PM} \approx S_i^* \, q + S_i \, (1 - q) \tag{3.17}$$

where S_i is given by (3.16) and S_i^* is defined by

$$S_i^* = \frac{c_i}{1 + \sum_{j=2}^{n} c_j}$$

$$c_i = \begin{cases} \frac{(n-1)!}{(n-i)!} \frac{1}{V_i} \, \overline{a}^{i-1} & i = 1, 2, \ldots, n \\ 0 & i = 0 \end{cases}$$

$$V_i = \prod_{j=1}^{i} v_j \quad (V_0 = 1)$$

$v_j =$ Expected number of components to be undergoing repair at the same time when one component is out for PM and j components are failed

The quantity S_i^* represents the expected portion of time that $i-1$ out of $n-1$ components are not functioning assuming that none of these $n-1$ components are out for PM. To establish the formula, we divide the time interval J into the time one component is out for PM and the time no components are out for PM. In the latter case we are in the same situation as described in the beginning of this section with no PM, and we can use S_i to approximate the expected portion of time i components are not functioning. In the former case the situation is described by one component out for PM and $n-1$ components normally operating. Thus a total of i components not functioning, means that $i-1$ of the $n-1$ normally operating components are not functioning. So, to compute the expected portion of time in this system state we can use the same type of formula as S_i, but with $n-1$ instead of n. Refer to Appendix D.5 for details.

The accuracy of the formula is demonstrated in Table 3.1 where the formula for S_2^{PM} has been computed for $n = 3$ under various assumptions and compared to the results obtained by Monte Carlo simulation. The program MIRIAM [56] has been used for the simulations. A time period of one year was considered. Two PM regimes were studied:

a) Each component is out for PM only once during the year

b) Each component is out for PM each month, for the same duration each month

Table 3.1: Comparisons of formula (3.17) for S_2^{PM} and simulated values (100-150 simulations were performed for each case giving a negligible estimated standard deviation)

Case	$\bar{a}$ (%)	$MTTF$ (hours)	$MTTR$ (hours)	q (%)	PM regime	Formula (3.17) for S_2^{PM} (%)	Simulated value for S_2^{PM} (%)
1	5	500	25.0	15	a	1.7	1.7
2	5	50	2.5	15	a	1.7	1.7
3	5	500	25.0	15	b	1.7	1.5
4	5	50	2.5	15	b	1.7	1.7
5	5	500	25.0	6	a	1.0	1.0
6	5	50	2.5	6	a	1.0	1.0
7	5	500	25.0	6	b	1.0	0.8
8	5	50	2.5	6	b	1.0	0.9
9	2	1250	25.0	15	a	0.6	0.7
10	2	125	2.5	15	a	0.6	0.6
11	2	1250	25.0	15	b	0.6	0.5
12	2	125	2.5	15	b	0.6	0.6
13	2	1250	25.0	6	a	0.3	0.3
14	2	125	2.5	6	a	0.3	0.3
15	2	1250	25.0	6	b	0.3	0.2
16	2	125	2.5	6	b	0.3	0.3

At each PM the three components are maintained sequentially. A PM is deferred if one or two units are under repair. The repair times are constant, $u_i = i$ and $v_i = i$ for all i.

The conclusion from the results in Table 3.1 and additional simulations is that the formula (3.17) gives very good approximations. In most cases (3.17) will overestimate S_2^{PM}. The largest differences in results are for cases with relatively large MTTR compared to the length of the PM, i.e. cases 3, 7, 11, 15. From the way the formula (3.17) was established this was to be expected. The S_i^* represents in this case a rough approximation of the expected portion of time that $i-1$ components of $n-1$ are not functioning, noticing that S_i^* represents a *limiting* availability.

As opposed to the type of repair distribution, the type of failure time distribution is critical for the accuracy of the formula. For non-exponential distributions the formula can give quite poor approximations depending on how the PM regimes affect the failure rates of the components.

Note that if q and $\overline{a}$ are small, then

$$\begin{array}{llllll} S_i^* & \approx 0 \text{ for } i > 2, & S_2^* & \approx \overline{a}\,(n-1)/v_1, & S_1^* & \approx 1 - S_2^* \\ S_i & \approx 0 \text{ for } i > 2, & S_2 & \approx \overline{a}^2\,(n-1)^2/u_2, & S_0 + S_1 & \approx 1 - S_2 \end{array}$$

Finally we note that a formula analogous to (3.17) can be established also for parallel systems without a standby component:

$$S_i^{PM} \approx S_i^*\, q + S_{i+1}(n)\,(1-q)$$

where $S_{i+1}^*(n)$ is defined as S_i^*, but with n instead of $n-1$ and U_i instead of V_i.

3.4.4 Periodical testing/replacement

We consider again a monotone system comprising n independent components. The components are put in operation at time $t = 0$. The system is tested and possibly repaired after time intervals of lengths τ. The state of the system can only be identified at a test. After a test (repair) the system is assumed to be good as new. For the time being we disregard test and repair times.

As an example, we can think about a fire detector system of a specific type. If the detector system fails in an interval between two tests, and a fire does not occur in the same interval, the failure will not be discovered before the next test.

An example of a state process for system/component is shown in Figure 3.7.

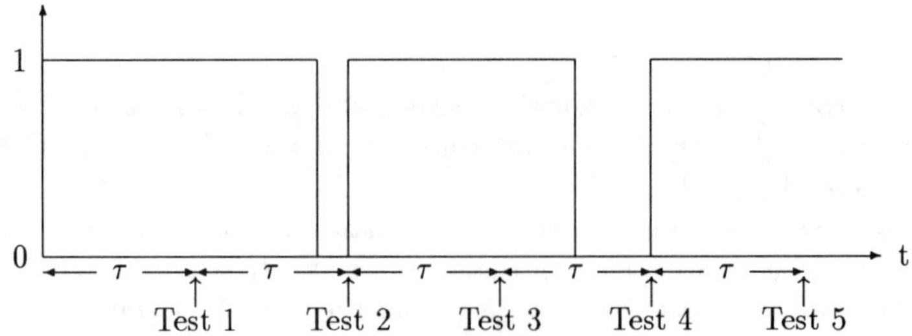

Figure 3.7: Example of a state process for system/component

Assume that component i has a lifetime distribution $F_i(t)$, $i = 1, 2, \ldots, n$. For each time period between two tests the availability of system/component will be identical as illustrated in Figure 3.8. It is therefore sufficient to consider the first time period.

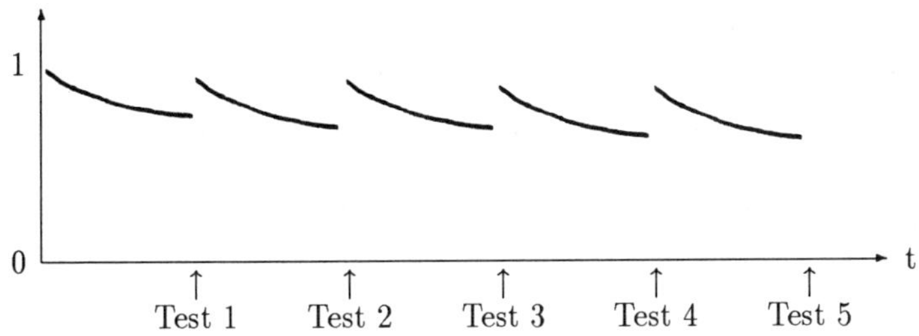

Figure 3.8: An example of availability as a function of time t

The availability of component i at time t, is given by $1 - F_i(t)$. To calculate system availability at time t we can proceed as explained in Section 3.2 and 3.3 with $q_i = F_i(t)$.

A measure that is often used to evaluate the performance of the system is the (expected) portion of time the system is not in the failure state, which is denoted by $MFDT$ (Mean Fractional Dead Time). It is easily seen from the Result E.6 of Appendix E that $MFDT$ can be expressed in the following way:

$$MFDT = \frac{1}{\tau} E \int_0^\tau [1 - \phi(\mathbf{X}(t))]\, dt = \frac{1}{\tau} \int_0^\tau E[1 - \phi(\mathbf{X}(t))]\, dt = \frac{1}{\tau} \int_0^\tau g(\mathbf{q}(t))\, dt \quad (3.18)$$

Note that $g(\mathbf{q}(t))$ represents the lifetime distribution of the system.

In the following we assume that the component lifetime distributions are exponential. Let λ_i denote the failure rate of component i. The integral in formula (3.18) can now in theory be calculated exactly. For example for $n = 1$ we find

$$MFDT = 1 - \frac{1}{\lambda\tau}(1 - e^{-\lambda\tau})$$

where $\lambda = \lambda_1$. In practice the following approximation formula is used for the computation of $MFDT$:

$$MFDT \approx \sum_{j=1}^{k} \frac{1}{|K_j| + 1} \prod_{i \in K_j} (\lambda_i \tau) \quad (3.19)$$

where K_j is jth minimal cut set, and $|K_j|$ equals the number of components in the set K_j, $j = 1, 2, \ldots, k$. This formula is established using (3.18) and the approximations

$$g(\mathbf{q}(t)) \approx \sum_{j=1}^{k} \prod_{i \in K_j} F_i(t) \approx \sum_{j=1}^{k} \prod_{i \in K_j} (\lambda_i t)$$

The above approximation symbols can be replaced by inequalities, $\leq$. The right hand side in (3.19) is therefore an upper bound on $MFDT$. For small values of $\lambda_i \tau$ the right hand side of (3.19) provides a good approximation of $MFDT$. It can be shown that if $\lambda_i\tau/2 \leq 1$, then

$$\sum_{j=1}^{k}\frac{1}{2}\frac{1}{(|K_j|+2)}\prod_{i\in K_j}\lambda_i\tau\,[\sum_{i\in K_j}\lambda_i\tau] + \sum_{j<l}\frac{1}{|K_j\bigcup K_l|+1}\prod_{i\in K_j\bigcup K_l}\lambda_i\tau$$

is an upper bound for the error term. To prove this, first note that by (3.5) we have

$$g(\mathbf{q}(t)) \geq w_1(t) - w_2(t) = \sum_{j=1}^{k}\prod_{i\in K_j}F_i(t) - \sum_{j<l}\prod_{i\in K_j\bigcup K_l}F_i(t)$$

Now, using

$$1-e^{-u}\leq u, \quad 1-e^{-u}\geq u-\frac{1}{2}u^2, \quad u\geq 0$$

and

$$\prod_i(1-a_i)\geq 1-\sum_i a_i, \quad 0\leq a_i\leq 1$$

we find that

$$\begin{aligned}
\prod_{i\in K_j}F_i(t) &= \prod_{i\in K_j}(1-e^{-\lambda_i t}) \geq \prod_{i\in K_j}\lambda_i t(1-\frac{1}{2}\lambda_i t)\\
&= (\prod_{i\in K_j}\lambda_i)\,t^{|K_j|}\prod_{i\in K_j}(1-\frac{1}{2}\lambda_i t) \geq (\prod_{i\in K_j}\lambda_i)\,t^{|K_j|}(1-\sum_{i\in K_j}\frac{1}{2}\lambda_i t)\\
&= (\prod_{i\in K_j}\lambda_i)\,t^{|K_j|} - \frac{1}{2}(\prod_{i\in K_j}\lambda_i)(\sum_{i\in K_j}\lambda_i)\,t^{|K_j+1|}\\
\prod_{i\in K_j\bigcup K_l}F_i(t) &= \prod_{i\in K_j\bigcup K_l}(1-e^{-\lambda_i t}) \leq (\prod_{i\in K_j\bigcup K_l}\lambda_i)\,t^{|K_j\bigcup K_l|}
\end{aligned}$$

Using the above inequalities and formula (3.18) the upper bound to the error term follows.
For an m-out-of-n system with $\lambda_i = \lambda$ for all i, the formula (3.19) can be written

$$MFDT \approx \binom{n}{r}\frac{(\lambda\tau)^r}{r+1} \tag{3.20}$$

where

$$r = n - m + 1$$

Noting that all the minimal cut sets of an m-out-of-n system comprise r components, and the number of minimal cut sets equals $\binom{n}{r}$, it is not difficult to see that an upper bound to the error term using the approximation (3.20) is given by

$$\binom{n}{r}\frac{1}{2}\frac{(\lambda\tau)^{r+1}}{r+2}\{r+\binom{n}{r}-1\}$$

for $\lambda\tau/2 \leq 1$. In Table 3.2 the formula (3.20) is computed for some common values of m and n.

Table 3.2: Formula (3.20) for $MFDT$ for some common m-out-of-n systems

m \ n	1	2	3	4
1	$\frac{\lambda\tau}{2}$	$\frac{(\lambda\tau)^2}{3}$	$\frac{(\lambda\tau)^3}{4}$	$\frac{(\lambda\tau)^4}{5}$
2		$\lambda\tau$	$(\lambda\tau)^2$	$(\lambda\tau)^3$
3			$\frac{3}{2}\lambda\tau$	$2(\lambda\tau)^2$
4				$2\lambda\tau$

Often $MFDT$ for a system is computed by $MFDT = g(\mathbf{q})$, where q_i equals $MFDT_i$ for component i when this component is considered in isolation, i.e.

$$MFDT_i \approx \frac{\lambda_i \tau}{2}$$

Using this approach the $MFDT$ of the system can be relatively roughly approximated, as this method means that the unavailabilities of the components are assumed to be equal to $MFDT_i$ independent of time. The information about the points in time of testing for the components are then disregarded.

Non-negligible test and repair times

Let us look at the case where the time it takes to test and repair the components is not negligible. Let T denote the expected test time, and let R_i denote the expected repair time for component i, given a failure of the component. We assume that the total repair time for the system equals the sum of the repair times for the failed components. The system is assumed not to operate under a test or a repair. Under these assumptions, $MFDT$ is computed using the following formula:

$$MFDT = \frac{\int_0^\tau F(t)\,dt + T + R}{\tau + T + R} = \frac{\frac{1}{\tau}\int_0^\tau F(t)\,dt + \frac{T+R}{\tau}}{1 + \frac{T+R}{\tau}} \tag{3.21}$$

where R equals the expected repair time for the system given by

$$R = \sum_{i=1}^{n} F_i(\tau)\,R_i$$

Example 3.10
Consider again the storage tank introduced in Section 2.1. As before we will study the system with respect to the failure mode "Overfilling of the tank", but we will now use a different approach. We assume that component i has a time to failure which is exponentially distributed with failure rate λ_i. If, for example, valve 1 fails at time t, this means that the first safety barrier will not function if the liquid level increases before the failure is discovered.

We assume that the components of the system are tested and repaired at intervals of length τ. The problem is to calculate the system $MFDT$ given that

$$\lambda_1 = \lambda_2 = \lambda_3 = 0.08,\ \lambda_4 = \lambda_5 = 0.04,\ \tau = 0.5$$

The time unit is one year. Notice that for these values the component i's $MFDT$ is approximately 2% for $i = 1$, 2 and 3, and 1% for $i = 4$ and 5, which is in accordance with the unreliabilities used earlier.

Now using (3.19) we find that

$$\begin{aligned}\text{MFDT} &\approx \tfrac{1}{3} \times 0.08 \times 0.04 \times 0.5^2 + \tfrac{1}{3} \times 0.04^2 \times 0.5^2 \\ &\quad + \tfrac{1}{4} \times 0.08^3 \times 0.5^3 + \tfrac{1}{4} \times 0.08^2 \times 0.04 \times 0.5^3 \\ &= 0.04\%\end{aligned}$$

The error term is negligible, so that we have $MFDT = 0.04\%$. Hence the expected portion of time the system does not function is slightly higher than the system unreliability found earlier (0.03%).

The above calculations assume that test and repair times are negligible. With negligible test times, but repair times of 3 hours for each component, we find by using (3.21) that

$$MFDT \approx 0.04\% + 0.01\% = 0.05\%$$

□

In many practical situations the test intervals are not constant, but vary. Consider a component with lifetime distribution $F(t) = 1 - e^{-\lambda t}$ and let τ be a stochastic variable representing the test interval for this component. Let $E\tau$ and $Var\tau$ denote expectation and variance of τ, respectively. Then

$$MFDT = \frac{E\{\int_0^\tau F(t)\,dt\}}{E\tau} \approx \frac{E\{\int_0^\tau \lambda t\,dt\}}{E\tau} = \frac{\lambda \frac{1}{2} E\tau^2}{E\tau} = \frac{C\lambda\, E\tau}{2}$$

where

$$C = 1 + \frac{Var\tau}{(E\tau)^2}$$

Thus, if we use the formula $MFDT \approx \lambda\tau/2$, where τ represents the mean test interval, we will underestimate $MFDT$. By a similar argument it follows that (3.19) holds with $\tau^{|K_j|}$

replaced by

$$\frac{E_T{}^{|K_j+1|}}{E_T}$$

For presentation of other similar models for analysing systems undergoing tests, see e.g. [16,59].

3.5 Reliability importance measures

An important objective of many reliability and risk analyses is to identify those components or events that are most critical from a safety point of view, and should be given priority with respect to improvements. To do this we need a *criticality measure*. A large number of such measures have been suggested [12,13,23,27,33,51,64,99]. Here we shall restrict ourselves to two measures, improvement potential [12] (Section 3.5.1) and Birnbaum's measure [33] (Section 3.5.2). In Section 3.5.3 some general comments on the use of criticality measures are given.

Example 3.11
Consider again Example 2.1. The unreliability of the system with respect to the failure mode "Overfilling of the tank" equals

$$\begin{aligned} g &= \{1 - p_1 p_4\}\{1 - p_5(p_2 + p_3 - p_2 p_3)\} \approx \\ w_1 &= q_1 q_5 + q_4 q_5 + q_1 q_2 q_3 + q_2 q_3 q_4 \\ &= 0.02 \times 0.01 + 0.01 \times 0.01 + 0.02 \times 0.02 \times 0.02 + 0.02 \times 0.02 \times 0.01 \\ &= 0.03 \times 10^{-2} \end{aligned}$$

cf. the Examples 2.1 and 3.4. If we look at the subsystems comprising the minimal cut sets, it is clear from the above expression that the subsystems $\{1, 5\}$ and $\{1, 4\}$ are most critical in the sense that they are contributing most to unreliability. To decide which components are most critical we must define more precisely what is meant by critical. For example, we might decide to let the component with the highest potential for increasing the system reliability be most critical (measure for reliability improvement potential) or the component that has the largest effect on system reliability by a small improvement of the component reliability (Birnbaum's measure).

3.5.1 Improvement potential

The following reliability importance measure for component i, I_i^A, is appropriate in a large number of situations, in particular during design:

$$I_i^A = h_i - h$$

where h is the reliability of the system and h_i is the reliability assuming that component i is in the best state. The measure I_i^A expresses the system reliability improvement potential of the component, in other words the unreliability that is caused by imperfect performance of component i. This measure can be used for all types of reliability definitions, and it can be used for repairable or non-repairable systems.

For a highly reliable monotone system with reliability interpreted as the probability that the system is functioning at a certain point in time, the measure I_i^A is equivalent to the well-known Vesely-Fussells importance measure [64]. In fact, in this case I_i^A is approximately equal to the sum of the unreliabilities of the minimal cut sets which include component i, i.e.

$$I_i^A \approx \sum_{j:i \in K_j} \prod_{l \in K_j} q_l \tag{3.22}$$

Notice that the measure I_i^A gives like the Vesely-Fussells measure the same importance to all the components of a parallel system, irrespective of component reliabilities. This is as it should be because each one of the components has the *potential* of making system's unreliability negligible, for example by introducing redundancy, cf. the discussion in Section 3.5.3.

Example 3.11 continued
Computation of I_i^A based on (3.22) gives

$$\begin{aligned} I_1^A &= 0.02 \times 10^{-2}, & I_2^A &= 0.00 \times 10^{-2}, & I_3^A &= 0.00 \times 10^{-2}, \\ I_4^A &= 0.01 \times 10^{-2}, & I_5^A &= 0.03 \times 10^{-2} \end{aligned}$$

Thus component 5 is the most critical component based on this measure. The components 1 and 4 follow in the second and third place, respectively.

3.5.2 Birnbaum's measure

Birnbaum's measure for the reliability importance of component i, I_i^B, is defined by

$$I_i^B = \frac{\partial h}{\partial p_i} \tag{3.23}$$

Hence Birnbaum's measure is found by partial differentiation of system reliability with respect to p_i. The approach is well known in classic sensitivity analysis. We see that if I_i^B is large, a small change in the reliability of component i will give a relatively large change in system reliability.

Birnbaum's measure might be appropriate for example in the operation phase where possible actions are related to operation and maintenance parameters. Before looking closer into specific improvement actions of the components, it will be informative to measure the sensitivity of the system reliability with respect to small changes in the reliability of the components.

To compute I_i^B the following formula is often used:

$$I_i^B = h(1_i, \mathbf{p}) - h(0_i, \mathbf{p}) \tag{3.24}$$

where $h(\cdot_i, \mathbf{p}) = h(p_1, p_2, \ldots, p_{i-1}, \cdot, p_{i+1}, \ldots, p_n)$. This formula is established using (3.4).

Example 3.11 continued
Using (3.24) we find that

$$\begin{aligned} I_1^B &= 1.03 \times 10^{-2} &= 1.0 \times 10^{-2}, \quad I_2^B &= I_3^B = 0.09 \times 10^{-2}, \\ I_4^B &= 1.02 \times 10^{-2}, &= 1.0 \times 10^{-2}, \quad I_5^B &= 3.0 \times 10^{-2} \end{aligned}$$

To demonstrate the calculations, let us look at the calculations of I_5^B.

First, notice that

$$h(1_i, \mathbf{p}) - h(0_i, \mathbf{p}) = g(1_i, \mathbf{q}) - g(0_i, \mathbf{q})$$

where $g(\mathbf{q}) = 1 - h(\mathbf{p})$. Using the approximation $g \approx w_1$ with $q_5 = 0$ we find that $g(0_5, \mathbf{q}) = 0.01 \times 10^{-3}$. To compute $g(1_5, \mathbf{q})$ we cannot use this approximation since q_5 is not small ($q_5 = 1$). Instead we consider a modified system where it is known that component 5 is not functioning, cf. Example 3.5. This modified system is a series system comprising the components 1 and 4. It follows that

$$g(1_5, \mathbf{q}) \approx q_1 + q_4 = 0.02 + 0.01 = 0.03 = 3.0 \times 10^{-2}$$

The error term is negligible. Thus $I_5^B = 3.0 \times 10^{-2}$.

Alternatively, I_i^B can be computed using (3.23). We then write

$$h(\mathbf{p}) = 1 - \{1 - p_1p_4\}\{1 - p_5(p_2 + p_3 - p_2p_3)\} = p_1p_4 + p_5(p_2 + p_3 - p_2p_3)(1 - p_1p_4)$$

so that

$$\begin{aligned} I_5^B = \frac{\partial h}{\partial p_5} &= (p_2 + p_3 - p_2p_3)(1 - p_1p_4) \\ &= (1 - 0.04 \times 10^{-2})(1 - 0.98 \times 0.99) = 3.0 \times 10^{-2} \end{aligned}$$

We see that for this example the Birnbaum measure gives the same ranking of the components as the measure I_i^A. This is, however, not true in general, see Problem 3.7.12.

3.5.3 Some comments on the use of importance measures

The two criticality measures presented in this section can be useful tools in the system optimization process/system improvement process. This process can be described as follows:

1. Identify the most critical units by means of the chosen criticality measure
2. Identify possible improvement actions/measures for these units
3. Estimate the effect on reliability by implementing the measure
4. Perform cost evaluations
5. Make an overall evaluation and take a decision

The criticality measure to be used in a particular case depends on the characteristics we want the measure to reflect. Professor Bo Bergman expresses this in [27] in the following way: "For the choice of a specific importance measure it is important to make clear in what sort of situation it is to be used. Undoubtedly, different situations call for different importance measures". In a design phase the system reliability improvement potential I_i^A might be the most informative measure, but for a system with frozen design, the Birnbaum measure might be more informative since this measure reflects how small component reliability improvements affect system reliability.

Some of the importance measures introduced are time dependent, as for example the Birnbaum measure, Vesely-Fussell's measure and the measure I_i^A for certain definitions of reliability. In practice, however, limiting availabilities are often used, so that the time dependency is eliminated.

When identifying the most critical components in a system it is important that the components are defined in an appropriate way. This is illustrated in Problem 3.7.13.

The two criticality measures, I_i^A and I_i^B, can be fitted to other models than the one presented above. For example, if we study the risk related to an activity, we can rank events or categories of events by calculating the risk reduction by disregarding the possibility that these events shall occur. In particular such a criticality ranking might be incorporated in a CCA/ETA, where e.g. the importance of the various tree branches are studied. Similarly, the effect on risk (e.g. expected loss) of small changes in probabilities/frequencies of certain events can be measured by partial differentiation of the risk with respect to the actual parameters.

3.6 Dependent components

Up to now we have assumed that the components are functioning/not functioning independently of each other. In many situations there will be some sort of dependency between the components. The main causes can be grouped in the following way:

1. A common external cause can result in failures of several components. Examples are common electric power, fire, flood, and erroneous maintenance.

2. Mutual influence. Components can affect each other through the internal environment. Examples are temperature, pressure and vibration.

3. Components supporting loads. If a set of components share a common load, failure of one component increases the load supported by the other components.

4. The occurrence of a specific failure event may preclude the possibility of another failure event. For example, after the failure of a fuse the electricity system is shut down and the bulb cannot burn out.

Normally, the dependency of category 4 can be disregarded since the time a component is down due to failures of other components is usually small compared to the total operating time. On the other hand side, frequent shut downs may in some cases result in increased failure rate due for example to increased start-up load.

If a component fails, the remaining components may

i) either reduce its tendency to failure

ii) or increase its tendency to failure.

If the situation is as described by *i*), we say that there is a negative dependency, and if the situation is as described by *ii*), we say that there is a positive dependency.

Failures as mentioned above, are often denoted "common-mode failures". In the following we will adopt this usage.

The fire at the nuclear power station Brown Ferry on 22 March 1975 is one of the most well-known accidents that were caused by common-mode failures. The fire started when two of the operators used a stearin candle to identify draught (air leakage) in the reactor building. The air was drawn out through a cable gate. The flames from the stearin candle were drawn out in the cable gate, and the polyurethene foam in the cable gate was ignited. The fire destroyed the insulation of approximately 2000 cables. Among these were all the cables of the automatic shut-down system, and also the cables of all the manually controlled valves - except 4 valves. With these four valves the reactors were closed so that a core melt was avoided.

After this accident new regulations were put in force, which stated that the cables of the various shut-down systems had to be in separate cable gates, and that no inflammable material must be used, such as polyurethene foam.

The most efficient defence against the consequence of randomly occurring component failures is to use redundancy, i.e. two or more units that can perform the same function, so that a failure of one component does not result in system failure. The fire at Brown Ferry shows that even redundancy is not sufficient. At Brown Ferry there was a number of redundant shut-down systems, but they were all made inactive by a small failure - the flame from a stearin candle.

Common-mode failures will often have a great effect on system reliability. In particular this is the case in highly reliable systems.

When working with common-mode failures, a detailed classification system has proved to be very useful. The grouping in Figure 3.9, which is taken from [45], gives an impression of the large number of causes that can result in common-mode failures.

ENGINEERING				OPERATIONS			
Design		Construction		Procedural		Environmental	
Functional deficiencies	Design realisation	Manufacture	Installation and commissioning	Maintenance and test	Operation	Normal extremes	Energetic events
Logical error	Channel dependency	Inadequate quality control	Inadequate quality control	Imperfect repair	Operator errors	Temperature	Fire
Inadequate measurement	Common operation & protection components	Inadequate standards	Inadequate standards	Imperfect testing	Inadequate procedures	Pressure	Flood
Inadequate control	Operational deficiencies	Inadequate inspection	Inadequate inspection	Imperfect calibration	Inadequate supervision	Humidity	Weather
Inadequate response	Inadequate components	Inadequate testing	Inadequate testing and commissioning	Imperfect procedures	Communication error	Vibration	Earthquake
	Design errors				Inadequate supervision	Acceleration	Explosion
	Design limitations					Stress	Missiles
						Corrosion	Electrical power
						Contamination	Radiation
						Interference	Chemical sources
						Radiation	
						Static charge	

Figure 3.9: Classification of common-mode failures [45]

3.6.1 Analysing dependent components (common-mode failures)

One of the most difficult tasks in a reliability analysis is to take into account common-mode failures. It is difficult to formulate the dependency in a mathematically stringent way and at the same time obtain a realistic model, and to provide data for the model.

Whether we succeed in incorporating a correct contribution from common-mode failures is very much dependent on the modelling ability of the analyst. By defining the components in a suitable way, it is often possible to preclude dependency. For example, common-mode failures that are caused by a common external cause can be identified and separated out so that the components can be considered as independent components. To demonstrate how this is done, let us consider a parallel system comprising n components. We assume that the causes of a component failure can be as follows:

1. The failure is due to factors only affecting the component itself, and it is independent of the states of the other components.

2. The failure is due to a common external event, e.g. failure of the electricity system, and when this event occurs all the components will fail simultaneously.

By modelling the system as shown in Figure 3.10, where the external event is considered as an "hypothetical" component F in series with the rest of the system, the dependency is eliminated from the analysis. The components of the parallel system are now assumed to fail only due to internal causes.

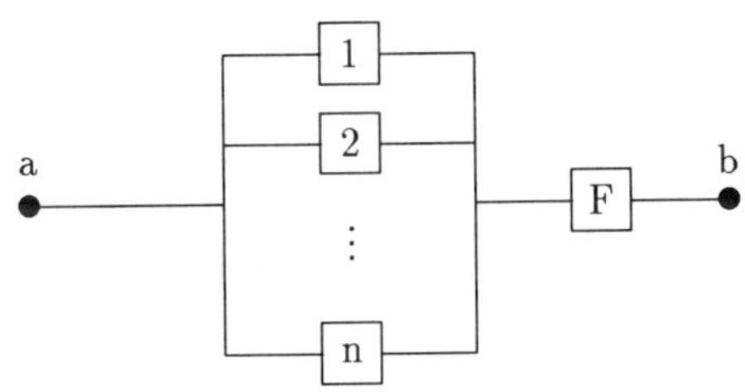

Figure 3.10: Parallel structure with common-mode failure source F

Another useful method for "elimination" of dependency is to redefine components. For example, instead of including a parallel structure of dependent components in the system, this structure be represented by one component. Of course, this does not remove the dependency, but it moves it to a lower level of the analysis. Special techniques, such as Markov modelling (see Appendix D), can then be used to analyse the parallel structure itself, or we can try to estimate reliability parameters directly for this new component.

Although it is often possible to "eliminate" dependency between components by proper modelling, it will in many cases be required to establish a model that explicitly take into account the dependency. We have already mentioned Markov models. There exist also other methods, such as the square root method [135] and the Heikkila model [63]. Care should, however, be taken when using these models in practice if in the given situation the choice of the particular common-mode failure model cannot be justified. Often it will be extremely difficult to estimate the parameters in these models.

Another way of taking into account dependency is to obtain bounds to the system reliability, assuming that the components are *associated* and not necessarily independent. Association is a type of positive dependency, for example as a result of components supporting loads. The precise mathematical definition is given in e.g. [24]. A number of results are established for associated components, for example the following inequalities:

$$\max_{1\leq j\leq s} \prod_{i\in S_j} p_i \leq h \leq 1 - \max_{1\leq j\leq k} \prod_{i\in K_j} q_i$$

where S_j equals the jth minimal path set, $j = 1, 2, \ldots, s$ and K_j equals the jth minimal cut set, $j = 1, 2, \ldots, k$. As far as applications are concerned, this approach does not seem to have won through. The main reason for this is the fact that this method usually leads to very wide intervals for reliability. If the problem is to compare different design alternatives or system modifications, or to identify the most critical components, such intervals are barely informative and represent a poor basis for making decisions. In the first phase of a reliability analysis or a risk analysis information about the dependency might be so vague that we are forced to apply bounds based on associated components. However, in order to arrive at results that can be used as a basis for making decisions, the dependencies must be looked more closely into and separated out from the top-level analysis, cf. the comments above.

3.7 Problems

1. Consider Problem 2.5.5. Calculate system unreliability given that the system is a parallel-series structure.

2. Consider the bridge structure as shown in Figure 3.5. Assume that the component unreliabilities q_i are given by:

   i) $q_i = 0.5$ for $i = 1, 2, \ldots, 5$

   ii) $q_1 = q_2 = 1 \times 10^{-2}$, $q_3 = 0.23$, $q_4 = q_5 = 0.75 \times 10^{-2}$

   Assuming independent components, calculate system unreliability.

3. Consider a system of 5 components and the minimal cut sets $\{1, 4, 5\}$ and $\{2, 3, 4, 5\}$. Assume that the lifetimes of the components are Weibull distributed with shape parameter $\beta = 2$ and scale parameter $\lambda = 1$, and that they are independent. Compute the system reliability at time 0.5 assuming that no repair/replacement is performed in the interval [0,0.5] and the components are "new" at time zero.

4. A system is defined by the following reliability block diagram:

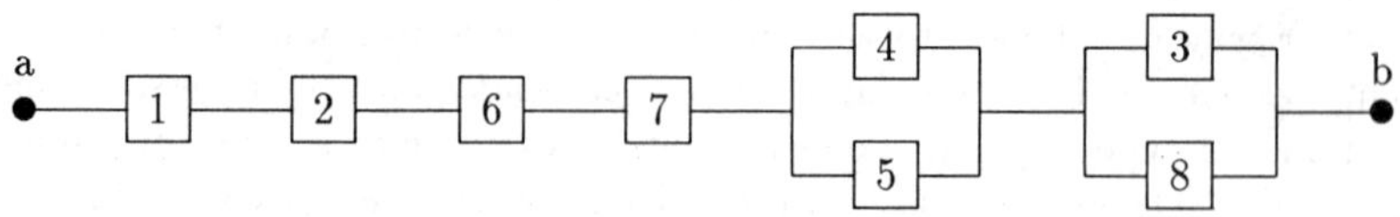

   Assume that we have given the following reliability data:

Component	MTTF (hours)	MTTR (hours)
1	1×10^5	50
2	2×10^4	50
3	1×10^2	5
4	1×10^2	5
5	1×10^2	5
6	2×10^4	50
7	2×10^4	50
8	1×10^2	5

A component failure is immediately detected, and the repair starts. Find the limiting availability for this system. State the assumptions you must make to be able to find this quantity.

5. A system comprises two identical stream paths A and B. Both stream paths must function to obtain full production. An alternative configuration is considered to increase the availability of the system: to install a third identical stream path such that the system gives full production if and only if at least two of the stream paths are functioning. Each stream path consists of one pump and one rotation filter, which must both function to secure a functioning stream path. It is assumed that pumps and filters are functioning and are repaired independently of each other. The pump unavailabilities equal 0.8%, whereas the filter unavailabilities equal 3.2%.

   a) Compute the availability of full production for the two systems. Explain what is meant by availability.

   b) Assume that the total cost associated with a pump equals 350 dollars per day (including installation, maintenance and operation), whereas the total cost associated with the rotation filters equals 1600 dollars per day. If the system cannot deliver from at least two stream paths, the company suffers a penalty of 50000 dollars per day. Which of the two systems should be installed if we want to minimize the expected costs?

6. Consider Example 3.9.

   a) Calculate S_2, i.e. the expected portion of time that both components are "down" when $u_i = 1$, $i = 1, 2$. Compare with the value found in Example 3.9.

   b) Assume that the two components can also be taken out of operation due to preventive maintenance (PM). We make the same assumptions as in Section 3.4.3. Compute S_2^{PM}, i.e. the expected portion of time that both components are "down" when each of the two components is out for PM approximately 5% of the time (438 hours per year). Set $v_1 = 1$.

7. Given a parallel structure of n independent components with failure rates $\lambda_i = 1 \times 10^{-5}$. The components are tested and possibly repaired at even intervals of length $\tau = 1000$. The test and repair times are ignored.

   a) Calculate the probability that the system does not function just before a test if $n = 2$.

   b) Find the $MFDT$ of the system for $n = 1$ and 2. Explain what is meant by $MFDT$.

   c) A requirement of a $MFDT$ of the system not exceeding 1×10^{-6} has been formulated. How large must n be to satisfy this requirement?

8. Let the situation be as described in Problem 7 above, but assume now that the system is a bridge structure. Compute the $MFDT$ of the system (approximate).

9. In Example 3.10, show that $MFDT \approx 0.05\%$ when the repair time equals 3 hours for each component.

10. In this exercise we study an alternative method for exact reliability calculations. As we will see, the method is only of practical interest for smaller system. To illustrate the method we look at a 2-out-of-3 system. The model is described in Section 3.1 and 3.2.

    a) Find the minimal cut sets of the system.

    b) Explain why

    $$\phi = (X_1 \coprod X_2) \cdot (X_1 \coprod X_3) \cdot (X_2 \coprod X_3)$$

    c) Show by multiplying out the right hand side in the above expression that

    $$\phi = X_1 \cdot X_2 + X_1 \cdot X_3 + X_2 \cdot X_3 - 2 \cdot X_1 \cdot X_2 \cdot X_3$$

    Hint: $X_i^r = X_i$ for $r = 1, 2, \ldots$.

    d) Use the expression in c) to show that

    $$h = p_1\, p_2 + p_1\, p_3 + p_2\, p_3 - 2\, p_1\, p_2\, p_3$$

    e) Generalize the method to an arbitrary monotone system. Discuss the efficiency of the method.

    f) The multiplication method described above can also be used based on the *minimal path sets*. Show how this can be done for the 2-out-of-3 system.

11. Perform the calculations of I_i^A and I_i^B in Example 3.11.

12. Consider the system shown below. We assume that the components are independent and that

$$p_1 = p_2 = 0.5, \quad p_3 = 0.9$$

Compute I_i^A and I_i^B and comment on the result.

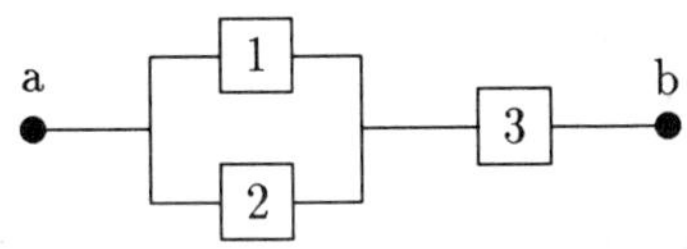

13. Let us assume that the system in Problem 12 comprises two components: component 3 and a new component representing the parallel system of the components 1 and 2. This new component will have a reliability equal to 0.75. Compute Birnbaum's measure and comment on the result.

14. This problem will look closer into Birnbaum's measure. The model is as described in Section 3.5.

   a) Show that

$$I_i^B = E[\phi(1_i, \mathbf{X}) - \phi(0_i, \mathbf{X})]$$

   b) Show by using the equality of a) that

$$I_i^B = P(\phi(1_i, \mathbf{X}) - \phi(0_i, \mathbf{X}) = 1) = P(\phi(1_i, \mathbf{X}) = 1, \ \phi(0_i, \mathbf{X}) = 0)$$

   If $\phi(1_i, \mathbf{x}) - \phi(0_i, \mathbf{x}) = 1$, we call $(1_i, \mathbf{x})$ a *critical path vector* and $(0_i, \mathbf{x})$ a *critical cut vector* for component i. For simplicity, we often say that component i is *critical* for the system.
   We have shown that I_i^B equals the probability that the system is in a state so that the component i is critical for the system. If the components are dependent, this probability can be used as a definition of Birnbaum's measure.

   c) Set $p_j = 1/2$ for all $j \neq i$. Show that

$$I_i^B = \frac{1}{2^{n-1}} \sum_{(\cdot_i, \mathbf{x})} [\phi(1_i, \mathbf{x}) - \phi(0_i, \mathbf{x})]$$

   This quantity is used as a measure of the *structural* importance of component i. Give an interpretation of this measure.

15. Consider the model described in Section 3.4.1. Use Problem 14 a) and b) to prove heuristically that the expected number of system failures per unit time is approximately given by

$$\sum_{i=1}^{n} [h(1_i, \mathbf{A}) - h(0_i, \mathbf{A})]/(MTTF_i + MTTR_i)$$

16. Consider Example 3.7. Estimate the average number of system failures for the two alternatives.

17. Consider the system shown in Problem 12. Assume that the components 1 and 2 are dependent, but independent of component 3. Furthermore, assume that $q_1 = q_2 = 0.10$ and $q_3 = 0.01$.

    a) Assume that the dependency between the components 1 and 2 is given by: $P(X_1 = 0 \mid X_2 = 0) = 0.5$. Find the system unreliability g. Try to comment on the result.

    b) Find bounds on the system unreliability g when we assume that the components are associated.

18. A system is said to be coherent if *a*) its structure function ϕ is non-decreasing in each argument and *b*) each component is relevant. Component i is relevant if there exists at least one vector $(\cdot_i, \mathbf{x})$ such that $\phi(1_i, \mathbf{x}) = 1$ and $\phi(0_i, \mathbf{x}) = 0$.

    Show that ϕ coherent implies that ϕ is monotone. Give an example of a system which is monotone but not coherent.

    The examples of monotone systems studied in this chapter are all coherent. Most textbooks on system reliability concentrate on coherent structures.

19. a) Let ϕ be a monotone structure function. Show that

$$i)\ \phi(\mathbf{x} \coprod \mathbf{y}) \geq \phi(\mathbf{x}) \coprod \phi(\mathbf{y})$$

    where $\mathbf{x} \coprod \mathbf{y} = (x_1 \coprod y_1, \ldots, x_n \coprod y_n)$.

    b) Let h be the reliability function of a monotone system. Show that

$$ii)\ h(\mathbf{p} \coprod \mathbf{p}') \geq h(\mathbf{p}) \coprod h(\mathbf{p}')$$

    for all $\mathbf{0} \leq \mathbf{p} \leq \mathbf{1}$ and $\mathbf{0} \leq \mathbf{p}' \leq \mathbf{1}$.

    These results state that redundancy at the component level is more effective than redundancy at system level. This principle is well known among design engineers. Note that if the system is a parallel system, then equality holds in *i*) and *ii*).

    c) Show that if the system is coherent (see Problem 18), then equality in *i*) and *ii*) holds if and only if the system is a parallel system.

    d) Formulate and prove similar results for $\phi(\mathbf{x} \cdot \mathbf{y})$ and $h(\mathbf{p} \cdot \mathbf{p}')$.

Chapter 4

MULTISTATE MODELS FOR ANALYSING SYSTEM RELIABILITY

In this chapter parts of the theory presented in Chapter 3 will be generalized to include multistate systems where component and system are allowed to have an arbitrary number of states/levels. It will focus on flow network models which are a significant special case of multistate systems. Flow network models are used to model e.g. production and transportation systems for oil and gas, and power transmission systems.

As in Chapter 3 the risk concept is not explicitly mentioned. Refer to the introduction of Chapter 3 for a comment on this.

The presentation of the multistate reliability theory in this chapter is based on [9,11, 13,18,100]. For some other interesting related papers, see [37,38,39].

4.1 Multistate monotone systems

We consider a system comprising n components, numbered consecutive from 1 to n. As in the binary case, x_i represents the state of component i, $i = 1, 2, \ldots, n$, but now x_i can be in one out of $M_i + 1$ states,

$x_{i0}, x_{i1}, x_{i2}, \ldots, x_{iM_i} \quad (x_{i0} < x_{i1} < x_{i2} < \cdots < x_{iM_i})$

The set comprising these states is denoted S_i. The states x_{ij} represent for example different levels of performance, from the worst x_{i0}, to the best x_{iM_i}. The states $x_{i0}, x_{i1}, \ldots, x_{i,M_i-1}$ are referred to as the failure states of the components.

Similarly, $\phi = \phi(\mathbf{x})$ denotes the state (level) of the system. The various values ϕ can take are denoted

$\phi_0, \phi_1, \ldots, \phi_M \quad (\phi_0 < \phi_1 < \cdots < \phi_M)$

We see that if $M_i = 1$, $i = 1, 2, \ldots, n$ and $M = 1$, then the model is identical with the binary model of Chapter 3.

Monotone structure
Analogous to the binary case we define:

A structure is said to be monotone if

1. $\phi(x_{10}, x_{20}, \ldots, x_{n0}) = \phi_0$ and $\phi(x_{1M_1}, x_{2M_2}, \ldots, x_{nM_n}) = \phi_M$
2. The structure function is non-decreasing in each argument

In the following we will restrict attention to monotone structures.

Minimal cut vectors
A vector $\mathbf{z}$ is a cut vector to level c iff $\phi(\mathbf{z}) < c$. A cut vector to level c, $\mathbf{z}$, is minimal iff $\phi(\mathbf{x}) \geq c$ for all $\mathbf{x} > \mathbf{z}$ ($(x_1, x_2, \ldots, x_n) > (z_1, z_2, \ldots, z_n)$ means that $x_i \geq z_i$, $i = 1, 2, \ldots, n$ and there exists at least one i such that $x_i > z_i$.)

Minimal path vector
A vector $\mathbf{y}$ is a path vector to level c iff $\phi(\mathbf{y}) \geq c$. A path vector to level c, $\mathbf{y}$, is minimal iff $\phi(\mathbf{x}) < c$ for all $\mathbf{x} < \mathbf{y}$.

Example 4.1
Figure 4.1 shows a simple example of a flow network model. The system comprises three components. Flow (gas/oil) is transmitted from a to b. The components 1 and 2 are binary, whereas component 3 can be in one out of three states: 0,1, or 2. The states of the components are interpreted as flow capacity rates for the components. The state/level of the system is defined as the maximum flow that can be transmitted from a to b, i.e.

$$\phi = \phi(\mathbf{x}) = minimum\{x_1 + x_2, x_3\}$$

If for example the component states are $x_1 = 0$, $x_2 = 1$ and $x_3 = 2$, then the flow throughput equals 1, i.e. $\phi = \phi(0, 1, 2) = 1$.

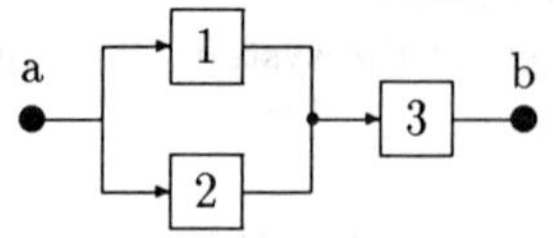

Figure 4.1: A simple example of a flow network

The possible system levels are 0, 1 and 2. We see that ϕ is a multistate monotone system. The minimal cut vectors and path vectors are as follows:

System level 2
Minimal cut vectors: (0,1,2),(1,0,2) and (1,1,1)
Minimal path vectors : (1,1,2)

System level 1
Minimal cut vectors: (0,0,2) and (1,1,0)
Minimal path vectors : (0,1,1) and (1,0,1)

4.2 Computing system reliability

Assume that the state X_i of the ith component is a stochastic variable, $i = 1, 2, \ldots, n$. Let

$$\begin{aligned} p_{ij} &= P(X_i = x_{ij}) \\ h_j &= P(\phi(\mathbf{X}) \geq \phi_j) \\ a &= E\phi(\mathbf{X})/\phi_M = \textstyle\sum_j \phi_j \, P(\phi(\mathbf{X}) = \phi_j)/\phi_M \end{aligned}$$

We call h_j the reliability of the system at system level j. Note that a represents the expected throughput (flow) relatively to the maximum throughput (flow) level. As in the binary case we write $X_i(t), \mathbf{X}(t)$, and so forth, if we refer to a specific point in time t.

The problem is to compute h_j for one or more values of j, and a, based on the p_{ij}s. We assume that the components are functioning/not functioning independently of each other.

Example 4.1 continued
Assume that

$$\begin{aligned} p_{i1} &= 1 - p_{i0} = 0.96, \; i = 1, 2 \\ p_{32} &= 0.97, \; p_{31} = 0.02, \; p_{30} = 0.01 \end{aligned}$$

Then by simple probability calculus we find that

$$\begin{aligned} h_2 &= P(X_1 = 1, X_2 = 1, X_3 = 2) = 0.96 \times 0.96 \times 0.97 = 0.894 \\ h_1 &= P(X_1 = 1 \cup X_2 = 1, X_3 \geq 1) = P(X_1 = 1 \cup X_2 = 1) \, P(X_3 \geq 1) \\ &= \{1 - P(X_1 = 0) \, P(X_2 = 0)\} \, P(X_3 \geq 1) = 0.9984 \times 0.99 = 0.988 \\ a &= (0.094 \times 1 + 0.894 \times 2)/2 = 0.941 \end{aligned}$$

□

For the above example it is a easy to calculate the system reliability directly by using elementary probability rules. For larger systems it will be very time-consuming (in some cases impossible) to perform these calculations if special techniques/algorithms are not used. If the minimal cut vectors or path vectors for a specific level are known, the system reliability for this level can be computed exactly, using for example the algorithm described in Appendix F. For highly reliable systems, which are most common in practice, simple approximations can be used as described in the following.

Analogous to the binary case, approximations can be established based on the inclusion-exclusion method:

$$1 - h_j \approx \sum_r \prod_{i=1}^{n} P(X_i \leq z_i^r) \tag{4.1}$$

where $(z_1^r, z_2^r, \ldots, z_n^r)$ represents the rth cut vector for level j. The right-hand side of (4.1) is always greater than or equal to the left-hand side. The error introduced by using (4.1) is less than or equal to

$$\sum_{r<l} \prod_{i=1}^{n} P(X_i \leq minimum\{z_i^r, z_i^l\})$$

Another simple calculation method for highly reliable systems is [11]:

$$h_j \approx 1 - \sum_{i,r:\phi(r_i,\mathbf{M})<\phi_j} p_{ir} \tag{4.2}$$

$$a \approx 1 - \sum_{i,r} \{1 - \phi(r_i,\mathbf{M})/\phi_M\}\, p_{ir} \tag{4.3}$$

where

$$\phi(r_i,\mathbf{M}) = \phi(x_{1M_1}, \ldots, x_{i-1,M_{i-1}}, x_{ir}, x_{i+1,M_{i+1}}, \ldots, x_{nM_n})$$

For each component i we sum all component states that give system failure, i.e. $\phi(r_i,\mathbf{M}) < \phi_j$. The possibility of two or more components being in a failure state at the same time is precluded. The error term by using (4.2) and (4.3) is less than or equal to:

$$\delta = \sum_{k=2}^{n} \sum_{(i_1,i_2,\ldots,i_k)\in L_k} \prod_{j=1}^{k} (1 - P_{i_j M_{i_j}})$$

where

$$L_k = \{(i_1, i_2, \ldots, i_k) : 1 \leq i_1 < i_2 < \cdots < i_k \leq n \ , \ i_j \in \{1,2,\ldots,n\}\}$$

Note that a term in the above sum equals the probability that two or more specific components are in a failure state. In Appendix A it is shown that δ is an error bound of (4.2) and (4.3).

Example 4.1 continued

If we use (4.1) to calculate h_j, we obtain

$$\begin{aligned} h_2 &\approx 1 - (0.04 \times 1 \times 1 + 1 \times 0.04 \times 1 + 1 \times 1 \times 0.03) &= 0.890 \\ h_1 &\approx 1 - (0.04 \times 0.04 \times 1 + 1 \times 1 \times 0.01) &= 0.988 \\ a &\approx (1 \times 0.098 + 2 \times 0.890)/2 &= 0.939 \end{aligned}$$

We can conclude that the approximations are quite good for this example.

If we use the formulae (4.2) and (4.3), we find

$$
\begin{aligned}
h_2 &\approx 1 - (0.04 + 0.04 + 0.03) &&= 0.890 \\
h_1 &\approx 1 - 0.01 &&= 0.990 \\
a &\approx 1 - (0.5 \times 0.04 + 0.5 \times 0.04 + 0.5 \times 0.02 + 1 \times 0.01) && \\
&= 1 - 0.06 &&= 0.940
\end{aligned}
$$

Also with the simple methods (4.2) and (4.3) we get good approximations for this example.

□

The above calculation methods may be applied to flow network systems as well as other multistate monotone systems. There exist also some algorithms designed especially for flow network systems. One of the most efficient of these seems to be the algorithm described in [44], see Appendix F.

The problem of determining the p_{ij}s, will as in the binary case, depend on the particular situation considered. Often it will be appropriate to define the p_{ij}s by the limiting availabilities of the components. These are described in Chapter 3 for the binary case. Generally, results from Markov theory and Semi-Markov theory are used, see Appendix D and [116].

Example 4.2

Consider a gas production and transportation system from a reservoir to the market (buyers). The production system comprises two installations, A and B, with maximum throughput rate equal to 75 and 50 per day. The maximum throughput rate for the pipelines is 100 per day. The system can be represented by the flow network shown in Figure 4.2.

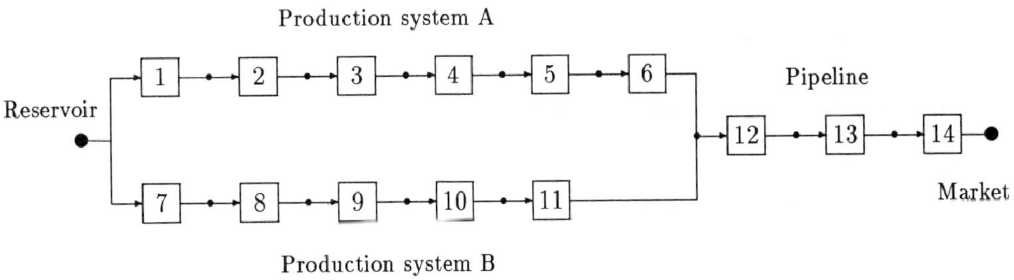

Figure 4.2: A flow network representation of a gas production and transportation system

The network system comprises 14 components representing different items of process equipment (such as separation system, dehydration system and compression system), pipelines, risers and valves. Most components have only two states, the functioning state and the not-functioning state. For equipment with parallel trains, a multistate representation has been used. For example, the dehydration system (component 8) comprises four identical trains in parallel, each with a maximum throughput rate of 12.5. The component representing this system, has five levels: 0, 12.5, 25, 37.5 and 50.

The gas demand is assumed to be 100. If all the components are 100% functioning, the demand will be satisfied. If, however, one or more components fail, the gas deliveries might be below the demand.

The problem is to calculate the reliability of the gas deliveries based on information about the expected sojourn times in the different states.

We make the following assumptions:

$$
\begin{array}{lll}
S_1 = \{0,75\} & S_2 = \{0,25,50,75\} & S_3 = \{0,75\} \\
S_4 = \{0,50,75\} & S_5 = \{0,75\} & S_6 = \{0,50,75\} \\
S_7 = \{0,50\} & S_8 = \{0,12.5,25,35.5,50\} & S_9 = \{0,50\} \\
S_{10} = \{0,50\} & S_{11} = \{0,25,50\} & S_{12} = \{0,100\} \\
S_{13} = \{0,100\} & S_{14} = \{0,100\} & \\
p_{10} = 0.2 \times 10^{-2} & & \\
p_{20} = 0.00 \times 10^{-2} & p_{21} = 0.1 \times 10^{-2} & p_{22} = 0.3 \times 10^{-2} \\
p_{30} = 0.2 \times 10^{-2} & & \\
p_{40} = 0.1 \times 10^{-2} & p_{41} = 0.5 \times 10^{-2} & \\
p_{50} = 0.2 \times 10^{-2} & & \\
p_{60} = 0.3 \times 10^{-2} & p_{61} = 1.0 \times 10^{-2} & \\
p_{70} = 0.3 \times 10^{-2} & & \\
p_{80} = 0.00 \times 10^{-2} & p_{81} = 0.00 \times 10^{-2} & p_{82} = 0.05 \times 10^{-2} \\
 & & p_{83} = 0.4 \times 10^{-2} \\
p_{90} = 0.5 \times 10^{-2} & & \\
p_{10,0} = 0.5 \times 10^{-2} & & \\
p_{11,0} = 0.04 \times 10^{-2} & p_{11,1} = 0.5 \times 10^{-2} & \\
p_{12,0} = 0.2 \times 10^{-2} & & \\
p_{13,0} = 0.2 \times 10^{-2} & & \\
p_{14,0} = 0.1 \times 10^{-2} & &
\end{array}
$$

For the two-state components these p_{ij} values are given by the limiting availabilities $MTTR/(MTTF + MTTR)$. For standby systems the p_{ij} values have been established using formulae similar to those presented in Section 3.4.3.

Based on this information we can now calculate h_j and a. Using the formulae (4.2) and

(4.3) we find that for example

$$\begin{aligned}
h_M &= P(\phi(\mathbf{X}) = 100) &\approx& \; 1 - 10^{-2}(0.2 + 0.1 + 0.2 + 0.1 + 0.2 \\
&&& +0.3 + 0.3 + 0.5 + 0.5 + 0.04 + 0.2 + 0.2 + 0.1) \\
&= 1 - 2.94 \times 10^{-2} &=& \; 97.1\% \\
h_1 &= P(\phi(\mathbf{X}) > 0) &\approx& \; 1 - 10^{-2}(0.2 + 0.2 + 0.1) \\
&= 1 - 0.5 \times 10^{-2} &=& \; 99.5\% \\
a &= E\phi(\mathbf{X})/\phi_M &\approx& \; 1 - 10^{-2}(0.5 \times 0.2 + 0.25 \times 0.1 + 0.5 \times 0.2 + 0.5 \times 0.1 \\
&&& +0.5 \times 0.2 + 0.5 \times 0.3 + 0.25 \times 0.3 + 0.25 \times 0.5 \\
&&& +0.25 \times 0.5 + 0.25 \times 0.04 + 1 \times 0.2 + 1 \times 0.2 + 1 \times 0.1) \\
&= 1 - 1.36 \times 10^{-2} &=& \; 98.6\%
\end{aligned}$$

The error terms by using these formulae are less than $\delta = 0.001$.

4.3 A time dependent model

We consider the system in a time interval $J = (u, u + s]$. We assume that there exists a reference level $D(t)$ at time t, $t \in J$, which expresses a desirable level of system performance at time t. The reference level $D(t)$ is a random variable, taking values in $\{d_0, d_1, \ldots, d_r\}$. For a flow network system we interpret $D(t)$ as the demand rate at time t. In the following we will use the word "demand rate" also in the general case. The state of the system at time t, which we in the following refer to as the throughput rate, is assumed to be a function of the states of the components and the demand rate, i.e.

$$\phi(t) = \phi(\mathbf{X}(t), D(t))$$

We assume that the $n + 1$ processes $X_i(t)$ and $D(t)$ are independent, when not stated otherwise. Furthermore, with probability one, these $n + 1$ processes make no transitions ("jumps") at the same time. All processes are assumed to be right-continuous.

Notation

$$\begin{aligned}
(r_i, \mathbf{X}(t), D(t)) &= (X_1(t), \ldots, X_{i-1}(t), x_{ir}, X_{i+1}(t), \ldots, X_n(t), D(t)) \\
\phi(r_i, \mathbf{M}, d) &= \phi(x_{1M_1}, \ldots, x_{i-1,M_{i-1}}, x_{ir}, x_{i+1,M_{i+1}}, \ldots, x_{nM_n}, d) \\
I(\cdot) &= \text{Indicator function, which equals 1 if the argument is true and 0 otherwise} \\
\phi^{(k)}(\mathbf{X}(t), D(t)) &= I(\phi(\mathbf{X}(t), D(t)) \geq \phi_k)
\end{aligned}$$

We see that $\phi^{(k)}$ equals 1 if the system is in state ϕ_k or better, and 0 otherwise. The length of an interval J^* is denoted $|J^*|$.

4.3.1 Performance measures

The performance of the system described above can be measured in a number of ways. Below we give an overview of some commonly used measures. The measures will be denoted $I_{1a}, I_{1b}, \ldots, I_{2a}, I_{2b}, \ldots$, etc.

1.
$$\begin{aligned} I_{1a} &= \text{Probability distribution of } \phi(t) \text{ given a demand rate } D(t) = d \\ I_{1b} &= E[\phi(t) \mid D(t) = d] \\ I_{1c} &= P(\phi(t) \geq D(t)) \end{aligned}$$

2. Let
$$Y = \int_J (D(t) - \phi(t))\, dt = \int_J D(t)\, dt \; - \; \int_J \phi(t)\, dt = Y_1 \; - \; Y_2$$

   We see that Y represents the lost throughput (volume) in J, i.e. the difference between the accumulated demand (volume) and the actual throughput (volume) in J.

$$\begin{aligned} I_{2a} &= \text{Probability distribution of } Y \\ I_{2b} &= EY = EY_1 - EY_2 \\ I_{2c} &= EY_2/EY_1 \end{aligned}$$

   The measure I_{2c} is called "throughput availability".

3. Let
$$Z = \frac{1}{|J|} \int_J I(\phi(t) = D(t))\, dt$$

   The random variable Z represents the portion of time the throughput rate equals the demand rate.

$$\begin{aligned} I_{3a} &= \text{Probability distribution of Z} \\ I_{3b} &= EZ \end{aligned}$$

   The measure I_{3b} is called "demand availability".

4. Let
$$V = \text{Number of times the process } \phi(t) \text{ is below } \phi_k \text{ in } J$$

$$\begin{aligned} I_{4a} &= \text{Probability distribution of V} \\ I_{4b} &= EV \\ I_{4c} &= P(V = 0) = P(\phi(t) \geq \phi_k,\ t \in J) \end{aligned}$$

   In the following we will assume that $D(t) \geq \phi_k$ for all t. Some closely related measures are obtained by replacing ϕ_k by $D(t)$.

4.3.2 Computation of the performance measures

The computation of the above measures can be an extremely difficult task in practice. The need for model simplifications and efficient computation methods is therefore vital. In this section we will discuss the computation problem for the various measures defined above. We will consider both analytical methods and the use of Monte Carlo simulation.

Analytical methods

1. **Measures of category 1.** There exists a number of methods for computing the probability distribution and the mean of the throughput rate at a fixed point in time, given a fixed demand rate $D(t) = d$, see Section 4.2 and [11,13] and the references therein. Computation of these measures represents the basic problem within multistate theory and has been given much attention in the reliability theory literature. In most applications the component probabilities, $p_{ij}(t)$, are assumed to be equal to the limiting (steady state) values as $t \to \infty$.

   Computation of the measure I_{1c} is trivial given that the measure I_{1a} has been computed for all d-values, noting that

   $$I_{1c} = P(\phi(t) \geq D(t)) = \sum_{i=0}^{r} P(\phi(t) \geq d_i \mid D(t) = d_i)\, P(D(t) = d_i)$$

2. **Measures of category 2.** From a practical point of view, it is not possible to find the exact distribution of Y, the lost throughput (volume) in J, using analytical methods. It is, however, possible to obtain an approximate distribution.

   Writing

   $$Y = \int_J (D(t) - \phi(t))\, dt = \sum_l \int_{a_l}^{b_l} (D(t) - \phi(t))\, dt \equiv \sum_l Y^{(l)}$$

   and assuming that the lost volumes in the intervals $(a_l, b_l]$, $Y^{(l)}$, are approximately independent and identically distributed, it follows by the Central Limit Theorem that Y has an approximate normal distribution. To guarantee identical distributions the process $D(t) - \phi(t)$ must be a regenerative process with cycles equal to the length of the subintervals.

   The asymptotical properties, as $s \to \infty$, of integrals/sums as above have been studied by e.g. [60,102,128,131]. Under natural assumptions it is proved that the integrals/sums are asymptotically (approximately) normally distributed. For the special case treated in Section 3.4.1 with only one component, it is shown in [131] that $\int_0^s \phi(t)\, dt/s$ is asymptotically normal as $s \to \infty$ with mean $A_i = MTTF_i/(MTTF_i + MTTR_i)$ and variance equal to

   $$(MTTF_i^2\, \sigma_i^2 + MTTR_i^2\, \tau_i^2)/(MTTF_i + MTTR_i)^3\, s$$

where σ_i^2 and τ_i^2 denote the variance of the lifetimes and and the repair times, respectively.

The mean of Y and Y_i can be calculated using Fubini's Theorem [32]:

$$E\int_J Z(t)\,dt = \int_J EZ(t)\,dt$$

where $Z(t)$ is one of the processes $D(t)-\phi(t)$, $D(t)$ and $\phi(t)$. The above right-hand integral can be computed with a desired level of accuracy by dividing the interval into subintervals:

$$\int_J EZ(t)\,dt \approx \sum_l EZ(t_l)\,h_l$$

where the intervals $(t_l, t_l+h_l]$ are sufficiently small. Hence the problem has been reduced to computing measures of category 1.

Computing the variance of Y and Y_i is more difficult. If we proceed as above for the mean using Fubini's Theorem, we find that

$$Var\int_J Z(t)\,dt = \int_J\int_J EZ(t)Z(s)\,ds\,dt - (\int_J EZ(t)\,dt)^2 \tag{4.4}$$

But in most cases we are not able to compute $EZ(t)Z(s)$ for $s \neq t$. By using Schwartz's inequality (or more general, Hölder's inequality) [32], an upper bound of the variance can be computed as explained for the mean, noting that

$$\int_J\int_J EZ(t)Z(s)\,ds\,dt \leq \int_J\int_J [E(Z(t))^2]^{1/2}\,[E(Z(s))^2]^{1/2}\,ds\,dt = (\int_J [E(Z(t))^2]^{1/2}\,dt)^2$$

Only in some simple cases it will be possible to compute the exact variance. For highly available systems it is, however, possible to calculate an approximation utilizing information about the failure and repair processes.

Let us look at $Z(t) = D(t) - \phi(t)$. Assume that it is possible to divide the time interval J into subintervals J_l as above (not necessarily the same intervals) and such that the probability that two or more component failures occur in the subinterval is small. Furthermore, we assume that the demand is a constant d in the whole subinterval. Then the variance of $\int_J Z(t)\,dt$ can be approximated by the sum of the variances of $\int_{J_l} Z(t)\,dt$, which in turn can be estimated by the sum over i of the variances of the lost volume in J_l due to failure of component i:

$$Var\int_{J_l} Z(t)\,dt \approx Var\sum_{i=1}^{n} C_i = \sum_{i=1}^{n} VarC_i \tag{4.5}$$

where C_i equals the lost volume due to a failure of component i. The equality of (4.5) follows from the assumption of independent components. Let C_{ij} equal the lost volume associated with a failure of component i to level x_{ij}, H_{ij} the sojourn time of component i in state x_{ij}, and let P_{ij} be the probability that component i fails to state x_{ij} in the subinterval J_l. Then it is not difficult to see that

$$\begin{aligned} VarC_i &= EC_i^2 - (EC_i)^2 = \sum_j EC_{ij}^2 P_{ij} - (\sum_j EC_{ij} P_{ij})^2 \\ &= \sum_j [d - \phi(j_i, \mathbf{M}, d)]^2 EH_{ij}^2 P_{ij} - \{\sum_j [d - \phi(j_i, \mathbf{M}, d)] EH_{ij} P_{ij}\}^2 \end{aligned}$$

3. **Measures of category 3**. With respect to computation, this category is similar to measures of category 2.

4. **Measures of category 4**. There does not exist any explicit formula for the distribution of V. Only in some simple cases is it possible to obtain practical computation formulae. For example, if all components have only two states and the lifetime distribution is exponential, it is possible to derive a simple approximation formula. Let $N_i(t)$ denote the number of failures of component i in the time interval $[0, t]$. Then if the repair times are small compared to the lifetimes and the lifetimes are exponentially distributed with parameter λ_i, it follows that the number of component i failures in the time interval $(u, u+s]$, $N_i(u+s) - N_i(u)$, is approximately Poisson distributed with parameter $\lambda_i s$. If the system is a series system, and we make the same assumptions as above, the number of system failures in the interval J is approximately Poisson distributed with parameter $\sum_{i=1}^n \lambda_i s$. The number of system failures in $[0, t]$, $N(t) = \sum_{i=1}^n N_i(t)$, is approximately governed by a Poisson process with intensity $\sum_{i=1}^n \lambda_i$. A more general approach is described in the following.

   Assume that the component sojourn times in the failure states are small compared to the sojourn time in the best states x_{iM_i} and assume that the failure counting process $N_{ij}(t)$, representing the number of failures of component i to level x_{ij} in $[0, t]$, has an intensity process $\lambda_{ij}(t) \approx \alpha_{ij}(t) I(X_i(t) = x_{iM_i})$, where $\int_0^t \lambda_{ij}(v)dv < \infty$ for all $t < \infty$ and the $\alpha_{ij}(t)$s are deterministic functions. The function $\alpha_{ij}(t)$ might for example represent the intensity of a (non-homogeneous) Poisson process, when we disregard the sojourn times in the failure states of component i. Then it is often possible to establish a good approximation using the following approach:

   (a) Divide the time interval J into subintervals J_l such that the probability that two or more component failures occur in the subinterval is small and the number of system failures in the consecutive subintervals, $N(J_1), N(J_2), \ldots, N(J_r)$, can be considered approximately independent.

   (b) Let
   $$g_{ij}(t) = \phi^{(k)}(M_i, \mathbf{X}(t), D(t)) - \phi^{(k)}(j_i, \mathbf{X}(t), D(t))$$
   We see that $g_{ij}(t)$ equals 1 or 0, and is equal to 1 if and only if a failure of component i to level j implies a system failure below state ϕ_k. Approximate the probability of a system failure in a subinterval J_l by the expected number

of system failures, i.e. the number of times the process $\phi(t)$ is below ϕ_k in J_l, given by

$$EN(J_l) \approx \sum_{i=1}^{n}\sum_{j=0}^{M_i-1}\int_{J_l} Eg_{ij}(t)\,\alpha_{ij}(t)p_{iM_i}(t)\,dt \tag{4.6}$$

This formula is justified by (4.8) below. Computing the right-hand expression of (4.6) is similar to computing measures of category 1 and I_{2c}.

(c) Approximate the probability distribution of the number of system failures in J by calculating the distribution of

$$V = N(I_1) + N(I_2) + \ldots + N(I_r) \tag{4.7}$$

where the $N(J_l)$'s are assumed to be binary random variables with the above expectations. When the $N(J_l)$'s are (approximately) independent and identically distributed, it follows that V has a Binomial distribution with parameters r and $EN(J_l)$. With r large and $EN(J_l)$ small we can approximate this Binomial distribution with a Poisson distribution with mean $rEN(J_l)$.

By summing over l in the expression for $EN(J_l)$ we obtain an approximation for EV, the expected number of failures in J.

It is also possible to generalize the binary results for the (expected) number of system failures presented in Section 3.4.1. Assume that the state process of component i is a regenerative process with a cycle length having finite mean μ_i, $i = 1, 2, \ldots, n$. Assume also that the distribution of the cycles is not periodic. Moreover assume that

$$E\gamma_{ilj}(t) \equiv E[\phi^{(k)}(l_i, \mathbf{X}(t), D(t)) - \phi^{(k)}(j_i, \mathbf{X}(t), D(t))]$$

converges to a limit, $\bar{\gamma}_{ilj}$, say, as $t \to \infty$. Let f_{ilj} denote the expected number of times component i visits state x_{il} and makes a transition to state x_{ij} during a cycle, and let $N_{ilj}(t)$ denote the number of transitions that component i makes from state x_{il} to state x_{ij} in the period $[0, t]$. Then it can be shown by arguing as in the proofs of the formulae (3.13) and (3.14) in Appendix E.5.1 that

$$\lim_{t\to\infty}\frac{EN(t)}{t} = \lim_{t\to\infty}\frac{1}{t}E\sum_{i=1}^{n}\sum_{j<l}\int_0^t \gamma_{ilj}(v)dN_{ilj}(v) = \sum_{i=1}^{n}\sum_{j<l}\frac{f_{ilj}\,\bar{\gamma}_{ilj}}{\mu_i}$$

$$\lim_{u\to\infty} E[N(u+s) - N(u)] = s\sum_{i=1}^{n}\sum_{j<l}\frac{f_{ilj}\,\bar{\gamma}_{ilj}}{\mu_i}$$

If the process ϕ is a regenerative process, then with probability one,

$$\lim_{t\to\infty}\frac{N(t)}{t} = \sum_{i=1}^{n}\sum_{j<l}\frac{f_{ilj}\,\bar{\gamma}_{ilj}}{\mu_i}$$

From these formulae approximations for $N(t)$, $EN(t)$ and $E[N(u+s)-N(u)]$ can be established.

If the counting process $N_{ilj}(t)$ has an intensity process $\lambda_{ilj}(t)$, then it can be shown by using similar arguments as in the proof of (E.7) of Appendix E.5.1, that the expected number of system failures in the period $(u, u+s]$ equals:

$$\begin{aligned} E[N(u+s)-N(u)] &= \sum_{i=1}^{n}\sum_{j<l}\int_{u}^{u+s} E\gamma_{ilj}(t)\lambda_{ilj}(t)dt \qquad (4.8) \\ &= \sum_{i=1}^{n}\sum_{j<l}\int_{u}^{u+s} E\gamma_{ilj}(t)E\lambda_{ilj}(t)dt \end{aligned}$$

Hence if the transition rate for component i from state x_{il} to state x_{ij} is a constant α_{ilj}, then

$$E[N(u+s)-N(u)] = \sum_{i=1}^{n}\sum_{j<l}\int_{u}^{u+s} E\gamma_{ilj}(t)\alpha_{ilj}p_{il}(t)dt$$

The result (4.8) also holds when the processes $X_1(t)$, $X_2(t), \ldots, X_n(t), D(t)$ are stochastically dependent. Then the intensity process of $N_{ilj}(t)$ may depend on the history of all the processes up to time t. A similar extension is possible for the approximation of $EN(J_l)$ given above.

It is also possible to use Markov models to compute the measures of category 4. But this approach is practical only for small systems since the state space normally will consist of all combination of component states: $S_1 \times S_2 \times \ldots \times S_n$, i.e. a number of states equal to $(M_1+1)\times(M_2+1)\times\ldots\times(M_n+1)$. Another important practical limitation of Markov models is the assumption of exponential repair times.

The measure I_{4c} is a subset of the measure I_{4a}, the distribution of V. The measure I_{4c} can therefore be computed (approximated) using the above approach. Clearly,

$$I_{4c} = P(V=0) \approx \prod_{l}(1-EN(J_l))$$

The accuracy of the above formulae is discussed in the following example. The measure I_{4c} has been studied in detail by Natvig [98,100,101]. He has, however, focused on finding *bounds* on I_{4c} under various assumptions. We will not look closer into this problem here.

The above calculation methods can also be used when V denotes the number of times $\phi(t)$ is below $D(t)$.

Example 4.1 continued
Consider now a time period J of 1 year. Assume that the demand rate equals 2 for the whole interval and that the above probabilities $p_{ij}(t)$ are the same for all t in the interval. Then we can calculate an approximate distribution of the lost throughput in this time interval by using an approximation to the normal distribution with mean equal to

$8760 \times (2 - 1.882) = 0.118 \times 8760 = 1033$. To estimate the variance we will use the formula (4.5). We must then specify information about the failure and repair processes. Let us assume that the performance process for the binary components 1 and 2 are determined by a constant failure rate $\lambda = 1/480$ and a Time To Repair (TTR) equal to 20 hours (the TTR is assumed to be deterministic). The component 3 is assumed to comprise two independent units: Unit $3a$ has a constant failure rate equal to $1/490$ and a $TTR = 10$ hours, whereas unit $3b$ has a constant failure rate equal to $1/990$ and a $TTR = 10$ hours. If the unit $3a$ fails, the system state equals 1, whereas if unit $3b$ fails, the system state equals 0. With these probabilities, the limiting availabilities $1/(1+\lambda TTR)$ equal the above component probabilities $p_{ij}(t)$. Using this information, we can compute the variance from (4.5). Let L denote the length in hours of the subintervals and let W denote the number of subintervals in J ($W \times L = 8760$). Then we obtain a variance estimate of

$$W\{2(1/480)L1^2 20^2[1-(1/480)L] + (1/490)L1^2 10^2[1-(1/490)L] + (1/990)L2^2 10^2[1-(1/990)L]\} \approx 8760 \times 2.3 \approx 2 \times 10^4$$

We see that the above variance estimate does not depend significantly on L. This is of course true as long as $P_{ij} \approx \lambda_i L$ is small.

Using this value for the variance we can easily calculate an approximate probability distribution for the lost throughput volume relative to the demand:

Lost throughput/ demand	3.00 – 4.99%	5.00 – 6.99%	7.00 – 8.99%
Probability	0.14	0.78	0.08

Finally, let us look at the computation of the measures of category 4. Assume the same failure and repair process as described above. Let us first consider the number of times the system state is below level 2. In this case the system can be viewed as a series system, hence the number of system failures is approximately governed by a Poisson process. The intensity of the process equals the sum of the failure rates: $1/480+1/480+1/490+1/990 = 7 \times 10^{-3}$.

Consider now the number of times the system state is below level 1. To compute the distribution of V in this case we can proceed as explained above, with subintervals of length L hours. It follows that V is approximate Poisson distributed with parameter $W\,EN(J)$, where

$$\begin{aligned} EN(J) &= L[P(X_2(t)=0)P(X_3(t)\geq 1)\lambda_1 p_{11}(t) + P(X_1(t)=0)P(X_3(t)\geq 1)\lambda_2 p_{21}(t) + \\ &\quad P(X_1(t)=1 \bigcup X_2(t)=1)\lambda_{3,0}p_{32}(t)] = L1.1 \times 10^{-3} \end{aligned}$$

Hence the average number of times the state of the system equals 0 is approximately 10 ($8760 \times 1.1 \times 10^{-3}$) per year.

Example 4.2 continued

Assume now that we consider the system in a time period of one year. In the winter period (duration 5 months) the demand rate is assumed to be 100, whereas the demand rate in the summer period (duration 7 months) equals 75. The problem is to calculate the throughput availability and demand availability. Consider first the winter period. Assuming that the p_{ij} values are constant (as an approximation), then we have already computed these measures and we can now interpret the results in the following way:

- The expected gas deliveries are about 98.6% of the gas demand in the winter period
- The expected portion of time the delivered gas satisfies the demand is about 97.1%

Let us assume that the throughput availability and the demand availability for the summer period have also been calculated, and that we have found that they are equal to 99.1% and 98.6%, respectively. Then it is not difficult to show that the throughput availability and demand availability for the whole interval equal 98.8% and 97.7%, respectively. To demonstrate the calculations, let us consider the calculations of the throughput availability:

$$\frac{0.986 \times 100 \times 7 + 0.991 \times 75 \times 5}{100 \times 7 + 75 \times 5} = 0.988$$

Monte Carlo simulation

Monte Carlo simulation represents an alternative to the analytical methods presented above. The technique is to generate a computer model of the system to be investigated, for example represented as a flow network, and then to simulate the operation of the system for the period J. Using the computer we generate realizations of the processes $\mathbf{X}(t)$ and $D(t)$. The sojourn times in the various states are determined by sampling from appropriate probability distributions. For example, for a two-state component the operating times (up-times) are sampled from the lifetime distribution and the downtimes are sampled from the repair time distribution. The system state, $\phi(t)$, is computed and logged, as the time t elapses. For each realization of the system, we can compute e.g. Y, Y_i, Z and V. Simulating the system performance a number of times, say v times, we can estimate the distribution, mean, etc. of the variables Y, Y_i, Z and V, i.e. we can estimate the performance measures I_{ij}. For example, the mean EY is estimated by the average value of the realizations of the variable Y. By increasing v, the estimation error can be made negligible.

It is often preferable to use a Monte Carlo simulation model to compute the performance measures of category 2-4. With a Monte Carlo simulation model the time aspect is more easily handled than with an analytical method. A Monte Carlo simulation model may be a fairly true representation of the real world. This is one of the great attractions of Monte Carlo simulation over analytical methods.

Simulation requires in general detailed input data. For example, the lifetime and repair time distributions must be known. Mean values as in many analytical models are not sufficient. On the other hand, the output from a Monte Carlo simulation model is very extensive and informative. The performance measures defined above are easy to estimate from a Monte Carlo simulation model.

The main disadvantage of the Monte Carlo simulation technique compared with an analytical approach is the time and expense involved in the development and execution of the model. To obtain accurate results using simulation a large number of trials is usually required, especially when the availabilities are high. The time and expense aspect is very important if the model is to be used to study effects of changes in system configurations, or if sensitivity analyses are to be performed.

With a complex Monte Carlo simulation model it is difficult to check if the program has been written correctly and, therefore, if the result can be relied upon.

A general Monte Carlo simulation model for analysing flow network system is MIRIAM [56].

Example 4.1 continued

The system has been studied by use of the Monte Carlo program MIRIAM. We have assumed the same failure and repair processes as described above. Thus the component/unit process is characterized by independent and identically distributed operating times (time to failure) and downtimes (time to repair). The operating times are governed by exponential distributions, whereas the downtimes are constant values. A run-in period of one year is assumed, that is, the process is simulated in a one year period before our observations begin. The process is simulated $v = 100$ times. The results from the simulations show very good correspondence with the analytical results obtained above. The throughput and demand availability estimates are identical with those obtained above. The distribution of the lost throughput estimated from the simulations is approximately the same as the one established using the analytical approach. In the simulations we obtain probabilities 0.15, 0.76 and 0.09 which are to be compared with the results from the analytical approximation method: 0.14, 0.78 and 0.08. The average number of times the system state is below 1 is 10 which is identical with the analytical estimate. Also for system level 2, the average number of failure times is the same as the previous estimate, i.e. 60 times. The simulations confirm that the distribution of the number of system failures is approximately Poisson distributed. The differences between the simulated values and the Poisson probabilities are quite small, the largest differences of course at the tails of the distribution since the number of simulations is too small to produce reliable estimates for these low probabilities. In the table opposite the estimated distribution of the number of system failures to level 1 is presented, based on the analytical approach and the simulation.

4.3.3 Including deferrable repairs

In this section it will be shown how the model can be generalized by considering two types of failure:

DEF-failure : A failure where the corresponding repair can be deferred to a more convenient time.

IMM-failure : Not a DEF-failure

	Estimate of $P(V \leq x)$	
x	Analytical approach	Simulation
3	0.01	0.00
4	0.03	0.04
5	0.07	0.05
6	0.13	0.12
7	0.22	0.23
8	0.33	0.32
9	0.46	0.47
10	0.58	0.58
11	0.70	0.67
12	0.79	0.77
13	0.86	0.84
14	0.92	0.92
15	0.95	0.95
16	0.97	0.98
17	0.99	0.99

This approach is appropriate when modelling for instance gas/oil transportation systems, where for example the repair of a leaking valve can be deferred. The following assumptions are made (some of these assumptions will be commented on later):

1. The deferrable repairs begin at fixed points $T_1, T_2, \ldots$. The repair of all DEF-failures that have occurred in the interval (T_{r-1}, T_r) begins at time T_r and is completed before T_{r+1}.
2. DEF-failures are specified only for level x_{i0} and $\phi(0_i, \mathbf{M}, D) = 0$, i.e. the repair of a DEF-failure implies that no flow is transmitted through the system.
3. For a DEF-failure, there is no flow reduction before the repair begins.
4. The repair time for a DEF-failure of component i, τ_{i0}, is a constant.
5. An arbitrary number of deferrable repairs can be carried out at the same time.
6. Given that no deferrable repair takes place at time t, the random variables $X_i(u)$, $i = 1, 2, \ldots, n$, are stochastically independent with distributions $p_{ij}(u)$, $u > t$.
7. The demand rate $D(t)$ is a deterministic function with $D(t) > 0$ for all t.

Assumptions 1 and 5 imply that if DEF-failures occur for components $i_1, i_2, \ldots, i_v$, then the total downtime of the system due to DEF-failures is:

$$max_{1 \leq j \leq v} \; \{\tau_{i_j 0}\}$$

The probabilities of DEF-failures need not be small. Such an assumption would simplify the calculations, but it would be very restrictive since the probabilities of DEF-failures are often quite high. DEF-failures are related to an interval (T_{r-1}, T_r) which might be very long, e.g. several years.

Consider now a fixed r. The following notation is needed:

B	Set of component indices for which DEF-failures are defined, $B \subseteq \{1, 2, \ldots, n\}$
R_{ij}	P(component i has a DEF-failure to level x_{ij} in (T_{r-1}, T_r))
a_l	lth smallest $\tau_{i0}, a_1 < a_2 < \ldots < a_s$, are the values the τ_{i0}s take, $a_0 \equiv 0$
A_l	Set of component indices (i values) such that $\tau_{i0} \geq a_l, A_l \subseteq B$
q_l	$\prod_{i \in A_l}(1 - R_{i0})$; probability that no DEF-failure to level 0 occurs with repair time $\geq a_l$

The problem is to compute some specific availability performance measures when taking into account the DEF-failures. We will restrict attention to the throughput availability I_{2c} and the demand availability I_{3b}.

Consider an interval $J^* = [T_r + a_{l-1}, T_r + a_l) \bigcap J$. Then conditional on whether a DEF-failure with longer repair time than a_{l-1} occurs or not and noting that the two availability measures are zero if such an event occurs, we find that for this interval,

$$I_{2c} = q_l \tilde{I}_{2c} \quad \text{and} \quad I_{3b} = q_l \tilde{I}_{3b} \tag{4.9}$$

where $\tilde{I}_{2c}$ and $\tilde{I}_{3b}$ are throughput availability and demand availability, respectively, for the interval J^* when ignoring DEF-failures. Thus the problem is reduced to the problem studied in Section 4.3.2 of computing the availabilities for intervals with no deferrable repairs.

Having established the availabilities for each interval J^*, the availabilities for the whole interval J are easily found as shown in Section 4.3.2, Example 4.2.

In practice, the following approximation method is often applied to highly available systems, when taking into account DEF-failures.

The throughput and demand availabilities are approximated by the sum of the unavailability contribution due to IMM-failures (found in Section 4.3.2) and a contribution due to DEF-failures. To establish the latter contribution, consider the set of intervals (J^*) defined above. For each such interval J^* the expected loss of throughput and the expected time of zero throughput due to the repair of a DEF-failure equals $(1 - q_l) \int_{J^*} D(t)dt$ and $(1 - q_l)L$, respectively, where L is the length of the interval. By summing over all intervals and dividing by the demand throughput, $\int_J D(t)dt$, and the length of the interval (J), we obtain the contribution due to DEF-failures.

To justify this approximation method, note that by (4.9) and the assumption that the system has a high availability, we have for each of the availability measures:

$$\begin{aligned} 1 - I &= 1 - q_l \tilde{I} = (1 - \tilde{I}) + (1 - q_l)\tilde{I} \\ &\approx (1 - \tilde{I}) + (1 - q_l) \end{aligned} \tag{4.10}$$

where I equals either I_{2c} or I_{3b}. By multiplying (4.10) by $\int_{J^*} D(t)dt$ for the throughput availability and by L for the demand availability, and summing over all intervals J^*, we see, using the above described calculation method, that approximate values for the unavailabilities are obtained. By (4.10) the error term using the above approximation method equals the unavailability contribution due to DEF-failures with $(1-q_l)$ replaced by $(1-q_l)(1-\tilde{I})$. Hence the error term is bounded by the unavailability contribution due to DEF-failures multiplied by the maximum unavailability $1-\tilde{I}$.

Example 4.2 continued

Assume that DEF-failures are specified for components 12-14, and assume that $R_{12,0} = 0.2$, $R_{13,0} = 0.3$, $R_{14,0} = 0.4$, $\tau_{12,0} = 4$ days, $\tau_{13,0} = 2$ days and $\tau_{14,0} = 1$ day. The deferrable repairs are carried out in the summer period, beginning at time zero.

To compute the throughput and demand availabilities when taking into account the DEF-failures, we consider the following intervals (in days): $(0,1], (1,2], (2,4]$, "remaining part of summer", and "winter". The last two intervals are not affected by deferrable repairs so that we can restrict attention to the three first intervals. When ignoring DEF-failures we have found that the throughput and demand availabilities for the summer period equal 99.1% and 98.6%, respectively. Using these results and that $q_1 = 0.8 \times 0.7 \times 0.6 = 0.336$, $q_2 = 0.8 \times 0.7 = 0.560$ and $q_3 = 0.800$ we obtain from (4.9) the following values for the throughput and demand availabilities:

$$\begin{aligned} \text{Interval } (0,1] &= 0.333,\ 0.331, \\ \text{Interval } (1,2] &= 0.555,\ 0.552, \\ \text{Interval } (2,4] &= 0.793,\ 0.789, \end{aligned}$$

Now combining the availabilities for the various intervals we find that the throughput and demand availabilities are 98.4% and 97.3%, respectively. We have set the interval "remaining part of summer" to 146 days and "winter" to 215 days. To demonstrate the calculations, look e.g. at the demand availability:

$$I_{3b} = (0.331 \times 1 + 0.552 \times 1 + 0.789 \times 2 + 0.986 \times 146 + 0.971 \times 215)/365 = 97.3\%$$

To obtain the availabilities using the approximation method described above, first note that the throughput and demand unavailability contributions due to IMM-failures are 1.2% (1.23%) and 2.3%, respectively. The throughput and demand unavailability contributions due to DEF-failures are:

$$(0.664 \times 1 \times 75 + 0.440 \times 1 \times 75 + 0.2 \times 2 \times 75)/(150 \times 75 + 215 \times 100) = 0.35\%$$

and

$$(0.664 \times 1 + 0.440 \times 1 + 0.200 \times 2)/365 = 0.4\%$$

respectively. Hence the total contributions are 1.6% and 2.7%, which is seen to be identical to the unavailabilities obtained with the exact method. As the error term by using the approximation method is bounded by $0.4\% \times 0.029 = 0.012\%$, we could have concluded that 1.6% and 2.7% are "correct" without comparing them with the results from the exact method.

4.4 Discussion

The traditional reliability theory based on a binary approach has recently been generalized by allowing components and system to have an arbitrary finite number of states. For most reliability applications, binary modelling should be sufficiently accurate, but for certain types of application such as gas and oil production and transportation systems, a multistate approach is usually required for the system and components. In a gas transportation system, for example, the state of the system is defined as the rate of delivered gas, and in most cases a binary model (100%, 0%) would be a poor representation of the system. A component in such a system may represent a compressor station comprising a certain number (M) of compressor units in parallel. The states of the component equal the capacity levels corresponding to M compressor units running, $M-1$ compressor units running, and so on.

For highly available systems, the most common situation in practice, multistate components can often be considered as binary (two-state) components without any significant loss of accuracy. For example the compressor station mentioned above may be represented by two states: M compressor units operating and $M-1$ compressor units operating. The possibility of two or more units being out of operation at the same time is then ignored. Often it is necessary to include also state 0 corresponding to a common-mode failure resulting in shut-down of all the parallel units.

There exists also a number of reliability importance measures for multistate systems, see e.g. [11,13,23,40,55,101] and the references therein. Many of these measures represent natural generalizations of importance measures of binary systems. We see e.g. that the measure I^A also applies to multistate models. For the Birnbaum measure, it is not so straightforward to generalize the measure. Several measures have been proposed, e.g. [40]:

The r,s reliability importance of component i, $I^{r,s}$, is given by

$$I^{r,s} = P(\phi(r_i, \mathbf{X}) \geq \phi_k) - P(\phi(s_i, \mathbf{X}) \geq \phi_k)$$

Refer to Section 3.5 for a discussion of the use of reliability importance measures.

In this chapter we have assumed that the components are stochastically independent. The problem of treating component dependencies as discussed in Section 3.6 for the binary case also applies for multistate systems. The approach presented in Section 3.6 is also recommended for the multistate case. For some reliability bounds for associated multistate components, see [50].

4.5 Problems

1. In the design process of a compression system for a gas and oil production platform two alternatives are considered:

   i) Two trains in parallel, each with a maximum throughput capacity of 60%

   ii) Three trains in parallel, each with a maximum throughput capacity of 40%

   Normal production is 100%. Each train comprises compressor and turbine, cooler and scrubber. A failure of one of these units will result in shut-down of the train. The following reliability data is given:

Unit	Failure rate per year	MTTR (hours)
Compressor and turbine	10	12
Cooler	2	50
Scrubber	1	20

   We assume that all the components in the system are independent.

   a) Show that the unavailability of a train, i.e. the (expected) portion of time in the long run that a train is not functioning, is approximately 2.7%.

   b) Under suitable assumptions, calculate the demand unavailability, i.e. the expected portion of time the production is lower than 100% for the two alternatives.

   c) Under suitable assumptions, calculate the throughput unavailability, i.e. the expected lost production relative to 100% for the two alternatives.

   d) The compression system represents a component in a flow network representation of the whole production system on the platform. For simplicity we assume that the network comprises only five components:

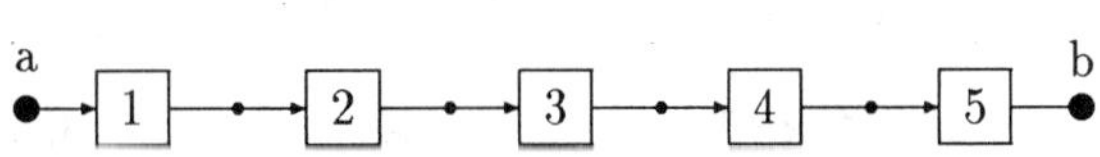

   Component 1 represents the compression system. Components 2, 3 and 4 are binary (0%,100%) with associated unavailabilities equal to 0.2%, 0.3% and 0.4%, respectively. Component 5 has three levels: 0%, 60% and 100%, with associated probabilities p_{ij} equal to 0.2%, 0.8% and 99.0%, respectively.

Calculate the demand and throughput unavailability for the production system for both design alternatives.

e) Discuss the criticality of the three equipment units in a compression train and the criticality of the components of the flow network.

2. Consider Example 4.2.

a) Find the minimal cut vectors for level 100.

b) Calculate the demand unavailability, i.e. $P(\phi(\mathbf{X}) < 100)$, using formula (4.1).

3. Establish approximation (4.1) and the expression for the corresponding error term.

4. The figure below shows a simple example of flow network system. The system comprises 6 independent components. Flow (oil/gas) is transmitted from a to b. All the components are two-state components: The components 1-4 have possible states 0 and 1, whereas the components 5 and 6 have possible states 0 and 2. The states are interpreted as flow capacity rates of the components. The state of the system, ϕ, equals the maximum flow that can be transmitted from a to b, and it follows that the system has the states $\{0, 1, 2\}$. Thus the demand equals 2.

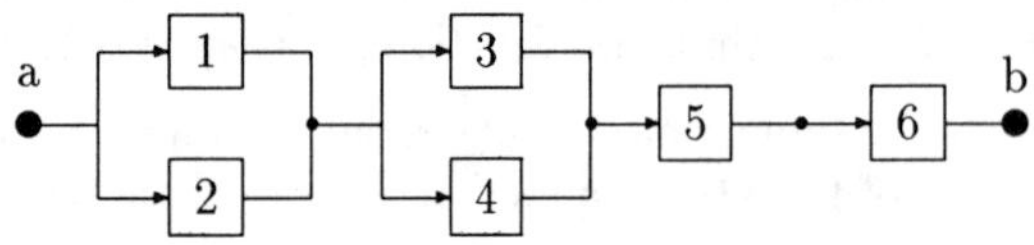

Components 1 and 2 have a $MTTF = 940$ hours and a $MTTR = 60$ hours, whereas components 3 and 4 have a $MTTF = 4900$ hours and a $MTTR = 100$ hours. Components 5 and 6 have two failure types: IMM-failures and DEF-failures. The following reliability data for these components are given:

Component	Failure rate (per year)	Repair time (hours)	Failure type
5	2×10^{-2}	250	IMM
6	4×10^{-2}	250	IMM
5	4×10^{-2}	200	DEF
6	5×10^{-2}	100	DEF

a) Explain what is meant by the two failure types.

b) Find the demand availability for the system. State possible assumptions you make.

c) Find the throughput availability for the system. Again state possible assumptions you make.

d) Indicate how the demand availability can be calculated in a generalized model where also component 3 has DEF-failures.

5. Consider the model described in Section 4.3 with the DEF-failures included. Assume that the probabilities R_{ij} are small and that the assumption 2 is not valid. Explain how we can use the approximation principle that is used in Section 4.2 (Formulae (4.2) and (4.3)) for the analysis of the deferrable repairs.

   Justify the following approximations for the throughput and demand unavailability contributions due to DEF-failures:

$$\sum_r \sum_i \sum_j U_{ij}(r)\, R_{ij}(r) \,/ \int_J D(t)\, dt$$

$$\sum_r \sum_i \sum_j V_{ij}(r)\, R_{ij}(r) \,/ |J|$$

   where

   $U_{ij}(r)$ = Lost throughput volume during the repair of a DEF-failure of component i to level value x_{ij}, when repair begins at T_r

   $V_{ij}(r)$ = Time the system throughput is less than demand during repair of a DEF-failure of component i to level value x_{ij}, when repair begins at T_r

   Calculate the above approximations for the data given in Example 4.2 and compare the results with the values computed in Section 4.3.3.

6. Consider again the DEF-failure model of Section 4.3. Generalize the results for the case that the restoration time for a DEF-failure of component i to level x_{i0} is a random variable which can take an arbitrary finite number of values.

Chapter 5

OPTIMAL REPLACEMENT POLICIES

5.1 Introduction

In Section 3.4.2 we considered a model for analysing the availability of units which are replaced under a preventive policy. The purpose of the maintenance policy is to increase the operating time of the units. Now, consider one unit (component) and suppose that the model analysed in Section 3.4.2 holds for all replacement times $T > 0$. Then a natural question to ask is the following. Which replacement time T maximizes the availability, or equivalently minimizes the unavailability? By replacement we mean either a physical replacement of the unit or a repair (overhaul) so that the unit is as good as new after the repair.

In this chapter, Section 5.2, we will look closer into the problem of optimizing the replacement times. The optimization criterion considered is the (expected) long run cost per unit of time. This criterion is more general than the availability criterion. This assertion follows directly from the theory of renewal reward processes, see Appendix E.2, by letting the reward (cost) process be the operating time of the unit.

In general we can write the optimization criterion in the following form, cf. Result R.6 of Appendix E.2:

$$B^T = \frac{M^T}{S^T} \tag{5.1}$$

where M^T represents the expected cost associated with a replacement cycle, and S^T represents the expected length of a replacement cycle.

As an example consider the model introduced in Section 3.4.2 and for simplicity assume that $\alpha(t) \equiv 1$, i.e. the model equals the classic age replacement model: the unit is replaced at failure or at time T, whichever occurs first. Let U^T represent the expected time to replacement, i.e.

$$U^T = \int_0^T R(t)\, dt$$

where $R(t)$ equals the survivor function of the unit given by

$$R(t) = e^{-\int_0^t \alpha(s)\lambda(s)\,ds} = e^{-\int_0^t \lambda(s)\,ds}$$

Furthermore, let D^T represent the expected downtime associated with the replacement, i.e.

$$D^T = v(1 - R(T)) + uR(T)$$

where v and u represent the expected downtime associated with the unplanned replacement and the planned replacement, respectively. Using (3.15) it is seen that the availability equals

$$\frac{U^T}{U^T + D^T}$$

and expressions for the nominator and denominator of (5.1) follow. Since

$$\frac{U^T}{U^T + D^T} = \frac{1}{1 + \frac{D^T}{U^T}}$$

it is seen that the maximization of the availability is equivalent to minimizing the ratio D^T/U^T, i.e. the expected downtime in a replacement cycle divided by the expected time to replacement. By interpreting the downtime as a cost, we have shown that the problem of maximizing the availability is equivalent to minimizing the ratio between the expected cost associated with a replacement cycle and the expected time to a replacement. In replacement models where the downtime is short compared to the operating time, the ratio D^T/U^T is approximately equal to the (expected) long run cost per unit of time. In most applications, the error introduced by this approximation is negligible.

In the following we will use the ratio between the expected cost associated with a replacement cycle and the expected time to replacement as the optimality criterion.

In Section 5.3 we study the block replacement policy. Such a policy stipulates preventive replacements at defined intervals of time, kT, $k = 1, 2, \ldots$, regardless of the unit age, and replacement at failure otherwise. The block replacement policy is not included in the general repair replacement model analysed in Section 5.2.

The optimality criterion considered in this chapter is the most common used criterion for replacement optimization. Another criterion that is sometimes used is the total expected discounted cost. This criterion also takes the form (5.1) under reasonable assumptions, and the results obtained in this chapter can easily be converted to this criterion, see Problem 5.5.6.

In the example above, as for most replacement models, the replacement time T is a fixed quantity. In situations where information is available on the underlying condition of the unit, for example by measurement of wear characteristics and damage inflicted on the unit, it is preferable to consider replacement times which are based on this information. Models for analysing such situations are presented in Section 5.4.

System deterioration is often reflected in higher production costs and lower product quality. To keep production costs down while maintaining good quality, preventive maintenance is usually performed on systems subject to deterioration. The growing importance of maintenance has generated an increasing interest in the development and implementation of preventive maintenance models for deteriorating systems.

There is extensive literature on models for optimization of repair, replacement and inspection, which is surveyed in the celebrated paper by Pierskalla and Voelker [112] and updated by Valdez-Flores and Feldman [136]. In the models presented in this chapter we seek an optimal replacement time. We are not dealing with the problem of optimizing repairs and inspections.

Our presentation covers the basic models, in addition to some new more advanced models. The replacement models presented in Sections 5.2 and 5.3 include the three basic replacement models: age replacement model, block replacement model and minimal repair replacement model.

In most of the models presented in this chapter we focus on one-unit systems. An exception is made in Section 5.4, where we optimize the replacement times of a monotone system comprising n components.

5.2 General minimal repair, age replacement model

As mentioned in Section 3.4.2 this model represents a modification of a model analysed in [34]. We have generalized the model slightly by allowing the cost of minimal repairs to be a stochastic process with a mean value function $r(t)$.

5.2.1 The optimization problem

Consider the general repair replacement model presented in Section 3.4.2. From (3.15) it is seen that the optimization function, B^T, the ratio between the expected cost (downtime) associated with a replacement cycle and the expected time to replacement, equals:

$$\begin{aligned}
B^T &= \frac{\int_0^T r(t)\beta(t)\lambda(t)R(t)\,dt \;+\; D^T}{U^T} \\
&= \frac{\int_0^T r(t)\beta(t)\lambda(t)R(t)\,dt \;+\; u+(v-u)(1-R(T))}{\int_0^T R(t)\,dt} \\
&= \frac{\int_0^T r(t)\beta(t)\lambda(t)R(t)\,dt \;+\; u+(v-u)\int_0^T \alpha(t)\lambda(t)R(t)\,dt}{\int_0^T R(t)\,dt} \\
&= \frac{u+\int_0^T [r(t)\beta(t)+(v-u)\alpha(t)]\lambda(t)R(t)\,dt}{\int_0^T R(t)\,dt} \\
&\equiv \frac{u+\int_0^T a(t)R(t)\,dt}{\int_0^T R(t)\,dt}
\end{aligned} \tag{5.2}$$

We assume that the limit $\lim B^T$ exists when $T \to \infty$, and denote the limit by B^∞. The policy in this case is interpreted as "perform no planned replacement".

The problem of finding an optimal replacement time is now reduced to that of minimizing the function (5.2) with respect to T.

We see from (5.2) that B^T is a continuous function in T. Thus a minimum exists in the interval $(0, \infty]$, since

$$\lim_{T \to 0} B^T = \infty$$

It is tacitly assumed that $u > 0$. By differentiating B^T we see that the derivative of B^T equals 0, is greater than 0, or is less than 0, if the function

$$h(T) \equiv a(T) \int_0^T R(t)\,dt - \int_0^T a(t)R(t)\,dt - u$$

equals 0, is greater than 0, or is less than 0, respectively. If $a(T)$ is continuous, then $h(T)$ is continuous and an optimal T^* satisfies $h(T^*) = 0$. It is not difficult to see that the function $h(T)$ is non-decreasing (strictly increasing) if the function $a(T)$ has this property. It follows that if $a(T)$ is non-decreasing, then B^T has the following characteristics:

> There exists an optimal value $T^* \in (0, \infty]$ such that B^T is non-increasing for $T < T^*$ and non-decreasing for $T > T^*$. If $a(T)$ is strictly increasing, then the value T^* is unique and B^T is strictly decreasing for $T < T^*$ and strictly increasing for $T > T^*$.
>
> Suppose that $a(\infty) = \lim_{T \to \infty} a(T)$ exists. Then observing that $\lim_{T \to 0} h(T) = -u < 0$ and that $h(T) = (a(T) - B^T)U^T$ we see that
>
> a) If $a(\infty) > B^\infty$, then there exists a finite T^* minimizing B^T.
>
> b) If $a(T)$ is strictly increasing and $a(\infty) > B^\infty$, then there exists a unique, finite minimum.
>
> c) If $a(T)$ is non-decreasing and $a(\infty) < B^\infty < \infty$, then "$T = \infty$" is optimal.

When the function $a(T)$ is non-decreasing we could alternatively have used the results of Appendix G. We then have to introduce the δ functions

$$B_\delta^T = M^T - \delta S^T = u + \int_0^T (a(t) - \delta)R(t)\,dt$$

We see that if $a(t)$ is non-decreasing, than $T_\delta = inf\{t : a(t) \geq \delta\}$ minimizes the B_δ^T function. The results of Appendix G follow.

Example 5.1 [80]

The unit is a pump used in offshore oil production. It is assumed that if the unit fails, it is replaced. This means that the model is the classic age replacement model. Based on a failure time analysis the lifetime distribution was assumed to be a Weibull distribution with shape parameter $\beta = 1.3$ and scale parameter $\lambda = 1/264$ days. The mean time to perform a replacement was estimated at 20 days for an unscheduled replacement and 1 day for a scheduled replacement. We seek a replacement time T minimizing the ratio between the (expected) long run cost (downtime) associated with a replacement cycle and the expected time to replacement. For this example $a(T) = (v - u)\lambda\beta(\lambda T)^{\beta-1}$, hence a(T) is strictly increasing. We see that $a(\infty) > B^\infty$ so that there exists a finite, unique value T^* minimizing B^T. Now solving the equation $h(T) = 0$ using a numerical method, we find that the optimum preventive replacement time $T^* = 72$ days. Under the optimal policy, the limiting availability of the unit is 0.941. Using the policy "Replace only at failures", the limiting availability is 0.924. Thus the increase in availability is 1.7 per cent.

5.2.2 An extension

The model presented above is quite general and includes several well-known models as special cases, see Section 3.4.2 and the above cited survey papers [112,136]. It is, however, possible to generalize the model further in different directions. As an example, we will generalize the cost structure for the replacement.

Let d^T represent the cost associated with a replacement and let D^T denote its expected value. Furthermore let L represent the replacement time of the unit. We assume the following cost structure:

$$E[d^T|L = t] = c_1(t) + c_2(t), \quad t < T$$

and

$$E[d^T|L = T] = c_2(T)$$

Thus the expected cost of a planned replacement at time T is $c_2(T)$, whereas the expected cost of an unplanned replacement at time t is $c_1(t) + c_2(t)$, $t < T$. We assume that $c_1(t)$ and $c_2(t)$ are bounded functions, and that $c_2(0) > 0$. Furthermore we assume that the function $c_2(t)$ is absolutely continuous, i.e. we can write

$$c_2(t) = \int_0^t c_2'(s)\,ds + c_2(0)$$

Letting $f(t)$ denote the derivative of $1 - R(t)$, it follows that

$$\begin{aligned} D^T &= \int_0^T (c_1(t) + c_2(t))f(t)\,dt + c_2(T)R(T) \\ &= \int_0^T [c_1(t)\alpha(t)\lambda(t) + c_2'(t)]R(t)\,dt + c_2(0) \end{aligned}$$

By using this expression for D^T we obtain an updated version of (5.2), with a new $a(t)$ function given by

$$a(t) = [r(t)\beta(t) + c_1(t)\alpha(t)]\lambda(t) + c_2'(t)$$

and $u = c_2(0)$. The optimization can be now be performed as described in the previous section.

5.3 Block replacement model

The block replacement policy is described as follows. A unit is installed at time $t = 0$. The unit is replaced at times $kT, k = 1, 2, \ldots$, and at failures.

Thus a block replacement policy is one in which preventive replacements occur at defined intervals of time. In order to preserve a fixed schedule for maintenance, replacements are carried out at these intervals, irrespective of when a failure last occurred. A block replacement policy is usually applied to several units. The policy is simple to administrate because the preventive replacements are fixed, in contrast to the age replacement policy considered in the previous section where the actual scheduled replacement times become randomised.

Let $F(t)$ denote the lifetime distribution of the unit and $\mu = MTTF$ the mean time to failure, and assume that the failure rate $z(t)$ exists. The basic model assumes a simple cost structure: a preventive replacement costs $K > 0$ and a replacement at failure costs $c > 0$.

The objective here is again to determine the optimal replacement time T, so that the long run (expected) cost per unit time is minimized.

Now, from the results from Appendix E.2 it follows that the optimality criterion, B^T, has the form

$$B^T = \frac{E \text{ cost in } [0,T]}{T} = \frac{cM(T) + K}{T}$$

where $M(T)$ is the renewal function generated by lifetimes with distribution $F(T)$. From the Result E.2 of Appendix E we see that

$$\lim_{T\to\infty} B^T = \frac{c}{\mu}$$

This quantity represents the long run (expected) cost per unit of time using the policy $T = \infty$, i.e. "Replace only at failures". If $F(T)$ is continuous, then $M(T)$ is continuous, and it follows that there exists a minimum on $(0, \infty]$. We see from Result E.5 of Appendix E that if the distribution $F(T)$ of the lifetimes is non-periodic with finite mean and variance equal to μ and σ^2, respectively, then

$$B^T = B^\infty + \frac{a}{T} + o_T(1) \qquad (5.3)$$

where

$$a = K + c(\sigma^2 - \mu^2)/2\mu^2$$

and $o_T(1)$ is a function such that $o_T(1) \to 0$ as $T \to \infty$. It follows that if $a < 0$, then B^T converges to B^∞ from below and thus there exists a finite minimum of B^T. If $a > 0$, then we can conclude that B^T converges to B^∞ from above. Now suppose that the derivative of $M(t)$ exists, and let us denote it by $m(t)$. Then the optimality criterion has the form (5.2) with $a(t) = cm(t)$, $R(t) = 1$ and $u = K$. The results from Section 5.2 apply. Using (5.3) it is seen that if $a > 0$ and $a(T)$ is non-decreasing, then $T = \infty$ minimizes B^T.

It can be shown [120], that if the probability density function $f(t) = \frac{d}{dt}F(t) \to 0$ as $t \to \infty$, and for some $p > 1$, $\int_0^\infty |f(t)|^p dt < \infty$, then

$$m(t) \to \frac{1}{\mu} \qquad \text{as } t \to \infty$$

It follows that if this limit exists, then $a(\infty) = B^\infty$. The results a)-c) in Section 5.2.1 are therefore not relevant for this model. Due to (5.3) above, we have, however, been able to establish similar results.

The renewal function has been extensively studied in the literature. Explicit formulae for the renewal function can be found for the gamma distribution and the normal distribution [25,52,111]. For example, if $F(T)$ has a gamma distribution with parameters λ and $n = 2$, then

$$M(T) = \frac{\lambda T}{2} - \frac{1}{4} + \frac{1}{4}e^{-2\lambda T}$$

The renewal function for the Weibull distribution has been calculated in [121].

The basic block replacement model presented in this section, can be extended in many directions, for example with respect to the cost structure. We will, however, not go further into this here.

Example 5.1 continued
Assume that a block replacement policy is used, and suppose that the mean downtimes are as under the age replacement policy. Then the optimal replacement time is found to be 70 days, resulting in an optimal availability equal to 0.941. Thus the results are almost identical with those previously obtained assuming an age replacement policy. It should be noted that if the replacement times for the two policies are approximately equal and the probability $F(T)$ is relatively small, then the optimality function for the two policies would be approximately equal. This is seen by observing that under the given conditions, we have

$$EN(T) \approx P(N(T) = 1) \approx P(N(T) \geq 1) = F(T), \quad \text{and} \quad \int_0^T R(t)dt \approx T$$

For an interesting relation between the optimality criterion for the age and block replacement, see [117].

5.4 On-condition replacement models

In this section we consider the problem of finding an optimal replacement time when there is available information about the underlying condition of the unit, for example through measurements of wear characteristics and damage inflicted on the system, and that the proneness of the unit to failure can be characterized by a failure rate, which is adapted to this information. In the basic model to be analysed, the unit is replaced at failure. For simplicity we will assume a very simple cost structure: a planned replacement costs K and an unplanned replacement costs $c + K$, $c, K > 0$. Let $X(t)$, $t \geq 0$, be an observable stochastic process, possibly a vector process, representing the condition of the unit at time t, assuming no replacement in the interval $[0, t]$. A planned replacement of the unit is scheduled at time T, which may depend on the condition of the unit, i.e. on the process $X(t)$. The replacement time T is a *stopping time* in the sense that the event $\{T \leq s\}$ depends on the process $X(t)$ up to time s, cf. Appendix E.4. We define $N(t)$ as the number of failures in $[0, t]$, assuming no replacements in this interval. Then $N(t)$ is a binary counting process. The failure time of the unit, i.e. the jump time of the counting process $N(t)$ is denoted η. The failure intensity of the process is denoted $\lambda(t)I(t < \eta)$, where $I(t < \eta)$ is an indicator function which equals 1 if $t < \eta$ and 0 if $t \geq \eta$. The intensity, $\lambda(t)$, may depend on $X(s), 0 \leq s \leq t$. Often we can formulate the relation in the following way:

$$\lambda(t) = v(X(t))$$

where $v(x)$ is a deterministic function. We assume that the intensity $\lambda(t)$ is non-decreasing up to the failure time of the unit. The problem is to determine a replacement time minimizing the average cost criterion.

This model is quite general and includes a number of interesting special cases. We shall look closer into three examples:

1. *r different type of units.* Assume that r different type of units may perform the same function. At an arbitrary replacement the new unit is of type i with probability p_i, independent of earlier replacements. At the replacement the type is immediately identified. We assume that the age-dependent failure rate function for a unit of type i is $z_i(t)$. We assume also that these functions are non-decreasing. By letting $X(t)$ be equal to the type of the chosen unit, we see that the intensity of the failure process $N(t)$ is given by

$$\lambda(t) = \sum_{i=1}^{r} I(X(0) = i) z_i(t)$$

   We see that $\lambda(t)$ is non-decreasing, and the model fits into our general set-up. A possible but naive replacement rule is *Replace each type of unit according to its optimal age replacement policy.* However, this rule is in general not optimal, as will be shown later in Problem 5.5.2.

2. *Shock model.* Assume that shocks occur to the unit at random times, each shock causes a random amount of damage and this damage is cumulative. At a shock the unit fails with a given probability. A unit failure can occur only at the occurrence of a shock.

   Let $V(t)$ denote the number of shocks in $[0, t]$, and let Y_i denote the amount of damage caused by the ith shock. We assume that $V(t)$ is a Poisson process with intensity ν, and that the Y_is are independent and identically distributed random variables with a distribution $H(y)$. Let $X(t)$ denote the accumulated damage in $[0,t]$, i.e.

$$X(t) = \sum_{i=1}^{V(t)} Y_i$$

Now, if the unit is active before time t, and the accumulated damage equals x, and a jump of size y occurs at t, then the probability of failure at this point is $p(x+y)$, where $p(x)$ is a non-decreasing function.

This model is in fact a special case of the one described above. Given the history of the unit up to time t, the probability of failure in $(t, t+h)$ is

$$\nu h \int_0^{\infty} p(X(t)+y)dH(y) + o(h)$$

where o(h) is a small term such that $o(h)/h \to 0$ as $h \to 0$. Thus the intensity of the failure counting process $N(t)$ equals

$$\nu \int_0^{\infty} p(X(t)+y)dH(y)I(t<\eta)$$

For a formal proof of this result, see [14]. We see that the intensity is non-decreasing up to the failure time of the unit.

3. *Multistate monotone system.* Consider a monotone system ϕ of n independent components. Let R_i be a random variable representing the length of life of the ith component, $i = 1, 2, \ldots, n$. Assume that component i has a lifetime distribution $F_i(t)$ and a failure rate equal to $z_i(t)$.
Introduce the following random variables:

$$\begin{aligned} X_i(t) &= 1 - N_i(t) = I(t < R_i) \\ \mathbf{X}(t) &= (X_1(t), X_2(t), \ldots, X_n(t)) \\ \lambda(t) &= \textstyle\sum_{i=1}^n z_i(t)X_i(t)(1-\phi(0_i, \mathbf{X}(t))) \end{aligned}$$

We see that $X_i(t)$ is the state of component i being equal to 1 if the component number i is functioning, and 0 otherwise. It follows that we have a special case of the general model, with the failure intensity of the system given by the above expression for $\lambda(t)$. To see that the intensity takes this form, consider a small time interval $(t, t+h)$. Given the history of the process up to time t, and assuming that the system functions at time t, then the system fails with a probability equal to

$$\sum_{i=1}^{n} z_i(t)hX_i(t)(1-\phi(0_i, \mathbf{X}(t))) + o(h)$$

observing that $X_i(t)(1-\phi(0_i, \mathbf{X}(t)))\phi(\mathbf{X}(t)) = X_i(t)(\phi(1_i, \mathbf{X}(t)) - \phi(0_i, \mathbf{X}(t)))$ is either 0 or 1, and equals 1 if and only if the system is functioning, component i is functioning and the system fails if component i fails. For a formal proof, see Appendix E.5.3. It is not difficult to see that $X_i(t)(1-\phi(0_i, \mathbf{X}(t)))$ is non-decreasing in t, $t < \eta$. Thus the failure intensity is non-decreasing up to the failure time of the unit.

We return to the general theory. It follows from the results obtained in Section 5.2 that the cost criterion can be written

$$B^T = \frac{M^T}{S^T} = \frac{cEN(T)+K}{Emin\{T,\eta\}} = \frac{cE\int_0^T \lambda(t)I(t<\eta)\,dt + K}{E\int_0^T I(t<\eta)dt} \tag{5.4}$$

The third equality follows from the definition of the intensity of a counting process, see Appendix E.4.

Now, instead of minimizing (5.4) directly we will use the δ technique presented in Appendix G. Define the replacement times T_δ, $\delta \in (-\infty, \infty)$, by

$$T_\delta = inf\{t \geq 0, a(t) \equiv c\lambda(t) \geq \delta\}$$

We see that T_δ minimizes

$$M_T - \delta S_T = E\int_0^T [c\lambda(t) - \delta]I(t<\eta)\,dt + K$$

The results of Appendix G follow. Below we summarize the main results. Let $B(\delta) = B^{T_\delta}$. Then

> The stopping time T_{δ^*}, where $\delta^* = infB^T$, minimizes B^T. The value δ^* is given as the unique solution of the equation $\delta = B(\delta)$. Moreover, if $\delta > \delta^*$, then $\delta > B(\delta)$, if $\delta < \delta^*$, then $\delta < B(\delta)$; $B(\delta)$ is non-increasing for $\delta \leq \delta^*$, non-decreasing for $\delta \geq \delta^*$, and $B(\delta)$ is left-continuous.
>
> Choose any δ_1 such that $P(T_{\delta_1} > 0) > 0$, and set iteratively
>
> $$\delta_{n+1} = B(\delta_n), \quad n = 1, 2, \ldots$$
>
> Then
>
> $$\lim_{n\to\infty} \delta_n = \delta^*$$

Figure 5.1 shows the principle features of the $B(\delta)$ function. Note that the $B(\delta)$ function is in general not continuous, but left-continuous.

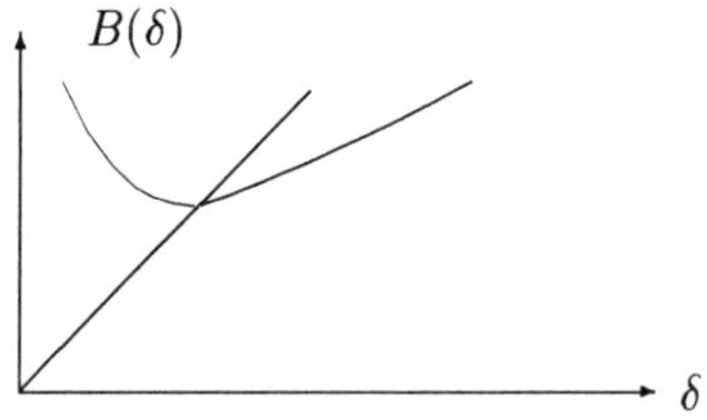

Figure 5.1: Principle features of $B(\delta)$.

Thus optimal replacement of the unit is at failure or when $a(t)$ reaches a given level δ, whichever occurs first.

Observe that

$$\frac{K}{E\eta} \leq \delta^* = inf_T B^T \leq B^\infty = \frac{c+K}{E\eta}$$

It can be difficult to compute $B(\delta)$ in a specific case. Below we carry out the calculations for a special case of the shock model presented above, see also Problems 5.5.2, 5.5.3 and 5.5.4.

Shock threshold model

Consider the shock model presented above. We make the following additional assumptions:

$$\begin{aligned} H(y) &= 1 - e^{-\theta y}, \quad \theta > 0, \quad y \geq 0 \\ p(x) &= I(x \geq M), \quad M > 0, \quad M \text{ is a constant} \end{aligned}$$

Hence the magnitude of a shock has an exponential distribution with expectation $1/\theta$ and failure occurs when the cumulative damage first exceeds a threshold M.

In this case we can write

$$\lambda(t) = \nu e^{-\theta(M - X(t))^+} \equiv v(X(t))$$

where $(M - X(t))^+ = max\{M - X(t), 0\}$. We shall prove that

$$B(\delta) = \begin{cases} \infty & \delta \leq c\nu e^{-\theta M} \\ \frac{K\nu+\delta}{1+\theta M + ln(\delta/c\nu)} & c\nu e^{-\theta M} < \delta \leq c\nu \\ \frac{\nu(c+K)}{1+\theta M} & \delta \geq c\nu \end{cases} \tag{5.5}$$

Using (5.5) it is easy to find δ^*. For $\delta \leq c\nu e^{-\theta M}$, we have $S^{T_\delta} = 0$ and $M^{T_\delta} = K > 0$, so $B(\delta) = \infty$. Now let $c\nu e^{-\theta M} < \delta \leq c\nu$. Define the renewal function $M(y)$ by

$$M(y) = \sum_{n=0}^{\infty} P(S_n \equiv \sum_{i=1}^{n} Y_i \leq y) \quad (S_0 = 0)$$

It can be shown that

$$M(y) = 1 + \theta y, \tag{5.6}$$

see Problem 5.5.3. Now, using the definition of T_δ, conditional on $V(t) = n$, defining

$g(\delta) = M + (1/\theta)ln(\delta/c\nu)$ and using (5.6), we obtain

$$\begin{aligned}
S^{T_\delta} &= E\int_0^\infty I(t < T_\delta)dt \\
&= E\int_0^\infty I(cv(X(t)) < \delta)dt = \int_0^\infty EI(cv(X(t)) < \delta)dt \\
&= \int_0^\infty \sum_{n=0}^\infty E[I(cv(X(t)) < \delta)|V(t) = n)P(V(t) = n)dt \\
&= \sum_{n=0}^\infty P(cv(S_n) < \delta)\int_0^\infty \frac{(\nu t)^n}{n!}e^{-\nu t}dt = \sum_{n=0}^\infty P(S_n < g(\delta))\frac{1}{\nu}\frac{1}{n!}\int_0^\infty t^n e^{-t}dt \\
&= \frac{1}{\nu}\sum_{n=0}^\infty P(S_n < g(\delta)) = \frac{1}{\nu}M(g(\delta)) = (1 + M\theta + ln(\delta/c\nu))/\nu
\end{aligned}$$

and

$$\begin{aligned}
EN(T_\delta) &= E\int_0^\infty v(X(t))I(t < T_\delta)dt \\
&= E\int_0^\infty v(X(t))I(cv(X(t)) < \delta)dt = \int_0^\infty Ev(X(t))I(cv(X(t)) < \delta)dt \\
&= \int_0^\infty \sum_{n=0}^\infty E[v(X(t))I(cv(X(t)) < \delta)|V(t) = n)P(V(t) = n)dt \\
&= \sum_{n=0}^\infty Ev(S_n)I(cv(S_n) < \delta)\int_0^\infty \frac{(\nu t)^n}{n!}e^{-\nu t}dt \\
&= \sum_{n=0}^\infty Ev(S_n)I(S_n < g(\delta))\frac{1}{\nu}\frac{1}{n!}\int_0^\infty t^n e^{-t}dt = \frac{1}{\nu}\sum_{n=0}^\infty Ev(S_n)I(S_n < g(\delta)) \\
&= \frac{1}{\nu}\int_{[0,g(\delta))} v(y)dM(y) = \frac{1}{\nu}[v(0)M(0) + \int_{(0,g(\delta))} v(y)dM(y)] \\
&= \frac{1}{\nu}[\nu e^{-\theta M} + \nu e^{-\theta M}\int_0^{g(\delta)} \theta e^{\theta y}dy] = e^{-\theta M} + e^{-\theta M}\{(\delta/\nu c)e^{\theta M} - 1\} = \delta/\nu c
\end{aligned}$$

The Formula (5.5) follows for $c\nu e^{-\theta M} < \delta < c\nu$. For $\delta \geq c\nu$ we have

$$B(\delta) = B(c\nu) = \nu(c + K)/(1 + \theta M)$$

This shock model has been analysed by a number of researcher, see e.g. [132]. A discrete time version of the model is presented in [21].

Finally in this section we shall briefly discuss the case where the unit is repaired at failure, and not replaced.

Assume the same model as described above, but assume now that the unit is repaired at failure. A repair costs c. Then $N(t)$, the failure counting process, takes values in $\{0, 1, 2, \ldots\}$. Letting $\lambda(t)$ denote the failure intensity of the process, the expression for the optimality criterion takes the form (5.4) with $I(t < \eta)$ replaced by 1. The results of Appendix G follow. The optimal policy takes the form: replace the unit as soon as the process $a(t)$ reaches the level δ.

If the intensity process depends only on the state process $X(t)$ and not on the failure process $N(t)$, we can interpret the repairs as minimal: a repair which changes neither the age of the unit nor the information about the condition of the unit. In this case, the running information about the condition of the unit can be thought to be related to a unit which is always functioning. The minimal repair assumption is reasonable when the repair does not materially change the condition of the unit from its condition immediately before failure. This might be true when, for instance, a small part in a complex system is repaired or replaced.

A large number of publications on on-condition based replacement models have appeared in recent years, see the survey paper [136].

The presentation in this section is to a large extent taken from [14,28].

5.5 Problems

1. Consider the model analysed in Section 5.2.1 and assume that the unit is minimally repaired at failure, i.e. $\beta(t) \equiv 1$ and $\alpha(t) \equiv 0$. Moreover assume that the mean value function $EN(t)$ is given by
$$EN(t) = (\lambda t)^{\beta}$$
where $\lambda = 0.12$ and $\beta = 1.75$, i.e. the model is based on the data given in Appendix C, Example C.4. The cost parameters are assumed to be as follows: $r(t) \equiv 1$ and $u = 2$.

   Compute the $h(T)$ function and find the optimal replacement time T^*. Sketch the optimality function B^T.

2. Consider the model introduced in Section 5.4: *r different type of units.* Suppose that $r = 2$ and $z_1(t) = at$, $z_2(t) = bt$, $a \neq b$.

   Determine the optimal replacement policy when $a = 1$, $b = 4$, $K = 0.1$, $c = 1$ and $p_1 = p_2 = 0.5$. Compute the cost criterion as a function of δ, i.e. compute $B(\delta)$. Sketch the function $B(\delta)$.

   Determine the optimal replacement time for each type of unit according to its optimal age replacement policy. Calculate the cost of this "naive" approach, and compare it with the one established above.

3. Consider the shock threshold model presented in Section 5.4. We shall show that
$$M(y) = 1 + \theta y$$

   a) Show that
$$P(S_n \leq y) = \sum_{i=0}^{\infty} \frac{(\theta y)^i}{i!} e^{-\theta y} I(i > n-1)$$

   b) Use a) to show that
$$M(y) = \sum_{i=0}^{\infty} (i+1) \frac{(\theta y)^i}{i!} e^{-\theta y}$$

c) Let

$$W(y) = ye^y = \sum_{i=0}^{\infty} \frac{y^{i+1}}{i!}$$

Show the desired result by using

$$\frac{d}{dy}W(y) = e^y(1+y) = \sum_{i=0}^{\infty}(i+1)\frac{y^i}{i!}$$

4. Consider the minimal repair model discussed briefly at the end of Section 5.4. Assume that $\lambda(t) = 1 + 3X(t)$, where $X(t)$ is a Poisson process with intensity 1. The cost of a replacement costs 6, whereas a minimal repair costs 1.

   Compute $B(\delta)$ and find the optimal δ^*.

5. Consider the model introduced in Section 5.4: Multistate monotone system. Let $v(t, \mathbf{x}) = \sum_{i=1}^{n} z_i(t)x_i(1 - \phi(0_i, \mathbf{x}))$ and

$$G(t, \mathbf{x}) = P(\mathbf{X}(t) = \mathbf{x}) = \prod_{i=1}^{n} [1 - F_i(t)]^{x_i} [F_i(t)]^{1-x_i}$$

   Show that

$$\begin{aligned} B(\delta) &= \frac{c\int_0^\infty \sum_{\mathbf{x}:\phi(\mathbf{x})=1} I(cv(t,\mathbf{x})<\delta)v(t,\mathbf{x})\phi(\mathbf{x})G(t,\mathbf{x})dt + K}{\int_0^\infty \sum_{\mathbf{x}:\phi(\mathbf{x})=1} I(cv(t,\mathbf{x})<\delta)\phi(\mathbf{x})G(t,\mathbf{x})dt} \\ &= \frac{c\sum_{\mathbf{x}:\phi(\mathbf{x})=1} \int_0^\infty I(cv(t,\mathbf{x})<\delta)v(t,\mathbf{x})G(t,\mathbf{x})dt + K}{\sum_{\mathbf{x}:\phi(\mathbf{x})=1} \int_0^\infty I(cv(t,\mathbf{x})<\delta)G(t,\mathbf{x})dt} \end{aligned}$$

   Simplify this expression for the case where the failure rates are constant.

6. Let $\alpha > 0$ be a positive discount factor; a cost c at time t has a value $ce^{-\alpha t}$ at time 0. Let T_i be the length of the ith replacement cycle, and let C_i be the total discounted costs associated with the ith replacement cycle. Then the total discounted cost incurred can be written

$$C_1 + e^{-\alpha T_1}C_2 + e^{-\alpha(T_1+T_2)}C_3 + \ldots$$

   Using this expression, show that the total expected discounted cost can be written

$$\frac{EC_1}{1 - Ee^{-\alpha T_1}}$$

   For the models treated in this chapter it is possible to write this criterion in the same form as the average cost criterion. Show this for the simple age replacement model, the minimal repair model, and the block replacement model.

Chapter 6

SAFETY MANAGEMENT

6.1 Introduction

This chapter gives an overview of the most important elements of safety management. By safety management we mean all systematic measures undertaken to attain and maintain a level of safety that conforms with defined policies, goals and acceptance criteria. Emphasis is placed on describing the role of reliability and risk analyses in safety management.

Safety and reliability can be viewed in line with other management tasks in the company, such as economic management, production management, etc. The same principles apply. We must establish overall policies, goals and acceptance criteria. Possible solutions that are expected to meet these policies, goals and acceptance criteria, must be developed. By means of different analysis methods we must find out how the solutions really conform to the defined policies, goals and acceptance criteria, and depending on the result of the analyses actions must be taken and specific solutions must be chosen. The process is iterative with increasing detail, to find the best possible solution. The principles of the process are shown in Figure 6.1.

To obtain efficient safety management, plans for safety measures must be developed and executed. The safety programme describing the safety policies, goals and acceptance criteria, and the measures, is to ensure that the safety management process is executed. Examples of such programmes are given in Section 6.2.6.

Safety management is a continuous process which runs in parallel with the other activities in the project or the company. It is included in the planning phase, the construction phase, the operational phase and the decommissioning and removal phase.

Internal control and *quality management* are two terms closely related to safety management. These terms can be defined as follows:

- Internal control comprises all systematic measures to ensure that the activity is planned, organized, executed and maintained in accordance with requirements laid down by law or regulations.

- Quality management comprises all systematic measures to ensure that quality (accordance with specified requirements) is planned and achieved. Often the term "quality assurance" is also used with this interpretation.

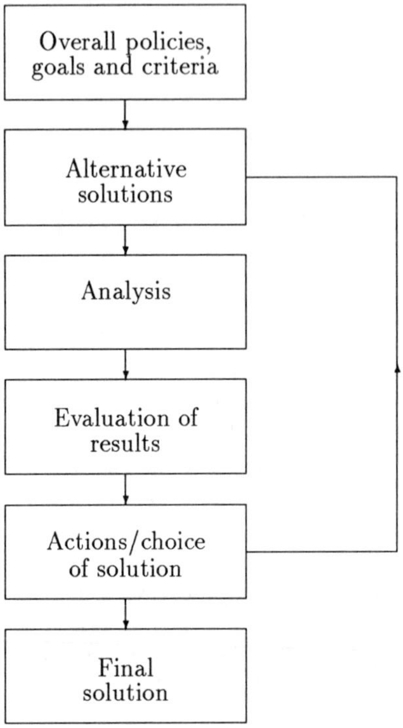

Figure 6.1: Safety management model

The term "internal control" has been used in connection with the change by the regulatory bodies from detailed control through inspection to system supervision and safety auditing. Internal control requires systematization in safety, with respect to planning, execution, follow-up and documentation. All parts of the organization must actively take part in this work. The responsibilities must be clearly defined. Internal control is also referred to in Section 7.4.

Companies which have developed quality management systems, may include safety in a total quality management system, to ensure accordance with the requirements specified by law and the company itself. This is particularly common in the petroleum industry, where the companies put strong emphasis on quality management. Safety is one of many items covered in a total quality management system.

Others prefer to consider quality management and safety management as two parallel activities, based on the same principles.

In the following the different elements of safety management will be described. The starting point of the presentation will be a company that produces and/or transports goods. Two different situations will be covered, with emphasis on the first:

1. The company is to manage the plant risk, where the risk is related to:
    - Human life and health
    - Environment
    - Material assets
    - Production
2. The company is to manage the reliability of a technical product

The first situation will in the following be referred to as "Safety management in a company", whereas the second situation will be referred to as "Reliability management of a product".

This chapter is based on the references [22,29,36,53,56,58,66,74,83,105,108,119,120,124]. A major part of the presentation of reliability management, including assurance of reliability through contract formulations, is based on [56].

6.2 Safety management in a company

6.2.1 Policies, goals and acceptance criteria

Definition
A policy is a statement of principle to be used as a broad guide or limitation to managerial action.

A policy may be formulated as for example:

- "The company shall be the best choice for international oil companies seeking safety and quality consulting services".

A policy may also be in the form of a "vision", such as that of the IBM President:

- "Let me give you my vision of the IBM of the future. It's very clear, It's very simple: I believe the IBM company must become the world's champion in meeting the needs of the customers".

Policy statements are seldom accurate and measurable. However, their mission is to point out the direction for the company, to limit the use of resources in areas which are not in line with the policy (e.g. the first safety and quality consultant company is probably not spending money on agriculture, since that is outside their policy). The policy should also be the main guideline for the employees, they should feel familiar with the policy and use it to guide their behaviour as professionals.

A good policy will describe and establish the real character of the company, and will form the foundation for:

- Delegation of authority and activities to lower levels of the organization.

- Ensuring common and co-ordinated behaviour through the whole organization. The policy forms the organization into a team, instead of a bunch of individuals acting completely in their own interests and priorities.

- Strengthening top management and making its work easier. Follow-up may concentrate on fulfilment of goals, instead of directing the organization in detail.

- Enabling middle and low level managers to make decisions, in line with the established policy.

In quality and safety matters clear policies are fundamental simply for the benefit of the last group above. To ensure safe activities and customer satisfaction, it will from time to time be necessary to make decisions which may be in conflict with the short-term profit. The policy will give first line managers backing for their behaviour and immediate action on such matters.

Some features are common to good policies:

- They are short and easy to understand

- They are related to the future, not the past

- The whole organization has been involved in the process of formulating the policy

- They are communicated to each and every employee, and to newcomers who joined the organization after the initial issue of information

- They are documented and approved by top management

Definition
A goal (objective) is the formulation of the organization's ambition.

The terms policies and goals are closely linked together, and many companies often combine policies and goals. If we are to distinguish between these terms, we can as a general rule say that a policy is the standing, long-term, "marching order" for an organization, while goals state the specific result that shall be achieved or is expected to be achieved.

An organization's goals define the common ambitions of the Board, the company management on all levels and the individual employee. The basis for such goals is often given as:

- They define what the organization shall strive to obtain.

- They define the norm by which success or failure can be measured.

Without goals, the organization cannot determine if progress is achieved or not, nor if the priorities are correct. Goals are guidelines for making priorities and strategies, which are the basis for selecting means.

However, goals are of limited value if the company does not enforce them through a process of systematic management, involving all parts of the organization and embracing a set of specific requirements and activities to identify necessary means and eliminate potential risk factors.

Goals should be [77]:

- *Optimal as to overall results.* Goals that suboptimize performance of various activities can easily damage overall performance.
- *All embracing.* Activities for which goals have been set tend to have high priority, but at the expense of the remaining activities.
- *Maintainable.* Goals should be designed in a modular fashion so that elements can be revised without extensive disassembly.
- *Economic.* The value of meeting the goals should be clearly greater than the cost of setting and administering them.

No less important is a list of criteria as perceived by those who are faced with meeting the goals. To those operating forces, goals should be:

- *Legitimate.* Goals should have undoubted official status.
- *Understandable.* They should be stated in clear, simple language. Ideally in the language of those who are faced with meeting the goals.
- *Applicable.* The goals should fit the conditions of use or should include the flexibility to adapt to conditions of use.
- *Worthwhile.* Meeting the goal should be regarded as benefiting those who do the added work as well as benefiting those who established the goal.
- *Attainable.* It should be possible for "ordinary" people to meet the goals by applying reasonable effort.
- *Equitable.* Since performance against goals is frequently used for merit rating of organizational units and individuals, the goals should be reasonably similar in difficulty of attainment.

Goals can be divided into economic goals, safety goals, goals related to quality factors, etc. We will focus on *safety goals.*

Safety goals are essential for many organizations. In order to avoid conflicts between such goals and others (e.g. economic goals), safety goals should be integrated into the general goals of the organization.

An example of a safety goal is "To reduce the lost working time caused by accidents by 25%".

Often safety goals describe an ideal situation. In the Norwegian regulations on the use of risk analyses in the petroleum industry with guidelines [108], the following statements are made:

> "The operator is required to define safety objectives for the total activity, cf. the scope of application of the regulations. These objectives express an ideal safety level. Thereby they ensure that the planning, maintaining and the further enhancement of safety in the activities becomes a dynamic and forward-looking process.
>
> The safety objectives are assumed to reflect the overall requirements of the applicable legislation with regard to full satisfactory safety for people, the environment, assets and financial interests.
>
> This means that:
>
> - Accidental events must be avoided
> - The level of risk in the activities is kept as low as possible
> - Attempts are made to achieve reduction of risk over time, e.g. in view of technological development and experience.
>
> The operator's safety objectives applicable to the activities may be expressed as long-term and short-term objectives.
>
> These safety objectives shall be used actively to initiate preventive safety measures such as system audit and motivation campaigns, based on the knowledge obtained through a risk analysis. This will contribute to a further enhancement of safety in the activity and will form the basis for, inter alia, revision of the operator's acceptance criteria".

Ideally safety goals are not very useful as a management tool. To be able to evaluate the risk related to an activity we must define acceptance criteria. They define the risk level which is considered acceptable with the available technology, economic limits and safety level in other parts of the industry and in society. A goal will, however, be to improve the acceptance criteria as available means makes such improvements possible. The need for risk reducing measures is assessed with reference to the acceptance criteria. The acceptance criteria express a standpoint with regard to risk connected with loss of human life, personal injury, damage to the environment, and assets and financial interests. For example of risk acceptance criteria, see Section 1.4.10 and Section 7.1. Refer to Section 7.4 for a thorough discussion of the use of acceptance criteria, based on experience in the petroleum industry.

In general, there are various bases for setting goals and acceptance criteria. Some of the most important are:

History. Historical performance is widely used as a basis for goals and acceptance criteria. With respect to safety, we can use statistics on the level of fatalities, injuries and accidents.

Engineering study. By engineering study we mean scientific collection and analysis of data. A risk analysis is an example of such a study. This approach is particularly useful in the planning phase of the system.

Benchmarking as a basis. The essential questions during a benchmarking process are:

- Which are the results obtained by the best one? What are the differences in my results?
- How is his performance to achieve these results? What are the differences in my performance?
- How can my organization obtain the same - or even better results?

In this process, we are able to continuously measure products, services and practices against the toughest competitors or those companies recognized as industry leaders.

If we systematize benchmarking in the various parts of the organizations, we will be surprised to see how much information we can obtain through "open" sources like company and product presentations, conference papers, visits, etc. In the petroleum industry most operators have a very open attitude regarding safety information, such as goals, acceptance criteria, practices and obtained safety results. Such information is free and available to us when we want to improve the performance of our activities.

Establishment of goals and acceptance criteria will depend upon:

- The requirements of regulatory authorities, etc., i.e. the minimum the company has to do to obtain necessary approval for its activities.
- What the organization desires in its own interests.
- What the organization is able to obtain given available resources and other limitations.

6.2.2 Alternative solutions

Alternative solutions that are expected to meet the overall policies, goals and acceptance criteria must be developed. To do this, different *means* will be used, both technical, operational, administrative and organizational. In the following we give an overview of the most important of these means. The means have been split into 11 main groups:

- Technical means
- Reliability management

- Procurement of goods and services
- Analysis
- Management involvement
- Organization
- Communication and motivation
- Personnel development and training
- Handling of accidents and contingency situations
- Transfer of experience
- Inspections, audits, etc.

These groups of means are listed in arbitrary order; they are *not* listed according to importance. Some of the groups are closely integrated and to some extent overlap.

Technical means

The safety level is very dependent on how the production plant is designed with respect to e.g.:

- *"Active protection":* Detection systems for gas and fire, shutdown system, fire fighting system, evacuation equipment, etc.
- *"Passive/inherent protection":* Area classification, fire protection, pressure relief panels, equipment redundancy, reliability of structure, equipment and components, etc.
- *Maintainability,* which is an expression of the ease with which maintenance can be conducted
- *Buffer and contingency storage.* Using such storage deliveries can be maintained for some time after production shutdown

Safe and economic operation is very much dependent on correct design and choice of system solutions and equipment.

Reliability management

The risk related to the activity of the company depends on the reliability of systems and equipment in several ways. Loss of revenue as a result of unplanned shutdowns, is a direct function of the reliability that is built into the system. The reliability of the production and transportation system will also have a direct effect on the investments and on the operation and maintenance costs. Furthermore, equipment reliability will affect safety of personnel and the environment, directly by means of safety equipment, and indirectly since failure and repair of equipment may result in higher risk of accident than normal operation. Large indirect losses can also occur if the company becomes known as an unreliable supplier.

Reliability management must ensure that the company meets its policies, goals and acceptance criteria. In particular, reliability management should ensure systematic evaluation of how much reliability is to be built into the production and transportation system, and how this can be performed in the most cost-effective way.

The principles for organizing reliability management are the same as for safety management. The model illustrated in Figure 6.1 also applies to reliability management.

Reliability clauses in connection with purchase contracts, and follow-up of these clauses, constitute an important part of reliability management. This aspect of reliability management is discussed under "Procurement of goods and services".

The reliability programme describing the reliability goals and criteria, and the planned reliability activities, is to ensure that the reliability management process is executed.

Procurement of goods and services

Means of securing procured goods and services of a sufficiently high standard with respect to safety and reliability, are e.g.:

- *Evaluation of contractors.* Goods and services are only purchased from contractors considered competent suppliers of safe products and services
- *Contract clauses with respect to contractor's safety management system.* Such clauses may for example be used when the contractor manages his own activity on the company's land
- *Assurance of reliability through contract clauses*

In the following we shall look closer into assurance of reliability through contract clauses. Choice of solutions is based on a number of assumptions of a technical and operational character. Many of these assumptions have a great impact on safety. If suppliers of critical equipment are not able to meet the expected reliability level, then key, technical decisions may be taken on insufficient grounds. In the worst case, this can mean that the actual risk related to the activity of the company, and in particular the economic risk related to production, can be unacceptably high. The wish to control the risk level is therefore a reason for specifying reliability requirements in contracts. Such requirements may also have their origin in cost/benefit evaluations. By improving the reliability of a particular item of equipment, it may be possible in some cases to avoid installing an extra, parallel unit, with all the costs this may incur.

Reliability clauses in purchase contracts, bid invitations and tender documentation can be an effective way of controlling the reliability level and thereby avoiding unwelcome variations. Purchase contracts include:

- Contracts for developing new types of equipment
- Contracts for purchasing equipment and equipment packages
- Contracts for studying and projecting tasks

The extent of the reliability requirements and the requirements for execution of reliability activities, must be adapted to the actual contract type and the ability of suppliers to meet the requirements and, for purchasing contracts, the type of equipment and criticality.

Reliability requirements and activities specified in bid invitations and tender documentation, and purchase contracts, may involve extra costs for the company, seen in isolation. Such requirements should therefore always arise from well-defined needs, supported by cost/benefit evaluations.

The reliability activities specified in a contract comprise reliability promoting measures and reliability verification measures. These two categories of measures are described in the following.

1. **Reliability promoting measures.** It should be a general objective for the company to get their equipment suppliers to improve the reliability of their equipment, even when quantitative reliability requirements are not part of reliability management. A number of reliability promoting measures can be used. Most of them will result in an initial cost for the company, whereas others will incur minimal costs. Just to mention the need for controlled reliability in a bid invitation or tender document may promote reliability, since the producer, supplier or project contractor will understand that the buyer is aware of reliability issues.

   The following topics can normally be referred to without financial compensation from the supplier:

   - *The supplier's competence in reliability technology*
   - *The supplier's experience with earlier deliveries of similar equipment*
   - *Documentation of performed and planned improvement measures.* As a part of the reliability promoting measures the supplier should be requested to describe the failure mechanisms for the critical failure modes of the equipment.
   - *Purchasing strategy.* A purchasing strategy that in itself can be reliability promoting is to use purchase contracts with options on additional deliveries. The purchase covers the immediate needs of the company, and the supplier is informed that importance will be attached to achieved operational reliability when deciding to take up the option or not. This strategy is normally only relevant for large projects.

- *Early identification of failures.* The supplier should describe his procedures for registration of failures during testing, installation and start-up of the equipment. Furthermore, the supplier should describe procedures for identification and execution of corrective and preventive actions as a result of registered failures.

The following measures will normally result in requirements for financial compensation, but they may still be cost-effective:

- *New reliability analyses.* The purpose of the analyses can be to identify earlier equipment problems, potential future problems, and to identify cost-effective, reliability improving measures related to the design or the operation of the equipment. Among the most relevant analysis methods are FMEA, FTA and CCA.
- *Maintenance evaluations.* In equipment purchase contracts the supplier is often requested to propose maintenance procedures that are in accordance with the maintenance philosophy and the safety procedures of the company. If the maintenance procedures are developed in close co-operation with experienced operating personnel at a sufficiently early stage, this activity may encourage reliability.

  Some maintenance data must be available at an early stage to ensure that reliability measures can be taken early enough. For reliability critical equipment the following information is required:

  - Intervals for execution of time-based, preventive maintenance which requires shutdown of the equipment, and assumed duration of the work
  - Average time between corrective maintenance for critical failure modes, and assumed duration of the repairs
  - Effect on other systems and possible alternative operation modes when the system is being repaired or maintained

2. **Reliability verification measures.** The purpose of reliability verification is to estimate one or more reliability parameters to be used in reliability analyses, or in relation to guarantee agreements. The verification can be performed by means of reliability analyses and tests, cf. Section 6.3.

   Quantitative reliability requirements are usually formulated as:

   - The minimum acceptable average time between critical failures
   - The lowest acceptable functional reliability for particular operation modes
   - The maximum duration of critical repair or maintenance activities

To ensure that these quantities have a unique and clear meaning, a definition of failure must be given, and the environmental conditions of the unit must be specified. Regardless of whether the verification is carried out by the supplier, the company, or by the parties jointly, it is essential that possible acceptance criteria are specified in detail in advance. The same applies to the consequences for the supplier if acceptance is not achieved. For the supplier to guarantee reliability, he will normally want to control the operation and maintain the equipment in the guarantee period. If the supplier himself carries out all the repairs, the planned maintenance, and purchases all spares, he will be motivated to minimize the failure rate and the unavailability to increase his profit. The customer will gain from not having to control the reliability programme or administer the maintenance work. When drawing up a guarantee contract which assumes that the supplier carries out and controls the maintenance of the equipment, the following guidelines should be followed:

- The contract should only apply to products that do not have a high development risk, and where the use will be relatively steady, so that both parties can agree about likely reliability improvements.
- The arrangements for payment must allow the supplier to make a large profit, while at the same time involving him in a reasonable amount of risk.
- Difficulties can arise in the administration of the contract in relation to other conventional repair policies. For example, equipment under such a contract needs special treatment and identifying to ensure that the unit is not opened or repaired by others than those approved.
- The supplier must be given freedom to modify the equipment to improve its reliability (at his own expense). But the customer must have a certain control of the modifications, since these can affect other equipment and systems.
- The contract should determine the way the equipment is operated and maintained, if there is a possibility that these factors can affect reliability.

Other arrangements that may have a positive affect on the reliability are bonuses related to achieved reliability after a certain operating period. Such bonus arrangements are normally to be preferred to penalties, such as a penalty where the achieved reliability level is not as high as assumed. It is important to motivate positively, instead of building in potential controversies and illegalities. It is easier to negotiate bonus agreements, since the offer is more easily accepted. Bonus agreements can represent a relatively small part of the customer's saving due to increased reliability, but a significant increase in the supplier's profit. For the supplier an achieved bonus will also have positive, indirect advantages, in strengthening morale and a point to refer to in future bids.

When drawing up contracts with bonus terms, it is necessary to ensure that other performance aspects are sufficiently specified, and possibly covered by financial arrangements, so that the supplier is not motivated by the bonus arrangement to achieve reliability at the expense of other important areas. Bonus contracts require thorough planning so that the supplier can meet the customer's requirements. The parameters chosen must represent a realistic challenge and the bonus must be high enough to warrant the additional effort.

Analyses

Risk analyses do not themselves improve safety, but they may indirectly affect it as the results from analyses may influence decisions on safety. Use of risk analyses in the various phases of the life cycle of the plant is described in Section 6.2.5.

Management involvement

To meet the defined policies, goals and acceptance criteria, management involvement is of key importance. The management can make its positive attitude to safety visible and contribute to create enthusiasm for safety in the company by for example:

- The development of the safety programme
- Safety inspections
- Making safety a permanent point on the agenda of relevant meetings
- Motivation to report on hazardous events and potential hazardous events
- Active follow-up of injuries, near misses, etc.
- Setting a good example
- Creating the possibilities for a good working environment
- Willingness to spend money on safety

The most effective means a manager has to influence the rest of the organization, is personal and direct involvement. This may be participation in relevant meetings, in safety inspections, and personally using safety protection equipment. The subjects that are important to the manager are given priority by the employees.

Organization

To obtain a high safety level, the organization of safety must be effective. Possible means that can be used are:

- *Safety as a line responsibility.* Safety, as other management tasks, is a line responsibility. This responsibility cannot be delegated. Hence, there is no way in which a safety manager can be responsible for safety. That responsibility rests with the line management, starting at the very top of the organization. However, top managers cannot themselves be expected to work out all the elements that go into the safety management system and co-ordinate their implementation. The most practical way to organize this is usually to have a staff safety position, reporting directly to top management (be it at corporate level, in an engineering project, or in an operating unit). This staff position will then work with management to establish a safety philosophy, establish and implement the safety management system, co-ordinate the safety efforts in the company and monitor safety performance.

  In major corporations, the corporate safety group works only on these tasks in order to maintain complete independence from the engineering and operating aspects.

  In engineering projects, safety is made up of two reasonably separate functions: safety management and safety engineering. The first does not work in line management, since it is a typical inter-disciplinary task which must report to project management. The question is then what to do with the second function, safety engineering. This is an engineering discipline, just like piping, process, instrumentation, etc. It is therefore preferable to have it as a line function, where it will be integrated with the other disciplines to which it relates. However, in a small project it might be a luxury to have two safety functions, and one might have to be limited to the staff function.

  The situation in an operating division is somewhat similar to that in engineering, except that safety engineering is replaced by a number of specialist functions such as:

  - Work environment (industrial hygiene, noise protection, lighting and ventilation, ergonomics, etc.)
  - External environment protection (pollution control)
  - Security (protection from criminal acts)
  - Emergency preparedness
  - Safe operating procedures (for vessel entry, hot work, etc.)
  - HAZOPs, Job Safety Analysis, and other safety studies
  - Accident reporting, investigation and analysis

  These functions are normally allocated to a safety department, which reports directly to top management in the operating division. It is not uncommon, however, to leave some tasks, like pollution control and the development of safe working procedures, to other departments, in which case they need to liaise with the safety department in order to ensure a high professional content and coherent approach.

  In matters like accident reporting and industrial hygiene, the safety department must co-operate with medical staff who normally belong to a separate medical department.

In respect of site safety, there are two lines of thought: safety officers are part of the safety department, or the production (operations) department.

If they belong in the safety department, they obviously need a "dotted line" on the organizational chart to the production management. The danger is that safety officers find themselves in a position of divided loyality: their boss is the safety manager, but from an operational point of view they must liaise very closely with the production supervisor.

It seems natural therefore to stick firmly to the principle of safety as a line responsibility, and have the safety officers reporting to the production supervisor, being employed in the production department. They will nevertheless find themselves in a dual role, since they must liaise closely with the safety department and therefore need the dotted line to the safety manager.

- *Safety committees.* Such committees, established at each main level of the organization, may represent an important element in organized safety work. Safety committees constitute a chain beginning with the Managing Director and ending at the very bottom. The leader of one committee is a member of the safety committee led by his/her superior.

- *Organization of work related to worker protection and the working environment.* In Norway the starting point is the general act of 4 February 1977 relating to worker protection and the working environment, according to which enterprises of a certain size shall have a working environment committee on which the employer, the employees and the safety and health personnel are represented. Safety delegates shall be elected at all enterprises covered by the act. The safety delegates shall safeguard the interests of employees in matters relating to the working environment.

- *Competence centre for safety.* Such a centre shall serve the whole organization. The establishment of such a centre will only be relevant for large companies.

Communication and motivation

Communication is a central factor for safety as well as for general management. Effective communication means e.g.:

- That it is open and works in both directions

- Exchange of relevant information and experience between the different organizational units in the company. The safety committees may contribute to information flow through the whole organization.

- Good co-operation between management, safety delegates, and trade unions.

Motivation is of great importance when it comes to safe and effective work performance. Important motivation factors are:

- Management attitude and involvement
- Acknowledgement and trust
- To influence one's own working situation
- To get feedback
- To have responsibility and authority
- To be informed

Personnel development and training

Many accidents occur as a result of inadequate qualifications. Prioritising personnel development and training can therefore improve safety. Relevant means in this context are e.g.:

- Personnel development by objectives. This means for example that specific goals are set for each person in the company, and plans are developed to ensure that these goals are achieved.
- Courses and up-grading training
- Site safety related training

Handling of accidents and emergency situations

An important feature of safety is warning, reporting and analysing accidents and near misses. Reporting may include computer systems for registration and presentation of statistics on personal injuries, discharges of chemicals, and loss of material assets. In order to obtain maximum learning from an accident it is essential for all personnel to be informed of what happened and why.

By emergency preparedness is meant all systematic measures to normalize an unstable and hazardous situation and minimize the consequences of a real accident. In particular, emergency preparedness includes measures to fight serious accidents such as fires and explosions, and to evacuate the personnel in a hazardous situation. Effective handling of accidents and emergencies can be achieved by e.g.:

- Identification of abnormal operating conditions and emergency situations, and planning for handling these
- Training and exercises

Transfer of experience

Transfer of experience from relevant units and plants is necessary to meet the defined goals and criteria in an effective way. Transfer of experience can be achieved through e.g.:

- Training
- Exchange of personnel
- Transfer of documentation and drawings
- Reliability and accident data

Inspections, audits, etc.

This group of means include:

- *Inspection,* which is a close look into the activity to check that it meets given requirements
- *Audit,* which is a planned, systematic and independent check of management systems to see that they are in accordance with the company's policies, goals and acceptance criteria.
- *Verification,* which is a study to confirm whether activity, product or service is in accordance with specified requirements
- *Self-monitoring,* which is a verification within one's own area of work to ensure quality in work

To obtain the desired effect by using the various means, *requirements* must be drawn up. These requirements may have different forms. Examples are acts and regulations, specifications, routines, guidelines and procedures. The requirements may be related to technical, operational, administrative or organizational conditions, depending on which means are considered.

Requirements are created for each type of activity, such as engineering, procurement, construction, operations, etc. In this way a uniform and co-ordinated set of requirements can be obtained, without lowering standards in any phase or activity.

Analyses are the means of linking the upper "policy and goal oriented" level and the lower "activity-oriented" level. Analyses are in many cases the only tools we have to verify that the chosen way of performing the activity meets the defined goals.

At an early stage of a system development, the means are more or less free variables, in the sense that one is free to select among them to obtain the desired goals and criteria. In practice, however, there will always be restrictions on the choice of the variables. These restrictions, which we refer to as constraints and boundary conditions, can be due to e.g.:

- Governmental requirements

- Political decisions
- Company requirements
- Sales agreements and product specifications
- Available technology
- Limited time and resources
- Limitations with respect to weight and space

When decisions are made, some of the free variables will become constraints in the optimization process. Other variables never become constraints since they can be changed in the operational phase. This is particularly true of operational and maintenance routines, but also of technical solutions if the actual changes are cost effective.

The constraints and boundary conditions also affect the choice of overall policies, goals and criteria, for example by governmental or company requirements.

The key elements of the development process are shown in Figure 6.2.

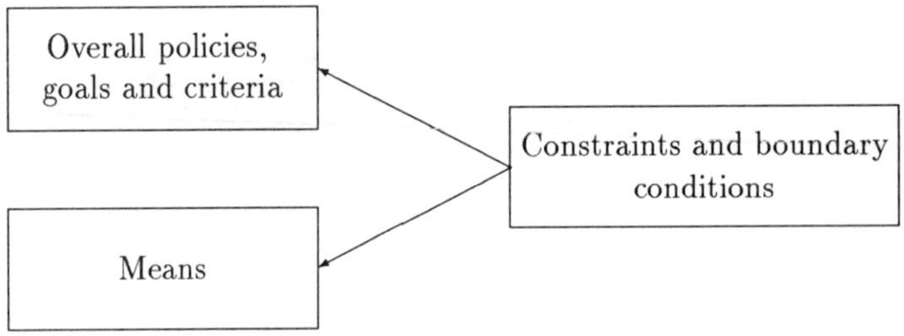

Figure 6.2: Main elements in a system development process

6.2.3 Analysis

By means of different analysis techniques we must find out how the alternative solutions meet the overall policies, goals and acceptance criteria. Refer to Chapter 2 for a description of some of the most common techniques.

6.2.4 Evaluation of results, actions/choice of solution

The results of the risk analysis must be evaluated in view of the defined policies, goals and acceptance criteria, and an economic analysis of the different alternatives must be carried out. These evaluations form the basis of choice of solution and possible actions/measures.

When we discuss safety, we meet a set of difficult priorities:

- How shall means directed at the (very seldom) fatal accidents, be balanced against means which may reduce frequent injuries?

- How shall preventive means (reduction of risk exposure) be balanced against means which reduce the consequence of accidents? The Norwegian Regulation concerning implementation and use of risk analyses in the petroleum activities states [108]:

  > "Probability reducing measures shall, to the extent possible, be given priority over consequence reducing measures."

  Most people will probably agree on this priority. But here we meet a psychological reaction. We feel safer when we know that there is an efficient emergency preparedness system ready to assist in case of an accident. Contingency measures can also themselves be probability reducing, for example emergency exercises might encourage realistic ideas of accidents and motivate safe behaviour.

- What principles should be adopted for the safety work strategy? How shall we balance:

  1. *Risk equality:* An equal level of risk is sought for all the employees in the company
  2. *Risk-cost effectiveness:* Measures are such that total risk reduction is maximized as a result of investments
  3. *Risk-compensation:* The benefit for the individual is judged to be equal to or greater than the risk. This might be achieved by using economic compensation, risk element of salary for particularly hazardous work, so-called "safety insurance" which gives financial compensation if a person is injured.

  Regardless of the principles we base our evaluation on, safety will be unevenly distributed in a company. Which policies, goals, acceptance criteria and strategies should be prioritized is in principle a political question. Therefore it is important that the priorities are set in a decision process where both the management and the employees are made responsible for the safety level and the priorities.

- Should personal safety or operational safety be the starting point for accident and loss prevention. Although these aspects of safety are usually closely linked, prioritizing areas of efforts can vary depending on whether emphasis is placed on personal injuries or pure economic losses.

Refer to Section 7.4 for a discussion of risk acceptance criteria.

Following up decisions on actions/measures and their implementation is extremely important. Safety is not improved by deciding to make changes; the changes must be carried out. Furthermore, it is essential that the results are controlled to see that they conform to the plans. In particular, it is important to check whether the measures introduce new risks.

6.2.5 Safety management in different phases

The life cycle of the plant can be divided into different phases:

- *Planning phase.* This phase includes the whole process from the screening and feasibility phases until the detailed engineering of the plant is complete
- *Building phase*
- *Start-up phase*
- *Operating phase*
- *Decommissioning and removal phase*

In the following we give a short description of safety management in the various phases.

Planning phase

The main purpose of safety management in the planning phase is to establish bases for decisions that affect safety, and to follow up the decisions. Results from risk analyses are an important part of this decision basis. This assumes, however, that the results of the risk analyses are available at the point of decision. Good time table planning is therefore essential in this phase.

Safe and economic operation is to a large extent a result of decisions made at an early stage in the planning phase. Factors that are not taken into account in this phase are difficult to incorporate later in an efficient manner. Improvised solutions in the operating phase will seldom be good ones. It is also more expensive to make modifications in the operation phase than in the planning phase.

Important safety activities in this phase are:

- Specify constraints and boundary conditions
- Establish policies, goals and criteria
- Establish safety programme
- Plan and execute analyses, identify and evaluate risk reducing measures
- Develop guidelines for following up deviations and identified problems
- Identify critical equipment and systems, contact supplier of critical equipment, define requirements for such equipment, implement requirements in bid invitations and tender documentation, execute improvement programmes for critical equipment
- Establish a list of assumptions

Building phase

The building phase is characterized by extensive activity, many contractors and subcontractors are involved. It is difficult to plan in detail since the development is dependent on a number of external factors. As a consequence of late deliveries, it may be necessary to change the work programme. Improvised solutions will often be the case.

A risk analysis of the building activity should be undertaken before the work starts, so that potential hazards can be identified and risk reducing measures introduced. The safety management in the building phase also includes control of the physical construction of the plant, to ensure that it conforms to approved drawings, requirements and assumptions.

Important aspects of safety management in the building phase are to:

- Secure that all relevant documentation from the planning phase is made available and applied by the persons involved
- Ensure transfer of knowledge from the planning phase to the building phase by letting personnel that have worked in the engineering phase work also on building the plant
- Evaluate thoroughly the consequences of introducing modifications, in particular with respect to the interaction between different systems
- Recruit operation and maintenance personnel with suitable education and experience.
- Ensure that recruiting of operation and maintenance personnel, and development of procedures, are carried out in good time before start-up
- Evaluate the procedures with respect to functionality and safety

Start-up

The start-up phase is quite similar to the building phase in the sense that many parties are involved. The activity is extensive and there is often great pressure to start up production on time. Machines and equipment are tested. It is often impossible to install safety equipment before the running-in period of the process has been completed. This must be given attention so as to avoid accidents.

Important elements of safety management in the start-up phase are:

- To make a plan for testing and test run.
- To evaluate the need for special safety precautions when testing and the test run
- To ensure that relevant information is transferred from personnel responsible for the planning and building of the plant to operational and maintenance personnel
- To ensure that detected failures are followed-up
- To ensure that competent persons have evaluated the following aspects before the plant is being used:

- Is the real plant identical with the final and approved drawings, specifications and procedures?
- Are modifications sufficiently checked?
- Have requirements on safety and working environment been implemented?
- Are the technical systems, procedures, qualifications of the personnel and organization of the work satisfactory?
- Has the interaction between technology, human beings and procedures been evaluated?

Operation

In general, the purpose of the safety work in the operation phase is to monitor and maintain the safety level that was assumed in the planning and building phase. In the operation phase risk analyses will have the following main objectives:

- Analyse the effects of modifications, and compare the up-dated risk picture with overall policies, goals and acceptance criteria
- Analyse the causes of safety problems (accidents or near misses)

Based on the risk analyses, technical, operational, administrative or organizational risk-reducing measures will be suggested. Furthermore the results from the risk analyses can be used in the following activities in the company:

- Training of operators and emergency response personnel (exercises on real scenarios)
- Development of procedures and safety instructions
- Documentation to the authorities

Examples of important elements in the safety management work in the operation phase are:

- Routines for safety inspections
- Routines for accident investigation and injury reporting
- Management involvement and support in the line organization
- Safety and emergency response training
- Routines to ensure that safety is considered in modifications
- Routines to ensure that safety is considered on a level with production availability in the daily operation

Decommissioning and removal

There will also be a need for risk analyses in the decommissioning and removal phase. This last phase of the life cycle of the plant will often be characterized by:

- Shortage of funds
- Reduced service function, e.g. preventive maintenance
- Reduced manning
- Reorganization
- The most competent and experienced personnel leave
- A feeling of defeat

Many companies experience a significant increase in the number of accidents in this phase. This is due to the above factors. The decommissioning and removal phase should be planned, and the effects of reduced manning and maintenance work on safety should be analysed in advance.

6.2.6 Examples of safety programmes

In this section we present two examples of safety programmes:

- Safety programme for a planned offshore production installation
- Safety programme for an operating offshore installation

The list of activities in the programmes is not complete; it includes a representative selection of all the activities.

Example 6.1 Safety programme for a planned offshore production installation

To design and build an offshore production installation in the North Sea company A has established a project organization B (project team). In the following we describe the safety programme for the unit B for the concept optimization phase. The purpose of this phase is to establish the final development concept, the basis for its construction and functional requirements, as well as its operational and maintenance philosophy.

1. INTRODUCTION

Purpose
The purpose of the safety programme is to describe all activities that will be executed by the project team to satisfy objectives and requirements for safety specified in the project. The safety programme specifies the safety objectives, safety organization and persons in charge of the various activities. The use of the programme shall ensure systematic identification, implementation, control and documentation of those safety activities required to obtain an acceptable level of safety concerning:

- Prevention of human injury and loss of life.
- Environment protection.
- Prevention of damage to equipment and production.

Limitations
This programme is developed for the concept optimization phase only. However, the outline of the programme is relevant to detailed engineering and subsequent project phases. A preliminary list of activities for the detailed engineering phase is included.

The safety programme is a part of the management system for the project.

Responsibility and approval
The project engineering manager has the overall responsibility for safety of the technical branches. This includes preparation, maintenance and implementation of the safety programme. The project safety unit will develop and maintain the safety programme on behalf of the engineering manager.

The safety programme shall be approved by the engineering manager subsequent to a review executed by the engineering and operational project staff, the company safety and quality assurance department, and by the company department for safety engineering. The programme shall be distributed to all project units in charge of planning, implementation and control of safety related activities.

Safety is a line management responsibility and each project organization unit and its personnel has a delegated responsibility to identify, plan, execute, and document safety related matters within their respective field of activity. The responsibility follows common principles for the organization and is an integrated part of the ordinary work routines.

2. GOALS AND MEANS
The principal objective of the safety work is to ensure that the platform and associated facilities are designed, constructed and operated with an acceptable level of safety, which means:

- Compliance with relevant rules, regulations, and requirements from the authorities and the company.
- The safety is to be optimized as far as practically possible within financial, technical, and other relevant constraints.

The safety related work shall include:

- Establishing built-in safety and acceptance criteria
- Implementation of the criteria
- Control and verification
- Documentation

A safety department is appointed to assist the project manager and project departments in their performance of safety related activities. The safety department will act as an integrated part of the project organization. The safety department will continuously review technical solutions, design changes and modifications, with regard to safety. The safety department will utilize risk analysis and reliability analysis as tools for evaluations. Thus the department will continuously provide input to concept optimization.

3. SAFETY REQUIREMENTS
The project shall comply with specific requirements, such as:

- Authority requirements
- Internal company requirements
  - Requirements stated in the company safety manual
  - Relevant specifications
  - Valid codes of design
- Specific project requirements
  (a) Acceptance criteria taken from one or more of the above requirements for personnel, environmental protection, investments and production
  (b) Design criteria, construction requirements which have to be met in order to satisfy acceptance criteria and to obtain cost effective solutions. Design criteria are related to accidental deaths as defined by risk analysis. Design criteria can e.g. be expressed as criteria for explosion pressure or thermal power.
  (c) Operational requirements

The requirements may be qualitative or quantitative. Acceptance criteria for personnel, internal as well as external staff, are given in the company safety manual. Other specific criteria have to be derived from general requirements, risk analysis, valid codes of construction, and operation philosophies. Establishing relevant design criteria is a separate activity which is divided into individual tasks according to the project time schedule. The main target in the concept optimization phase will be to identify all authority and company requirements and establish acceptance criteria. The risk analysis performed at this stage will identify preliminary design criteria. The list of design criteria has to be completed and verified during the detailed engineering phase.

4. SAFETY ACTIVITIES

General

The safety work shall be organized and performed as tasks integrated with the other activities in the project. The safety activities are divided into general activities, and system specific activities related to optimization and engineering work. In this section the general activity list is shown. A complete concept optimization activity schedule and a preliminary list for the detailed engineering is given in Section 8.

A safety plan and a safety activity list will be prepared before start up of each phase of the project.

The safety programme and working environment programme

Initiate revisions of the safety programme before usage in the detailed engineering and construction phases. Create a working environment programme and make revisions whenever required.
Responsible: Safety department.

Safety requirements and criteria

Identify laws and regulations, develop specific acceptance criteria, develop design criteria in accordance with the description given in this programme.
Responsible: Safety department.

Safety design activities

The activities within this field cover:

- Review and evaluate arrangements and systems
- Review built-in safety for alternative concepts, and optimize chosen concepts, regarding:
  - Area classification
  - Safety zones
  - Escape routes and evacuation systems
  - Specification and evaluation of safety and security systems

- Participation in design reviews, i.e. evaluations of the construction to ensure that it will behave as intended, that it is possible to construct it in a cost effective manner, and that the maintenance needed will be sufficiently limited
- Verification and implementation of design criteria

Responsible: Safety department. (Note! Each department has a responsibility of safety within its field of activities)

Risk analyses
Risk analyses shall be performed for the entire platform system including the oil and gas transportation system. During the optimization of concept design, technical solutions are continuously evaluated on the basis of risk analyses. In addition risk analyses will be performed when required in case specific problems arise.
Responsible: The safety department is responsible for specification, analysis and review of results. The safety department will also co-ordinate the implementation. The individual project departments are responsible for implementation of the risk analysis results.

Safety reviews
Safety reviews of the construction will be performed to control and verify project documentation to ensure that safety related matters are properly treated and design criteria are complied with.
Responsible: Relevant department executives.

Audits
System audits of the project are a part of the quality assurance system and co-ordinated by the project quality assurance department. External audits will be carried out on consultants, suppliers, etc. Such auditing is a part of the quality assurance system and further description is given in the quality assurance programme.

5. ORGANIZATION OF THE SAFETY WORK

   In this section the organization of the safety work in the concept optimization phase is described. This section will be modified for the detailed engineering and construction phases of the project. For further information references are made to the project manual.

   The project engineering manager is responsible for the organization and performance of the project safety activities. The safety department, reporting to the project engineering manager, has delegated responsibility for specification, performance and control of the safety activities as described in Section 4. The project engineering manager is responsible for ensuring that the project safety department is properly manned to ensure that they are able to plan and perform the safety work in a acceptable manner. The company safety department will provide personnel to the project safety department as requested by the project engineering manager.

6. TRANSFER OF EXPERIENCE
The project will employ relevant experience necessary to achieve safe construction. The engineering and construction experience will be incorporated in the concept optimization process with the aim of developing an installation with built-in safety meeting all relevant requirements. Operational experience will as far as practically possible be utilized to prevent problems relating to operation, maintenance, inspection and the working environment.

Transfer of experience will be documented and filed according to the filing system as described in the project manual.

Transfer of experience will be included in the safety activities by adopting experienced personnel into the project team, by review of written material and by direct communication with other organizations. The transfer of experience will be documented according to established project procedures.

7. DOCUMENTATION - SAFETY REPORTS
All safety activities will be documented and filed in accordance with the filing system as described in the project manual.

By the end of each project phase a safety report shall be prepared. This report will give a summary of safety results and give guidelines for work to be performed in the next project phase. The safety report shall contain the following:

- Objectives and results of the performed safety work.
- A list containing performed risk analyses and appraisals.
- Summary of the results from the risk analyses and appraisals.
- Recommendations for further work.
- An up-to-date list of safety acceptance criteria.
- Relevant notes concerning safety for construction, commissioning and operation.

8. SAFETY PLAN - LIST OF SAFETY ACTIVITIES
The safety plan is a time schedule for activities as described in Section 4. These activities are divided into specific tasks for each phase of the project. The safety scheme is developed as a part of the total project plans. For detailed information of the safety activities reference is made to the project manual.

A detailed list of safety activities will be worked out during start-up of each project phase. In the rest of this section the following lists are presented:

- A list of activities performed at the concept definition stage. This list is included for completeness

- A list of scheduled activities in the concept optimization phase
- A preliminary list of activities in the detailed engineering phase

List of Safety Activities

I Concept definition phase

- Work out safety programme
- Built-in safety: Evaluation of safety aspects concerning construction of the installation and connected pipelines
- Analysis: Coarse risk analysis of the installation and connected pipelines, coarse availability analysis of the production and transportation system

II Concept optimization phase

- Co-ordination and administration
- Make a revision of the safety programme, create a working environment programme.
- Built-in safety: Continuous evaluation of system changes
- Safety criteria: Identify requirements, develop general acceptance criteria, express preliminary design criteria for the detailed engineering and construction phases.
- Analyses: Concept risk analysis of the installation (this analysis is required by regulation of the Norwegian Petroleum Directorate [106]), study of installation, ship collision, risk analysis of the pipelines, risk analysis of the total installation system, reliability analysis of critical systems, availability analysis of production and transportation systems
- Evaluation of the working environment and preliminary philosophy for contingencies. The contingency policy will be developed in co-operation with the project operational staff
- Safety report: Prepare a safety report at the end of the phase

III Detailed engineering phase

- Update the safety programme
- Built-in safety: Continuous evaluation of system changes
- Safety criteria: Final criteria for detailed engineering and construction phases. Work out criteria for installation operation. Specify requirements in bid documents.
- Analysis: Total risk analysis, to obtain a picture as complete and correct as practically possible of the risk conditions at the installation, detailed HAZOP, analyses of the working environment, analysis related to construction, update reliability and availability analyses.

Example 6.2 Safety programme for an operating installation

In company A the production division is responsible for the operation of an oil and gas producing installation located in the North Sea. The production division prepares annual safety programmes for the installation. Such a programme is presented in the following.

1. INTRODUCTION
   The purpose of the programme is to ensure that the safety related work is taken care of in a systematic and controlled manner.

   The safety programme is issued and maintained by the production division safety and quality assurance department. Deviation from the programme shall be explained and documented by the organizational unit in charge.

   The terminology is in accordance with definitions given in the company safety manual.

2. SCOPE
   The production division safety management system is described in the safety management manual - production division, where requirements for a safety programme are stated. The manual reflects the legal and company requirements for safety management and safety in general.

3. SAFETY GOALS AND CRITERIA

   Safety goals
   The company's safety goals and policy are given in the company safety manual. The production division has derived its own safety goals based on the overall company safety goals:

   > Production of oil and gas without human injury, and without damage to environment and property

   By the expression "without injury" in this context is meant:

   - **For humans**
     - No working days lost due to accidents at work
     - No disease or damage to health shall occur as a result of long term exposure to a hazardous working environment
   - **For the environment**
     - Damage to the natural environment shall be as limited as technically and economically possible
     - No discharge/pollution beyond the levels permitted.

- **For Property**
  - No damage shall occur to equipment or construction leading to a reduction of safety for humans, the environment and production.

Acceptance Criteria
The frequency of lost working time due to injuries shall be less than 10 in 1 000 000 work hours.

4. SAFETY ACTIVITIES
This section presents a list of activities showing the major tasks to be performed by the production division during the year within the field of safety. The listing is in subject order:

- Common Activities (A)
- Technical and operational safety (T)
- Pollution Prevention (P)
- Work environment (W)
- Emergency planning (E)
- Security (S)

The responsible unit, time schedule, estimated hours of work within each activity are not included in this list. For this information reference is made to attachment 1 (not included here). For the year emphasis is put on inspection of fire/fire incidents and gas leakages and incidents.

ACT. NO.	DESCRIPTION OF ACTIVITY	OBJECTIVE
1.A	Work out statistics for accidents and near misses	Watch the risk level
2.A	Update the total risk analysis for the installation	Appraise effects of previous performed modifications
3.A	Work out the general part of a safety introduction package for new contractors	Motivation/information
4.A	Arrange safety seminars and safety meetings	Motivate for good safety attitudes
1.T	Improve access for flexible escape hoses	Personnel safety
2.T	Establish quality assurance system control of procedures and work instructions	Improve safety of work performance
3.T	Create a system for hotwork permits and control	Avoid fires/incidents caused by hot work
4.T	Perform safety reviews	Observe critical situations, motivate
5.T	Establish priority system for inspection and maintenance of critical equipment containing gas and for fire prevention equipment	Improved inspection programme
6.T	Establish a system for registration, assessment, and monitoring of operational safety related problems as component failures, gas alarms, etc.	Systematized routines

ACT. NO.	DESCRIPTION OF ACTIVITY	OBJECTIVE
1.P	Establish plans and information on environmental protection	Improve attitude
2.P	Discuss possibilities of reducing the amount of flared gas	Reduce noise level and prevent pollution
3.P	Create methods for analysing injection chemicals and perform such analyses on produced water	Detect chemical discharge from produced water
1.W	Basic survey of the working environment	Improve working environment
2.W	Hearing protection campaign	Improve use of personal protection equipment
3.W	Evaluate dangerous situations in cargo handling and use of chemicals	No lost time accidents caused by such situations
4.W	Identify reasons for sick leave by catering personnel	Eliminate reasons for sick leave
5.W	Work out long term programme for working environment	Establish strategy, find areas for activities and establish plans
1.E	Emergency drill	Train the emergency response team
2.B	Prepare a proposal for revised drill schedule	Vary the drills and make them more effective
1.S	Perform a campaign for security related activities	Secure the company's interests

6.3 Reliability management of a product

6.3.1 Policies, goals and criteria

For industry, the long-term profit-making aspect will always be the paramount issue, even though other requirements must also be taken into account. A high level of reliability is not an objective in itself, but merely a means to gain benefit. It is the total economy that is important, and to evaluate this, one has to take into account capital expenditure, operational expenditure, productivity, etc. The reliability level of the product/system should be a result of the economic optimization process.

But there is a strong connection between high reliability and profitability. Loss of revenue due to production shutdowns is a direct function of the reliability that is built into systems and equipment. But unreliability also produces indirect costs. Quality and reliability is today to a large extent used in marketing. Documented quality and reliability give a competitive edge and are important in determining market penetration and retention. What quality and reliability mean with respect to customer goodwill and product reputation are difficult to quantify, but are obviously very important factors.

A cornerstone of a successful quality and reliability strategy is constant effort towards quality and reliability *improvement*. "There will always be a way of improving quality and reliability at reduced cost". This rule for quality and reliability improvement is surprisingly often valid. In many cases minor actions can have dramatic effects on quality, reliability and costs.

Developments of the last 10-20 years have led to more of the quality and reliability work being carried out earlier in the design and production process. Immediately after the second world war, control and verification activities predominated. Products were controlled and defective items removed. This defensive technique is replaced by *management* of the design and production phase. Already in the development process one seeks to identify items that will be defective. The trend now is to put effort into increasing the quality before the production process begins. The emphasis is now on optimizing reliability. Reliability techniques are used to obtain cost-effective solutions. Management of reliability is a key to successful implementation of reliability technology.

Awareness that people want reliable products, and that reliability is a saleable property, has been most developed in Japan. The Japanese have the knack, apparently not prevalent in other countries, of deciding at a high level what needs to be done and then executing the decision right down to where the action is. They decided on quality and reliability as national priorities, and we can all see the results.

Public safety consciousness has been a powerful spur to quality and reliability in some areas, such as nuclear power and air travel, where government agencies set up and monitor reliability and safety standards. Product liability legislation is a new factor which is striking fear, if not dread, into manufacturers' hearts.

No one disputes the need for products to be reliable. People are hardly ever satisfied with less than perfect reliability, and they are acutely aware of the problem of less than perfect reliability in domestic products such as TV sets and cars. Organizations such as airlines, the military, and public utilities are aware of the costs of unreliability, and manufacturers often bear high costs due to failures occurring in the warranty period. But achieving high reliability is often costly, in particular when the product is complex and involves relatively new technology. All the economic aspects related to improved reliability should therefore be considered.

The bases for setting goals mentioned in Section 6.2.1 also apply to the reliability management of a product. In addition the market is an important basis. As shoppers we make price comparisons and thereby use the market price as a basis for buying. Industrial companies and other organs of society for centuries have similarly done comparison shopping by getting competitive quotations or bids as a basis for buying. There are some persuasive reasons for using the market as one of the bases for setting goals:

- The market goal is attainable under operating conditions. The proof is that others are already attaining it.
- If we are not meeting the market requirements, our clients in due course will discover that fact. (We may already be in trouble.)
- We should have early warning of lack of competitiveness so we will have time to take remedial action.

As a result of e.g. clauses in purchase contracts, requirements with respect to the reliability level can be specified. Such levels are usually expressed as Mean Time To Failure (MTTF), unavailability or Mean Time To Repair (MTTR), cf. Section 6.2.2, Procurement of goods and services.

6.3.2 Alternative solutions

To meet the defined policies, goals and acceptance criteria, different means will be used - technical, administrative and organizational. To avoid repetition from Section 6.2.2 we will in this section limit ourselves to giving an overview of the most important *technical means*, and briefly discuss means related to management and organization. Analysis and testing is covered in Section 6.3.3.

Technical means

Below we present a list of some important technical means that can affect reliability:

- *Choice of technology.* Often alternative technological solutions are available to perform a given function. Examples are subsea-completed instead of platform-completed oil and gas wells, new valve concepts that do not give production loss at maintenance, instead of traditional valve concepts, and cyclone systems instead of the usual separator technology. There can be large differences between the alternative solutions with respect to reliability, investment and operation costs, weight, etc.

- *Redundancy:* Introduction of several components in a system, which can perform the same function, so that failure in one component does not imply a system failure. It is common to distinguish between active redundancy, where all parallel units are operative under normal operating conditions, and passive redundancy, where only the necessary number of units are in operation at any time. Redundancy is an effective way of increasing reliability. There are, however, some disadvantages: increased weight, space, operational costs, etc.

- *Production capacities.* At failures of redundant systems, the production loss will depend on the capacity of the remaining intact units.

- *Instrumentation philosophy.* Failures in the warning and control systems can result in unnecessary shut-down of the process. A result of such failures can also be that hazardous situations are not detected and controlled. The reliability of the warning and control systems depend on e.g.

  - The size of the instrumentation system
  - Use of computer based monitoring and control systems
  - Choice of suppliers and models

- *Size and complexity of equipment and components.* Robust construction will give few failures that are caused by high or possibly varying operational loads. On the other hand, increased complexity in the construction may increase the unreliability. Highly complex construction may also increase the probability of common mode failures, i.e. failures that can put several redundant units out of function simultaneously.

- *Choice of materials.* The probability that critical failure mechanisms, such as corrosion will develop is very much dependent on the choice of materials. Factors that should be considered in this context are the environmental and operational loads that the unit can be exposed to, and the lifetime of the unit.

- *Environmental protection.* Many failure mechanisms are a result of exposure to environmental factors such as humidity, salinity, temperature variations, vibrations, etc. If the choice of materials and the design are insufficient to ensure that the equipment withstands the environmental loads, the probability of failures and problems can be reduced by protecting the equipment against the relevant environmental impacts.

- *By-pass:* For some types of equipment the consequences of failure can be reduced by having the production stream bypass the unit that has failed.

- *Early elimination of failures:* Defective components are removed before production starts by allowing the components to be exposed to higher loads than the expected operational loads. The method is particularly suitable for components that are cheap to produce, and it is used as routine in the production of electronic components.

- *Maintenance.* A high level of maintainability will reduce repair times to a minimum, and improve safety. Other maintenance aspects that affect availability, are the level of maintenance resources, such as personnel and spare parts.

Useful tools in the construction process, in particular relating to the determination of construction parameters, are Taguchi methods and statistical methods for planning experiments, see [29,77].

Management and organization
High reliability can be achieved only in an organization where the achievement of high reliability is recognized as part of corporate policy, and is given top management attention. If these conditions are not fulfilled, reliability will be cut whenever pressures of cost or time arise. Morale of reliability staff will be low and they will not be accepted in project teams. Therefore quality and reliability awareness and direction must start at the top and must permeate all functions and levels where reliability can be affected.

Because several different factors contribute to the reliability of a product, it is difficult to be categorical about the best way to ensure effective management of reliability. An effective reliability organization is, however, categorized by leadership and control. To be effective, the reliability organization must be executive, not merely advisory. On the other hand, it must not usurp the specialist design and production functions. The reliability team does not design, produce or maintain. However it should be able to give or withhold approval, and decide how the product should be tested to ensure reliability.

Reliability management must be integrated with other project management functions, to ensure that it is given appropriate attention and resources alongside all the other project requirements and constraints.

6.3.3 Analysis and testing

By means of various analysis methods and tests we must find out how alternative solutions really meet the goals and acceptance criteria. The following methods can be used for this purpose:

1. *Direct use of operational experience.* If systemized operational experience from similar equipment is available, this may be sufficient, provided that the operation and maintenance conditions are comparable.

2. *Use of theoretical models for analysis of reliability.* Examples of such analyses are presented in Chapters 3 and 4.

3. *Testing of equipment.* Examples of tests are:

    - Qualification testing. This test is common when introducing new technology, and it is performed to find out whether equipment or components are suitable for a specific operational situation. The qualification testing can also be carried out as lifetime testing or extended functional testing, possibly in accelerated form. The purpose of such extended testing is to establish an accurate estimate of the lifetime distribution or the survival probability for specific operational modi.

- Acceptance tests. Such tests are performed to find out whether equipment or components satisfy the constructive and functional requirements that are specified in the purchase contract.
- Integration testing (or system testing). The purpose of such testing is to reveal problems and deviations in the borderline between subsystems and equipment packages from different suppliers before the system is put into normal operation.

6.3.4 Evaluation of results, actions/choice of solution

The results from the reliability analyses and tests must be evaluated in view of the defined policies, goals and acceptance criteria. These evaluations form the basis for choice of solution and possible action. Such action can be:

- Modification of unreliable equipment and components
- New requirements, or change of subcontractors
- Systematic measures to reduce the vulnerability of equipment to environmental and operational loads
- Changes in routine to achieve a general or specific improvement of the product quality

6.3.5 Reliability programme - Reliability management in different phases

The reliability programme is to ensure that the reliability management process is performed in an effective way. More specifically, the reliability programme must:

- Ensure systematic planning of the reliability work
- Emphasize project milestones which are important to reliability
- Ensure that the priority of the reliability goals is recognized and maintained throughout the project
- Help to create awareness of, and understanding for, reliability issues
- Describe which reliability activities are to be carried out, how they will be carried out, by whom, and when
- Ensure that proper consideration is given to the interface of reliability work with other project activities

- Provide a means for effort and achievements in reliability to be reviewed for their effectiveness

The reliability programme is to be established prior to, or early in, each development phase, and otherwise updated as required.

The content of a reliability programme will vary, dependent on, for example, the complexity of the product and limitations with respect to time and costs. Below we present the content of a reliability programme with the main headings Introduction, Policies, goals and criteria, Organization and responsibility, Reliability activities, Activity plan and Documentation.

1. **Introduction.** This part of the programme may include the following items:

    - The purpose and scope of the reliability programme
    - Necessary definitions
    - Reference to relevant documentation

2. **Policies, goals and criteria.** See Section 6.3.1.

3. **Organization and responsibility.** The programme should describe the responsibility given to:

    - Managers and other key personnel
    - Functions, disciplines, sub-projects and other organizational units
    - Major engineering contractors

    See also Section 6.3.2 "Management and organization".

4. **Reliability activities.** Here the activities included in the reliability management are listed. The activities are related to e.g.:

    - Establishment and approval of policies, goals and criteria
    - Reliability analysis, evaluation, including identification of critical equipment and components
    - Reliability data
    - Procurement of parts
    - Design review, which means an evaluation of the design to ensure that it will function in operation, can be produced cost effectively, and is sufficiently maintainable
    - Recommendations with respect to actions and design proposals
    - Verification of reliability, testing

5. **Activity plan.** The reliability programme must include a plan for the activities. Correct timing is essential to ensure that the reliability work is functioning efficiently.

6. **Documentation.** Here requirements for documentation from the reliability activities are formulated. Normally the results are summarized at the end of each phase.

Below we present an overview of the most important reliability activities that are carried out within the various phases of the life cycle of the product.

Screening/feasibility phase

- Define framework conditions (constraints)
- Establish reliability programme for the phase
- Establish policies, goals and criteria
- Plan and undertake analyses of the reliability for main alternatives
- Identify critical items (systems)
- Evaluate risk by new technology
- Define assumptions and suppositions

Concept and engineering phase

- Control/update the framework conditions
- Establish/update the reliability programme
- Update policies, goals and criteria
- Undertake analyses of the reliability for optimization of chosen concept
- Update and develop a list of critical parts (systems, equipment, components) and operational conditions
- Make contact with potential suppliers of critical parts
- Specify reliability requirements, framework conditions and assumptions for critical parts, and incorporate necessary requirements in technical specifications
- Implement requirements with respect to reliability in invitation to bidders and tender documentation for detailed engineering and for parts with long delivery times
- Initiate improvement programmes for critical parts
- Update assumptions and suppositions

Construction

- Control/update the framework conditions
- Establish/update reliability programme
- Update policies, goals and criteria
- Select components
- Undertake improvement programmes for critical parts
- Test critical parts and the product
- Perform quality control
- Perform data collection and analysis

Operation

- Control/update framework conditions
- Establish/update reliability programme
- Update policies, goals and criteria
- Perform data collection and analysis
- Perform reliability analyses of modifications
- Undertake maintenance optimization

6.4 Problems

1. Describe the position of the reliability analysis/risk analysis in safety management.
2. Why is it so important to establish overall policies, goals and acceptance criteria?
3.
    a) What is the objective of the safety programme?
    b) List the most important elements of the safety programme.
    c) Give some examples of some typical safety activities in a safety programme. Distinguish between the design phase and the operation phase.
4. As Problem 3, but substitute reliability programme for safety programme.

Chapter 7

SOME APPLICATIONS

In this chapter three real-life examples of reliability and risk analyses are presented. The examples are taken from the petroleum activity in the North Sea. Due to confidentiality, all numerical values are changed for the first example. Some adjustments have been made for teaching purposes.

7.1 Risk analysis of an offshore oil and gas producing platform

The platform analysed is situated on the Norwegian Continental Shelf, and the platform had been in operation for many years when the analysis was performed. The analysis formed part of the company safety programme.

7.1.1 Objectives of the analysis

The following objectives of the study were established:

1. By conducting a risk analysis of the platform, provide, to the extent possible, a true and objective picture of existing risk elements. Primary focus shall be on the risk of loss of life.

   The analysis shall identify elements making significant contributions to risk, and discuss, possibly assess, impacts on safety inherent in changes/modifications of these elements. This shall provide the basis for an evaluation of risk level as well as possible measures that may serve to enhance safety.

2. Through engagement of personnel involved in the operation of the platform, contribute to increased vigilance and insight by application of analytical methods in safety work.

7.1.2 Study approach

The risk assessment methods developed and applied on offshore development projects since the 1980s were adopted for the analysis. Some elements were further developed in the analysis. They include, for example, the development of the event tree technique which allows for variation in the terminating event over several consequence categories, see Section 7.1.7.

The study was carried out as an in-house project in the company, with personnel from several organizational units. The project team included personnel with competence in risk analysis and personnel with operational experience from the platform. The project team was assisted in special tasks by platform personnel and engineers from specialist disciplines. Such tasks included for example explosion calculations, fire load analyses and structural load analyses.

The analysis itself was carried out over a period of five months, and about 2000 man-hours were budgeted.

As part of the analysis a HAZOP study was conducted. The HAZOP study was applied in two ways: comments/proposals were passed directly for decision and action, and results from the study were used as input to the risk assessments.

7.1.3 Risk

In the analysis the risk was calculated with respect to

- loss of life
- loss of material assets,

with emphasis on the first category. For process leaks, the risk was also expressed in terms of the Norwegian Petroleum Directorate measures (NPD measures) [108]. Below we present these three types of risk, as defined in the analysis:

Loss of life
The risk related to loss of life is expressed by accident frequencies (f_1, f_2, f_3, f_4, f_5) for different loss of life categories:

Loss of life (number of persons)	$1-2$	$3-5$	$6-20$	$21-100$	>100
Accident frequency	f_1	f_2	f_3	f_4	f_5

An accident frequency of for example $f_2 = 2 \times 10^{-4}$ per year, means that, in the course of 10000 operating years, there will be an average of 2 accidents with between 3-5 deaths; expressed differently, the probability of there being an accident involving 3-5 deaths in the course of one year, is 0.02%. The sum of all the f_is indicates the frequency of accidents causing loss of life (at least one death).

In addition to the above frequencies, the FAR value is computed. In order to calculate the number of hours, it was assumed that there are on average 240 persons on board the platform, 24 hours a day, which gives about 2.1×10^6 (= $240 \times 24 \times 365$) hours per operating year. A FAR value of 10 thus corresponds to an average of one death for every five operating years. To calculate the FAR value relating to occupational accidents, the number of working hours is used.

Based on the f_is the FAR value can be calculated using the following formula:

$$FAR = 1.5f_1 + 4f_2 + 13f_3 + 60f_4 + 170f_5$$

where the f_is are related to a period of 100 million hours. Here 1.5, 4, 13, 60 and 170 are the mean values in the loss of life intervals (categories).

Material loss

The risk related to material loss is expressed by accident frequencies (g_1, g_2, g_3) for different loss categories:

Damage (% of total value)	Limited 0-10%	Large 10-50%	Very large 50-100%
Accident frequency	g_1	g_2	g_3

The material losses are related to investment costs, and not to the real costs associated with the restoration following an accident.

Production losses have not been taken into account.

NPD measures

The starting point for these risk measures is the risk potential of the personnel outside the immediate vicinity of the accident, for example outside the module where the accident was initiated. This personnel can be considered safe if the following two functions are satisfied:

1. At least one escape route is intact until a safe escape has been completed

2. The shelter area is intact until a safe evacuation from the platform has been completed

These conditions assume that the structure maintains its load-bearing capacity to a sufficient degree during the escape and evacuation.

The risk is expressed by the frequencies (h_1, h_2), where h_i $(i = 1, 2)$ denotes the frequency of accidents where the above safety function is not satisfied. The NPD measure will be further discussed in Section 7.1.10 and 7.4.1.

7.1.4 Uncertainty

In the analysis the following philosophy was adopted for the treatment of uncertainty:

> In the risk analysis, probabilities and frequencies have been used for describing uncertainty as to whether certain incidents, such as leaks, might occur. These probabilities and frequencies are parameters in a model in which an attempt has been made to account for the most important factors affecting safety. Based on operating experience, statistics, etc., the probabilities and frequencies in the model have been estimated. The estimates used reflect the analysis group's best evaluation of the uncertainty of whether the incidents concerned will occur or not. Significance of the most important parameters of the model, such as leak frequency and ignition probability, has been assessed by looking at changes in the risk in the parameters. In this way the sensitivity associated with the estimates has been assessed.
>
> In this study no attempt has been made to quantify explicitly the uncertainty of the risk values derived. Such quantification is associated with considerable methodological difficulties and in our view contributes to complicating the risk picture quite unnecessarily. This is because quantification of uncertainty inherent in the estimates might easily be characterized by randomness and thus introduce a new element of uncertainty. Besides, quantification of uncertainty would only reflect uncertainty in the estimates of the model. Uncertainty associated with the model would be impossible to establish. This form of uncertainty might be at least as high as the uncertainty bound up with the estimates used in the model.

7.1.5 Assumptions

The analysis was based on a number of assumptions, including the following:

- Transportation to/from the platform and the risk related to transportation is not included
- For helicopter traffic, the risk related to helicopter landings and departures is included, also with respect to the passengers

- Shipping is included to the extent it presents a risk for the platform, in practice limited to ship collisions

- Pipelines to/from the platform are included to the extent they affect the risk related to the platform

A specific year was chosen as a reference year. Activities and activity levels, manning, manning distribution, etc., were estimated for this reference year.

7.1.6 Identification of undesirable events

A systematic identification of undesirable events was carried out based on operational experience, the HAZOP study and experience from earlier analyses and studies. The undesirable events were divided into the following main groups:

1. Process leaks, fire and explosion

2. Blowout

3. External influence (helicopter crash, ship collisions, etc.)

4. Occupational accidents

5. Others

In the following two sections we will look closer at the accident groups 1 and 4.

7.1.7 Process leaks, fire and explosion

To obtain reliable estimates of the risk related to process leaks, and evaluate the importance of factors/risk reducing measures, an event tree technique was used. Event trees are simplified models of the possible accidents and the sequence of events, but they are expected to give a reasonably correct picture of the risk that follows from the undesirable event described by the tree.

Figure 7.1 presents the standard event tree that was developed for the analysis. This tree was used for both gas and oil leakages. As an example of an accident scenario in the event tree, consider the path from the initiating, undesirable event on the top of the tree to the terminating event 14:

- A leakage has occurred in the module

- An ignition source was present and the gas was ignited

- Explosion has occurred

- The fire fighting system is functioning

- There is no accident propagation

For each tree a risk contribution is calculated for the various risk measures described in Section 7.1.3. The sum of these risk contributions gives the total risk related to process leaks. Notice that with this method no scenario is included more than once. Spread sheets (Lotus 1-2-3) were used to perform the extensive multiplication and summation operations in the risk calculations.

In the following we will look closer at the undesirable events (leakages), the branches of the tree and the terminating events. In the risk analysis, special substudies were performed to estimate the various frequencies/probabilities in the trees.

Leakage frequency

To be able to estimate the possible outcomes of the identified events it was necessary to split each undesirable event (leakage) into a number of categories. The following categories were used:

Minor	< 0.05	kg/s
Medium	$0.05 - 1$	kg/s
Large	$1 - 50$	kg/s
Very large	> 50	kg/s

These rates refer to initial leakage rates. The same categories were used for both oil and gas leaks. Minor leaks (< 0.05 kg/s) were not analysed further as it was concluded that such rates would make an insignificant contribution to the risk. To get an impression of the size of the leakage rates, examples of initial flame lengths of a jet fire from the leak point are presented in the table below. The figures presuppose idealized jet fires not encountering any obstructions.

Leakage rate (kg/s)	Flame lengths (m)
0.05	4
1	15
30	60
50	75

To estimate the leakage frequencies (leakage probabilities) within each module, the so-called "Parts count" method was used, a method that is usually adopted in such analyses. This method will be sketched in the following.

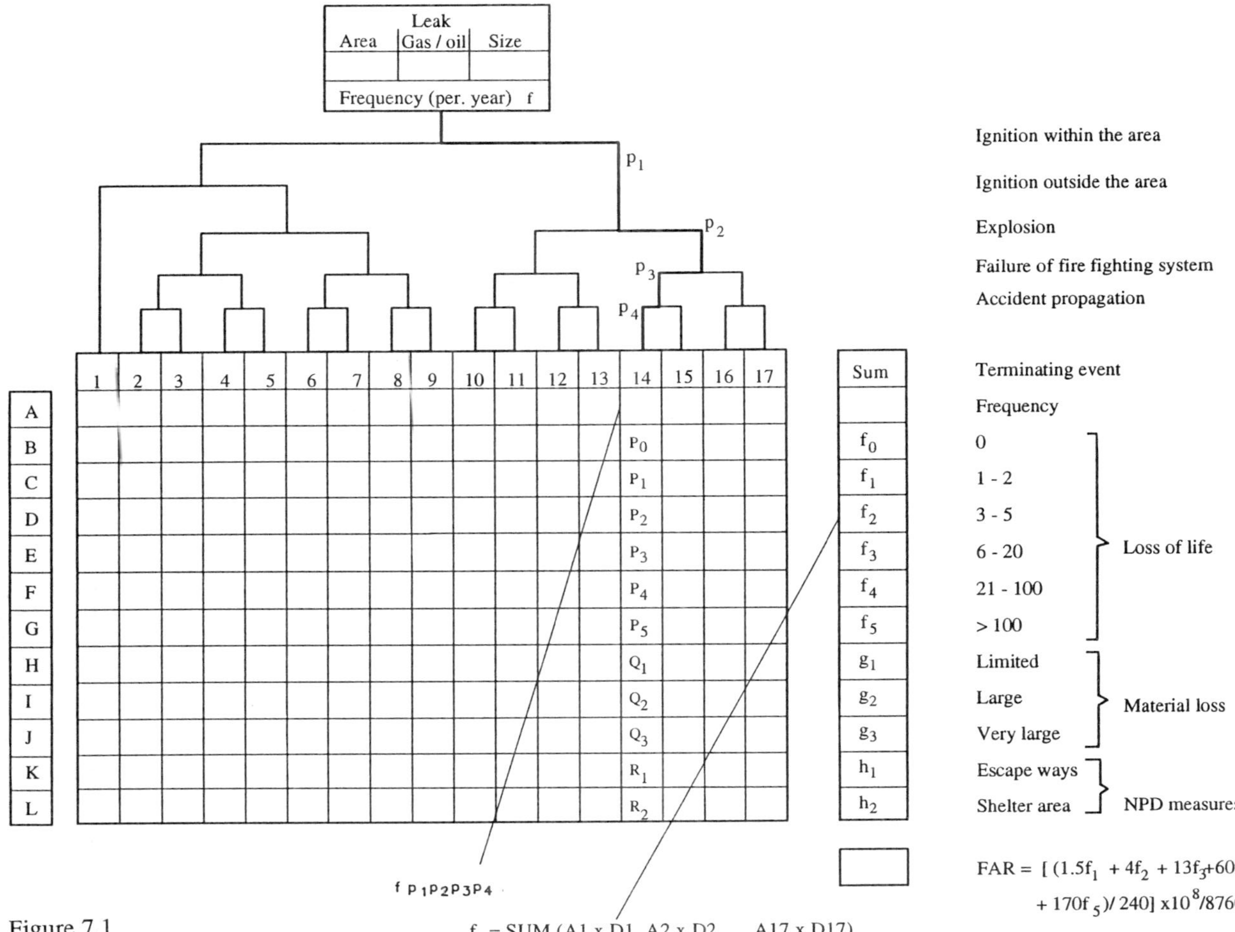
Leak
Area
Gas / oil
Size
Frequency (per. year) f
p1
p2
p3
p4
1 2 3 4 5 6 7 8 9 10 11 12 13 14 15 16 17
A B C D E F G H I J K L
P0 P1 P2 P3 P4 P5 Q1 Q2 Q3 R1 R2
Sum
f0 f1 f2 f3 f4 f5 g1 g2 g3 h1 h2
f p1p2p3p4
f2 = SUM (A1 x D1, A2 x D2, ..., A17 x D17)
Ignition within the area
Ignition outside the area
Explosion
Failure of fire fighting system
Accident propagation
Terminating event
Frequency
0
1 - 2
3 - 5
6 - 20
21 - 100
> 100
Loss of life
Limited
Large
Very large
Material loss
Escape ways
Shelter area
NPD measures
FAR = [(1.5f1 + 4f2 + 13f3+60f4 + 170f5)/ 240] x10^8/8760

Figure 7.1

A potential leakage could come from several sources, e.g. valves, flanges, vessels, compressors, etc. As an example, consider leakages from valves. The first step will be to count the number of each type and size of the valves in the area of consideration, and to multiply with the leakage frequency for each valve type. This leakage frequency is estimated from a large statistical data basis, e.g. data banks. Then we do the same for all other equipment categories, and finally we add them together to obtain the total leakage frequency in the area.

This frequency assessment is complicated by the fact that we usually have to take into account the release rate. To do this we estimate a hole size distribution for each type of equipment in question. An example is shown below. In the table "D" refers to the diameter of the pipe and "d" to the diameter of the hole.

Normalised hole size (d/D)	Portion of holes
0.1	0.60
0.2	0.25
0.4	0.10
1.0	0.05

Hence we need to introduce an additional step in the above calculation. After having counted the number of items of a particular piece of equipment and multiplied this by the leakage rate of the equipment, we split the resulting leakage frequency into a set of frequencies, each specific to a given range of release rate. Having done this calculation for all equipment categories involved, we add all the frequencies which belong to a certain release range.

To determine the different leakage rates, standard formulae for releases of hydrocarbons from a hole in pressurized equipment were used. Among others the following simple formula for gas release was used:

$$Q = U \times A \times P \times \sqrt{5.35 \times 10^{-5} \times M/T}$$

where

Q = Initial release rate (kg/s)
U = Discharge coefficient = 0.8
A = Area of release plane (m^2)
P = Absolute pressure inside equipment (Pascal)
T = Absolute temperature inside equipment (Kelvin)
M = Molecular weight of gas = 16

Some leakages had been registered on the platform in the operational phase. But there was too little data to produce estimates of the frequencies for the different modules (areas) on the platform. The total registered leakage frequency for the platform gave, however, an indication of the total leakage frequency of the platform.

Ignition

The probability of ignition depends on:

a) The probability that a combustible mixture occurs

b) The probability that the combustible mixture reaches a source of ignition

For each module and leakage category, an ignition probability that took into account a) and b) was estimated. The estimates were determined based on special models and expert evaluations. As a basis for the estimation, calculations were carried out to determine the development of the discharge concentration, i.e. when and where we could get a combustible mixture. The ventilation is a key factor in these calculations.

Explosion

An ignition will result in a fire or an explosion, depending on the available volume of combustible gas at the point in time of the ignition. A group of experts estimated a distribution between fire and explosion for each leakage category, where considerations were given to factors such as the place of the leakage, spreading pattern from the leakage, and the position of the source of ignition.

The expert group also estimated a probability distribution for the pressure build up in some typical modules. It is an extremely complex task to predict the explosion development in a closed or partly closed area that is full of equipment and has limited vent areas. The reason is that the mechanisms determining the process are large in number, complex and coupled. The most important parameters are quantity of gas; kind of gas; concentration and distribution of gas; source of ignition (place, strength); the total volume of the area; the length, width, height ratio of the area; number, size, shape and location of equipment; vent areas (location, size). By varying these parameters a gas explosion can give pressures in the range from a few millibars to 20 bars.

Explosion differs from fire with respect to the loads that equipment and personnel are exposed to. In addition to the pressure loads, objects that are carried away by the pressure wave can be a threat to the personnel in the area and rupture new pipes and escalate the accident. The scenario following an explosion is likely to include a fire.

Failure of the fire fighting system

Failure of the fire fighting system means in this context insufficient water/foam as a result of failure in the automatic fire fighting system, including damages due to an explosion.

For each leakage category and module a probability of failure of the fire fighting system was estimated. A fault tree analysis of the automatic fire fighting system was carried out as a part of the work of determining this probability.

Accident propagation
To estimate the probability of accident propagation to other areas/modules consideration was given to:

- The duration and intensity of the fire (volume of hydrocarbons, leakage rate, successful shutdown or not, effect of the fire fighting system and the blowdown system)
- Explosion effects (damage to equipment and pipes, and/or structure)
- Place of ignition (inside or outside the area)

Separate studies were performed to estimate the volumes of hydrocarbons between shutdown valves, and to evaluate the reliability of the shutdown system and the use and the effect of the blowdown system. For typical explosion pressures respond analyses were performed to estimate the damage both locally in the accident module and globally for the structure of the installation.

Distribution - loss of life and material assets
For each terminating event of the tree (each accident scenario) a probability distribution $(P_0, P_1, P_2, P_3, P_4, P_5)$ over loss of life is determined; P_0 represents the probability that the terminating event will result in 0 fatalities, P_1 represents the probability that the terminating event will result in 1-2 fatalities, etc. The sum of the Ps equals one. Based on the frequencies of the terminating events and their associated probability distribution over loss of life, the total accident frequencies for the categories 1-2 fatalities, 3-5 fatalities, and so on, i.e. f_1, f_2, f_3, f_4 and f_5, can be calculated. For example, for the category 3-5 fatalities this frequency (f_2) is found by multiplying the terminating frequency and its associated P_2 value and then summing over all terminating events in the tree, see Figure 7.1.

The accident frequencies g_i (i=1,2,3) for material losses are determined analogously. The method used to determine the Ps and Qs is described in the following.

Each terminating event in the tree may produce a number of different accident scenarios, depending on e.g. the duration of the fire and explosion effects. In this analysis ten different categories of accident scenarios are assumed, as is shown in Table 7.1. Each terminating event in the tree is assumed to result in one of these ten categories; which is shown in the table. For medium-sized gas leakages and all oil leakages the possibility of ignition outside the module has been disregarded. The analysis of these leakages therefore covers only the terminating events 1 and 10-17 in the event tree. For medium-sized leakages it is also assumed that the terminating event 10 results in the first category of accident scenarios in Table 7.1. The number of categories of accident scenarios has been set to a minimum. The definitions of small fire, large fire and very large fire are as follows:

Small fire: Includes only one area/module. A fire resulting from a medium leakage rate with an effective fire fighting and blowdown system will normally cause a small fire. The material losses will be limited. If evacuation is carried out, helicopters are used.

Large fire: Limited spread to adjacent areas/modules. A fire resulting from a large or very large leakage rate with an effective fire fighting and blowdown system will normally produce a large fire. Limited or large material losses. If evacuation is performed, helicopters are used.

Very large fire: Extensive spread to other areas. Normally very large material losses. Evacuation from the platform (if possible).

To determine the loss of life distribution for a given accident scenario, the following three distributions of loss of life are estimated (cf. Table 7.1):

A. Number of fatalities in the accident module (denoted by T_1)

B. Number of fatalities in other areas, as a result of e.g. explosion effects, no possibilities for escape, etc. (denoted by T_2)

C. Number of fatalities in connection with the evacuation from the platform (denoted by T_3)

These three distributions are then summed, i.e. the distribution of $T_1+T_2+T_3$ is computed, to establish a total distribution. To illustrate the calculations, let us consider a scenario where the portion of fatalities in the module is estimated to be 50% and the distribution B is estimated as follows:

Number of fatalities (T_2)	0	1-2
Distribution	0.6	0.4

The personnel distribution in the module is estimated to:

Number of persons	0	1-2	3-5
Distribution	0.2	0.4	0.4

Thus there is a probability of 20% that there will be no persons in the accident module at the time the accident occurs, and so on.

If e.g. 3-5 persons stay in the module, a 50% loss portion corresponds to approximately 2 persons, which means that the category 1-2 gets a contribution of 0.4. Similar reasoning for the other personnel categories gives the following distribution (distribution A):

Number of fatalities (T_1)	0	1-2
Distribution	0.2	0.8

Then the total distribution for A and B, i.e. for $T_1 + T_2$, can be calculated:

Number of fatalities ($T_1 + T_2$)	0	1-2	3-5
Distribution	0.2×0.6 $= 0.12$	0.2×0.4 $+0.8 \times 0.6$ $= 0.56$	0.8×0.4 $= 0.32$

Now the total distribution for A, B and C is found by summing the distributions for A and B above with C, similar to the summing of the distribution for A and B.

For the evacuation phase, only the very large fire scenarios will give a risk contribution. With a small or a large fire, a negligible risk is assumed of loss of life related to the evacuation. With a very large fire it is assumed that the evacuation of personnel from the platform is primarily performed by lifeboats. Due to heat radiation and smoke, it is assumed that only a small number of persons can be evacuated by helicopter. To determine the loss of life distribution related to evacuation by lifeboats, a detailed study was carried out. A special method was used to determine the distribution of the number of fatalities related to the evacuation. An important element of this method is to identify the scenarios where the evacuation could be difficult or impossible due to the accident, for example as a result of fire on the sea and fire radiation or smoke on the lifeboat deck. The method can briefly be described as follows:

1. Estimate a loss of life distribution when the evacuation is *not* made difficult or impossible by the accident (this distribution is denoted C_1)

2. Estimate for each accident scenario a factor D that represents the probability that the evacuation is made difficult or impossible by the accident

3. Estimate a loss of life distribution given that the event "evacuation by lifeboats is difficult or impossible" occurs (the distribution is denoted C_2)

4. For the given accident scenario, calculate the loss of life distribution (C) by $(1 - D) \times C_1 + D \times C_2$

Table 7.1

Area: Terminating event	Gas / Oil: Ignition place	Gas / Oil: Initial explosion	Size: Fire	A: Portion of fatalities in the area	B: Probability distribution over the number of fatalities not included in A and C. 0	1-2	3-5	6-20	21-100	> 100	C: Probability distribution over the number of fatalities in the evacuation 0	1-20	21-100	> 100	D: Probability that the evacuation via the lifeboats is made difficult/impossible due to the accident	Distribution material loss: Limit.	Large	Very large	NPD - measures: Escape ways	Schelter area
	Inside the area	No	Small													1				
10,12			Large														1			
11,13			Very large															1		
14,16		Yes	Small/ Large														1			
15,17			Very large															1		
	Outside the area	No	Small													1				
2,4			Large														1			
3,5			Very large															1		
6,8		Yes	Small/ large														1			
7,9			Very large															1		

7.1.8 Occupational risk

In this section we describe the method used for estimating the FAR value and the accident frequency for occupational accidents.

Occupational accidents typically include:

- Falls (to the same level, to a lower level, to the sea, etc.)
- Dropped objects, impacts, blows
- Poisoning, lack of oxygen
- Damage caused by tools, machinery

Data basis

The Norwegian Petroleum Directorate registers accidental injuries and deaths on the Norwegian Continental Shelf. From the Annual Report 1988 it is seen that in the period 1976-1988 there had been 7 deaths and 6575 injuries as a results of occupational accidents on fixed installations. The number of working hours in the same period was 186 million.

On the platform that was analysed, there had not been any occupational accidents involving death. The number of accidental injuries per million working hours was approximately 30 for the later years. The number of working hours was approximately 1.5 million.

Estimated risk contribution

To estimate the risk of loss of life related to occupational accidents, we can use different approaches:

1. If we assume that the platform is at the same level with respect to occupational accidents as the average of the fixed installation on the Norwegian Continental Shelf in the period 1976-88, we get

$$FAR = 7/1.86 = 3.8$$

2. The personnel injury frequency can be a good indicator of the risk compared to other platforms or in relation to the average on the Norwegian sector. One starting point is the relation between the number of accidental deaths and injuries registered in accordance with the Norwegian Petroleum Directorate's criteria for injury registration. For the period 1976-88 this relation is $7/6582 = 1/940$. Now, if we assume that this is a typical relationship, then we can use this to calculate a FAR value for the platform with the registered injury frequency as the starting point. Calculations give

$$FAR = 100 \times 30/940 = 3.2$$

The second method takes into account specific experience from the platform, and it was decided in the analysis to use this method.

With a FAR value equal to 3.2 we obtain a accident frequency per year in the category 1-2 equal to

$$1.5 \times 10^6 \times 3.2 \times 10^{-8} = 4.8 \times 10^{-2}$$

7.1.9 Risk picture

Tables 7.2-7.4 show the estimated, total accident frequencies and FAR values for the platform. In the analysis a number of graphic illustrations of the results were made. Two examples are shown in Figures 7.2 and 7.3. Thus the total FAR value is estimated at 35. Process leakages and blowouts give the largest risk contributions, approximately 80% if we look at the FAR value. It is the large-scale accidents that contribute most to the FAR value. There is a probability of 1% that an accident will occur during a year with very large material losses.

Table 7.2: Estimated accident frequencies - loss of life and FAR values

Category loss of life	$1-2$	$3-5$	$6-20$	$21-100$	> 100	*Total* ≥ 1
Accident frequency per 10^4 year	680	186	82	34	15	1000
FAR	4.6	3.5	5.1	9.7	12	35

Table 7.3: Estimated accident frequencies - material loss

Damage (% of total value)	Limited 0-10%	Large 10-50%	Very large 50-100%
Accident frequency per 10^4 year	1500	120	110

Table 7.4: Distribution of total FAR values by accident types

Accident type	FAR	Portion (%)
Process leaks	17	50
Blowout	9.6	27
Occupational accidents	3.2	9
External impacts	2.7	8
Others	2.2	6

The accident types, process leaks and blowout, contributed most to the risk and were therefore given most attention in the analysis. In addition to the results presented above, the risk description related to these accident types covers the following aspects:

- FAR value and accident frequencies distributed by leakage categories
- FAR value and accident frequencies distributed by areas/modules (leakages occurred in the module M1 contributes with 10% of the total FAR value, and so on)
- Distribution of the FAR value over the initial accident and the accident propagation
- The FAR value contribution from evacuation
- FAR values for personnel working in the most hazardous areas

In addition, the importance of some factors, such as the leakage frequency, ignition probability, reliability of the shutdown system and the fire fighting system, were discussed. The factors leakage frequency and ignition probability were given most attention, since the risk related to a particular leakage category is proportional to these factors. For example a halving of the leakage frequency will give a halving of the risk.

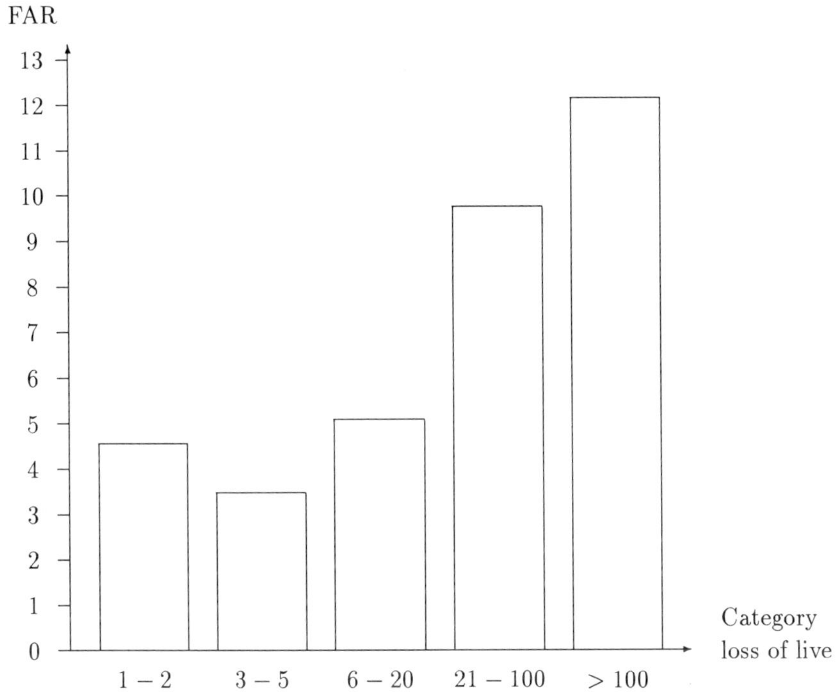

Figure 7.2: Estimated FAR values for the platform

7.1.10 Risk evaluation

In the foregoing the risk picture is described as it appears in the analysis. In the following this risk picture is discussed in relation to other comparable activities, and in relation to requirements and acceptance criteria which are relevant in our context, especially the company's own and those imposed by Norwegian authorities. The discussion applies to risk related to loss of life and material values.

Risk related to loss of life
The total FAR value for the platform has been estimated at 35. This value is directly comparable with the acceptance criterion given in the company's safety manual, which is FAR = 10. The Norwegian authorities have not expressed their safety requirements in FAR values.

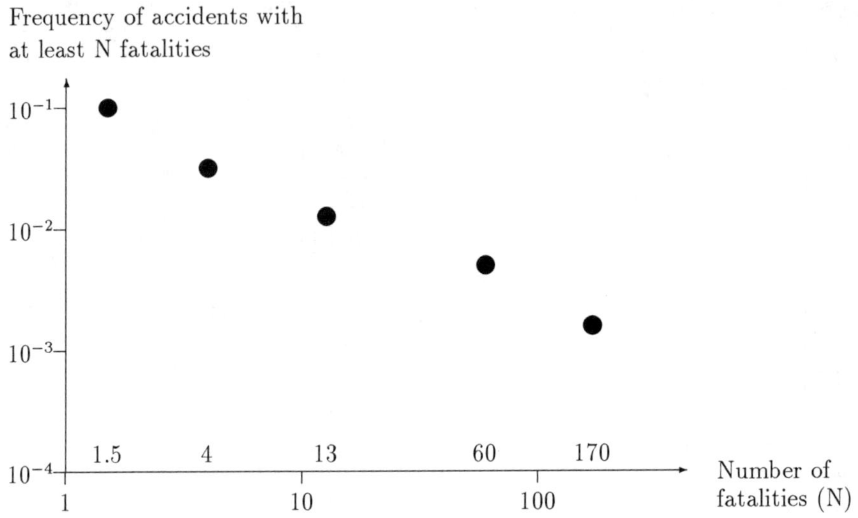

Figure 7.3: Connection between accident frequency (F) and loss of life (N) (F-N curve)

To get an impression of magnitude with respect to FAR values, a list of other relevant FAR values is presented below for various platforms and activities:

- Risk analyses of the platforms A1 and A2 have calculated FAR values equal to 8 and 11, respectively.
- If we include all accidents on the Norwegian Continental Shelf related to fixed installations for the period 1976-1988, then we obtain an observed FAR in the range 5-6 [107]. (In this period there was no accident with a large number of fatalities.)
- The Norwegian Continental Shelf (including 1988) has an observed FAR in the range 35-40 [104]. The Alexander Kielland accident contributes approximately half of this. On the English Sector the corresponding value is approximately 30 (including the Piper Alpha accident) [104].
- Of onshore industry, the chemical industry may present a relevant comparison. Here the observed FAR value is typical in the range 4-10 [89].

The conclusion from this is that the risk to personnel on the platform is quite high, and a special programme should be initiated to reduce the level of risk. There is great potential for improvement, see Section 7.1.11.

Obviously, the average FAR value (35) will be unevenly distributed, depending on which activity each individual performs, and where. Measures to improve conditions in "the high risk areas" will favourably affect the total risk, at the same time reducing the risk of the most exposed groups.

A particular characteristic feature of the platform is the relatively great contribution to the FAR value from major accidents. There are no specific requirements/acceptance criteria in the company or from the regulatory authorities associated with risk analysed by the various categories of accident size. However, a company will normally prefer a distribution showing a significantly lower FAR contribution for the large scale accident than for accidents with few fatalities. Expressed in a different way, the F-N curve should show an increasing fall for higher values of the number of fatalities (N) (concave form). When giving weight to accident factors, an aversion factor is commonly incorporated for major accidents. This may not be rationally justifiable, but can be explained by the additional strain on society, on the branch of industry involved - the oil industry - as well as on the company in the wake of the major accidents.

Planning requirements for new installations specify that accidents which the platform cannot withstand shall not exceed 10^{-4} per year [108]. This requirement applies to fire, blowout, etc., together about 10 accident types (cf. Section 7.4). This has in some cases been interpreted to mean that the total frequency/probability of all major accidents that do not meet the requirement for safe evacuation, escape ways, and structure, shall be lower than 10^{-3} per year. The requirement is directed towards design in the conceptual phase. If the criterion is interpreted as a requirement relating to large scale accidents with many fatalities, a number of platforms in operation, including this one, would have problems with the categories process leaks and blowout, which have accident frequencies that are significantly higher.

Risk related to material losses

For each incident, the consequence in the form of material losses is estimated on a very rough scale. For the category "limited", the uncertainty is greatest, since a number of incidents which would fall into this category were excluded from the analysis (when the risk to life was insignificant or nil).

As to the category "very large", the nature of the accident is usually such that full evacuation will take place, the situation has developed/escalated to new areas and attempts to get it under control will have been given up. As an example, a fire will then continue until all available combustible materials have been consumed. This category of consequence will therefore to a considerable extent have to be termed a total loss. In addition to loss of platform and production, the risk of oil pollution will be great. Obviously, this is true when a blowout is the initial incident, but also other fire/explosion incidents could develop in such a way that they might affect the wellhead areas and cause blowouts.

The probability of an accident with this consequence, in the category "very large", is estimated at 1.0%. This is a high figure. Based on strict cost/benefit criteria (considerations on strictly material/financial values) it is demonstrable that measures to improve this state of affairs should be implemented.

Risk acceptance
The use of risk acceptance criteria does not mean that accidents are accepted. The company safety objective is to avoid accidents and losses. Therefore we must continuously work to find measures that can avoid accidents. The results from the risk analysis may be used as a basis for determining the type and extent of such measures.

In the next subsection some possible areas of effort are discussed in view of the estimated risk and evaluations made in this section.

7.1.11 Risk factors and risk reducing measures

In the analysis possible areas of effort were discussed to reduce the risk. These areas include process leaks, blowouts, escape routes and evacuation. As an example we present some of the discussion on process leaks.

Fire and explosion in conjunction with hydrocarbon leaks contribute most to the level of risk on the platform. A reduction in the number of hydrocarbon leaks would mean substantial enhancement of safety on the platform. This applies in particular to large leaks. Measures aimed at lowering the leak frequency should therefore be introduced.

It is characteristic of leakages that we cannot point at a few measures that will reduce the leakage frequency. A wide spectrum of factors cause leakages. If we are to influence the leakage frequencies to any degree, we must succeed by measures that are broad and far-reaching. The individual's attitude, vigilance and knowledge cannot be overrated.

Leakages have always received attention on the platform, in particular in recent years. The importance of this risk factor is, however, so great that additional efforts are recommended. In the following we briefly review some possible areas of effort.

A considerable number of recorded major leaks are caused by human error. During non-routine operation, attention must be paid to the possibility of error, especially with regard to opening/closing of equipment (maintenance and modifications) and subsequent start-up. The focus should be on good planning (safe job analysis), co-ordination of activities and good communication on hand-over of equipment during shift changes. Operating routines/procedures to prevent excess pressure on the drainage system should be prepared, as well as measures aimed at detecting leakages, eliminating ignition sources, contingency plans for rapid shutdown and pressure relief, etc. Procedures for inspection of work executed prior to equipment start-up should be developed.

First-class workmanship and use of the right materials must be ensured to reduce the risk of leakages. As a case in point, one of the largest leaks on board the platform was due to incorrect rubber specification in a component's sealing material.

Qualifications and attitude need to be emphasized. In the company, great emphasis is placed on correct attitude and management involvement, reporting and investigation of accidents and near misses as well as consideration of safety issues in safety committees involving everyone. This has no doubt brought results and should therefore be continued but with greater emphasis on conditions that may lead to major incidents, such as hydrocarbon leaks. Leaks and fires should be reported systematically and should be reviewed in safety meetings/committees in a similar manner to work accidents.

The frequency of leaks is related to the qualifications and ability of operating and maintenance personnel. All inspection ranging from operator inspections to special condition monitoring is very important in identifying abnormal conditions. Personnel training should therefore have high priority. Process-related training is considered important in this context to achieve a full understanding of the processes on the platform. With regard to contractor personnel, it is important that they receive sufficient safety training to cope with the assignments allocated.

Production of oil will over time lead to changes in operating conditions, such as increased production of water, H_2S and CO_2 content, scaling, bacteria growth, emulsions, etc.; problems that to a large extent need to be solved by the addition of chemicals. These are all factors causing increased risk of corrosion, material brittleness and other conditions that may cause leakages. Experience gained on offshore installations indicates that unexpected problems occur. Knowledge about these conditions, identification of critical areas and development of effective inspection techniques are important areas of activity which serve to reduce the risk of hydrocarbon leaks.

Other causes of leaks may be breakdown of rotating equipment; turbines, compressors and pumps or liquid slugs in compressors. Mechanical strain, such as dropped objects and impacts on hydrocarbon systems, may result in damage and ruptures involving a risk of hydrocarbon discharges. Procedures call for special measures in connection with heavy lifts over the wellheads. In planning and preparation of safe job analysis for heavy lifts in process areas, special measures should also be considered. Minor mechanical loads may also cause serious leakages. Rupture of an instrument connection would, in a 100 bar gas system, cause a leakage in the order of 1-1.5 kg/s.

Some of the identified, possible measures were evaluated to estimate the risk reducing effect. They included:

- Reduced presence of personnel in the most highly exposed modules
- New escape routes from a module
- Installation of a new type of lifeboats
- Installation of explosion panels
- Improvement of the fire-fighting system

7.1.12 Follow-up

The study was presented and discussed with management, senior technical staff and union representatives, and a strategy and plan for follow-up activities were established in line with safety management principles. The main actions and conclusions were:

Information and communication

- Immediately, a brief description of the study and its results were sent by the Senior Vice President to each platform employee at home.
- Information tours were arranged to the platform with participation from operations management to give detailed information to each crew about the study, its results and action planned to improve safety on the platform.
- An in-depth presentation was given to the Norwegian Petroleum Directorate.

Future work

Objective:

- The safety level on the platform is to be increased to meet the safety goal for fixed installations (FAR = 10)
- Any improvements related to change in procedures and safe working instructions are to be implemented continuously
- Evaluations, engineering and approvals on risk reducing measures which require modifications are to be completed by January 1990
- Implementation of approved risk reducing measures will have high priority.

Strategy:

- Risk reducing measures that qualify for implementation will contribute to reduce the FAR value or be cost effective directly or indirectly in reducing material losses.

- The safety study will be updated by July 1991 to verify that the objectives of the follow-up work have been met.

Organization of follow-up work:

- All activities required to evaluate and perform engineering studies shall be handled in-house in accordance with established practice and procedures.

- Platform personnel will be involved in the work to the greatest possible extent.

- Planning and reporting is to be treated as a separate activity in the established planning system.

- Follow-up work is to be monitored by a "reference group" consisting of representatives from management and unions and supervised by the safety and quality assurance manager. The reference group will also be a forum for information on the progress of the work.

A list of possible risk-reducing measures was proposed to be considered further in 6 areas of effort:

1. Reduce process leaks
2. Avoid sources of ignition
3. Limit volume of hydrocarbons
4. Reduce consequences of fire and explosions
5. Improve evacuation and escape
6. General safety promoting measures

Each measure was evaluated in respect of risk-reducing effects and costs. The task of performing the studies and project activities was given to those functions in the organization according to their responsibility and in the following months a considerable amount of review and engineering study work was carried out.

7.2 Availability and life cycle costing in a gas project

This section shows how reliability and availability analyses are used in the Troll Gas Project. The presentation is based on [81].

Troll is a major offshore gas field situated about 80 kilometres north-west of Bergen. The field was discovered in 1979. In 1986 part of the recoverable gas reserves was sold together with gas from the Sleipner field to a consortium of European buyers. Deliveries are scheduled to start in 1996 and there will be a build-up in contract quantities until the year 2002, from which point they will remain fairly constant until the end of the contract period in 2022. At the the time the paper [81] was written, the base case plans for the Troll field development featured a four-legged, gravity base substructure, production, drilling and quarters platform located in 305 meters of water. The plan was to export the dry gas in two pipelines, one leading via the Sleipner platform to Zeebrugge in Belgium, the other connected to the Statpipe/Norpipe system to Emden in Germany. Condensate was to be stabilised and transported by pipeline to shore. A gas storage facility to be built at Etzel near Emden provides some stand-by capacity in case of disruptions to gas production.

7.2.1 Approach

The Troll platform is expected to be in operation for up to 70 years. This lifetime, considered very long for an offshore production platform, has induced the Troll Project to place emphasis on operational requirements and long-term cost effective solutions.

The current agreements for gas delivery from Troll require the contracted amounts of gas to be delivered to the buyers with reasonable reliability. However, it is not an objective to attain high levels of reliability at any price. If additional expenditure is proposed on grounds of increased reliability, the expenditure must be properly justified, by using for example reliability and cost analysis.

The Troll Project's approach to reliability engineering and life cycle costing is to use the available methods as a part of the overall evaluation process to screen alternative designs and equipment configurations. The purpose is to achieve a sound balance between capital and operational expenditure and gas sales income.

7.2.2 Cost elements

The main cost elements considered in the Troll life cycle cost evaluations are:

- Capital expenditure
- Operational expenditure
- Production output

Capital expenditure. When alternative solutions are to be compared with capital expenditure, it is important that not only the equipment itself but also the associated pipes, cables, instruments and supporting structure are considered. Also the effect of potential differences in the capital expenditure for living quarters and other supporting facilities, as a result of differences in manning, should be included. To allow for these elements, a parametric model was developed. This model estimates the difference in capital expenditure for the different alternatives. The primary equipment parameters used in the model are type, weight and footprint area (deck area taken up by the equipment). The model allows for the cost of

- Engineering and project management
- Equipment purchase and installation
- Bulk and structural materials/fabrication
- Freight
- Hook-up and commissioning
- Insurance

Operational expenditure. This expenditure will generally increase with the amount and complexity of equipment installed on a platform. However, improved equipment reliability and/or maintainability could reduce the demand for support, in addition to having a positive effect on production. Operational expenditure is typically divided into:

- Pure operational costs, such as energy consumption, operator labour and transportation of people and goods
- Maintenance costs, such as maintenance crew labour, materials and spare part holding

Where alternatives with different amounts of equipment of similar type, weight and size are involved, the difference in operational expenditure is often assumed to be proportional to the capital expenditures for course comparisons.

Gas/condensate transportation and storage tariffs are included separately in connection with the production.

Production. An appropriate establishment of overall system boundaries is essential when assessing the performance of production systems. In general, all significant factors affecting product delivery down to the point where the custody of the product is transferred to the buyer should be taken into account.

Before reaching the European buyers, the Troll gas will be routed through an offshore pipeline transportation network. The large internal volume of the pipelines and the compressibility of gas will enable the pipeline system to function as a buffer storage. Shortfalls in platform production can thus be fully or partially covered at the points of delivery by using the buffer gas volumes, which are recovered later by boosting platform output. Moreover, the gas storage facility planned near Emden will be able to meet the demand at this delivery point for many days.

To analyse the planned production and transportation system a flow network model of the system has been developed, and the reliability (availability) of the system is analysed by means of the methods described in Chapter 4, cf. Example 4.2 and Section 4.3.

Many previous offshore projects have allowed for an extensive use of spares with the aim of attaining high production availability. Evaluation procedures have had a tendency to underestimate the full cost of redundant equipment by taking insufficient account of all the additional capital and operational expenditures involved. Furthermore, the benefits of spares are often overestimated. Operational flexibility such as by-passes, line-pack, storage and overcapacity often reduce the impact of unreliability, whilst the increase in complexity introduced by spares generates new possibilities of both technical and human error.

7.2.3 Reliability data

Reliability data are defined in terms of failure rates and restoration times for Troll equipment on the basis of piping and instrumentation diagrams and equipment descriptions. In addition to the main items, associated equipment for process control, monitoring, blowdown, relief and shutdown are included.

Failure rates are derived by pooling data from several sources. Where failure modes are not precisely defined, judgement is used to estimate the proportion of the total number of failures for a component which can be expected to have an immediate effect on production. Other failures are assumed to be repaired during planned shutdowns or as opportunity maintenance during unplanned shutdowns.

The restoration data used are based on parametric models and expert judgements, mainly by experienced personnel on the Troll Project Team. Available experience data bases are not considered adequate for this type of information.

The estimated restoration times represent the time required to recover a function affected by failure of an item either by

a) restoring the item to full function, or

b) restoring the affected function by reversion to a safe alternative mode of operation pending repair of the failed item

In case a), the restoration times will usually include allowances for:

- Fault diagnosis
- Preparation
- Repair
- Test and production restart

In case b), the following elements are considered:

- Fault diagnosis
- Switch-over
- Production restart

7.2.4 Economic analysis

The definition of the economic implications of non-delivery of Troll gas caused by equipment unreliability is of major significance in this context. In addition to the short-term effects on cash flow of such non-delivery, the long-term implications must be considered. In principle, products not delivered from a given oil/gas production system are not lost (unless of course they are vented or burned), but the realisation of their income potential is deferred. The actual degree and effect of deferral will depend on the technical and contractual possibilities of utilising any overcapacity in the system to make up the shortfall. This is in principle also true of Troll. However, the viability of the Troll Project rests on the delivery of defined quantities of gas, within defined time-frames and at a defined unit price. Apart from these terms, nothing specific is known of the future market for Troll gas. The uncertainty is emphasized by the long contract period and the fact that the current sales agreements cover only one-third of the recoverable reserves. Therefore, any gas which is agreed delivered within the framework of the current sales agreements, and which is subsequently not delivered due to equipment unreliability, is assumed lost as far as project economics are concerned.

Other economic assumptions developed for Troll life cycle cost analyses include

- Gas and condensate prices as functions of crude oil price
- Gas and condensate transportation and storage tariffs
- Tax rates for cost and income
- Depreciation rules

- Inflation rates
- Exchange rate NOK/US$
- Crude oil prices

A range of discount rates is used to show the dependency of the decision on the criteria.

There are potentially great uncertainties in much of the input data used in the life cycle cost evaluations. Performance of sensitivity analyses of critical parameters and assumptions is therefore important prior to making a decision based on the results.

7.2.5 Example

The condensate system receives untreated condensate from the gas processing trains and removes vapour and water in two stages of separation, a flush drum and a stripper column and reboiler. Processed condensate is cooled, boosted to export pressure and metered before being transferred to the export pipeline.

Several options have been considered for the condensate system. The two alternatives evaluated in this example differ mainly in the level of equipment spares held:

Alternative 1: This configuration incorporates a single 1 × 100% separation train and condensate boiler. The condensate pump consists of 2 × 100% units. One unit is running, while the other is normally idle, but pressurised.

Alternative 2: 2 × 100% units are also included for the separation train and the cooler. The stand-by separation train consists of two successive flash drums with interstage heating. All standby equipment is normally idle but pressurised.

A shutdown or trip of the condensate system will, unless a quick restart or start of a standby unit is possible, cause the main gas trains to shut down. If the failure lies downstream of the condensate flush drum, however, gas production can be resumed with condensate being routed to the burner system pending repair of the failed item. Limited condensate burning is considered allowable as condensate represents only a minor portion of the hydrocarbons produced on Troll.

The differences in capital expenditure and operational expenditure are estimated as described previously. The calculation of differences in production includes differences between the options in terms of:

- Gas production
- Condensate production
- Condensate burning
- Gas venting

The differences in condensate production and gas venting are in this case marginal compared to the two other items.

An example of the reliability data selected for this analysis is shown in Table 7.5.

Table 7.5: Example of reliability data, equipment condensate reboiler

Item	Number	Failure rate	Failure mode	Mode fraction	Net failure rate	MTTR (hours)
Tube bundle	1	13	TL	1	13.0	48
48" reboiler flange	2	3.2	EL	1	6.4	58
2" valve	2	3.6	EL	0.11	0.8	3
2" flange	2	0.16	EL	1	0.3	9
10" flange	6	0.6	EL	1	3.6	34
8" control valve	1	12	ST	0.14	1.7	20
Control valve actuator	1	10	FC	0.54	5.4	9
Total					31.2	38.6

TL: Tube leak
EL: External leak
ST: Stuck
FC: Failed closed

The failure rates are per 10^6 hours.

Results

The net present value difference for the two alternatives have been calculated. The results for the pre-tax position are presented in Figure 7.4 as a function of crude oil price. Net present values are presented for discount rates equal to 5%, 8% and 12%. Depending on the chosen criteria, the breakeven point will be in the area of 20-35 US$/barrel, indicating that an oil price equivalent to these levels or higher is required in order to justify the extra expenditure.

A sensitivity test of the assumption that gas which due to unreliability is not delivered according to contract will not be sold, was carried out, alternatively assuming that the undelivered gas is sold at current price terms in 2023, i.e. the year following the contract termination. The results indicate a lower preference for equipment spares than in the base case: crude oil prices in excess of 25 US$/barrels are necessary to make the additional investment attractive.

On these grounds, it was decided not to change the condensate system from alternative 1 to alternative 2.

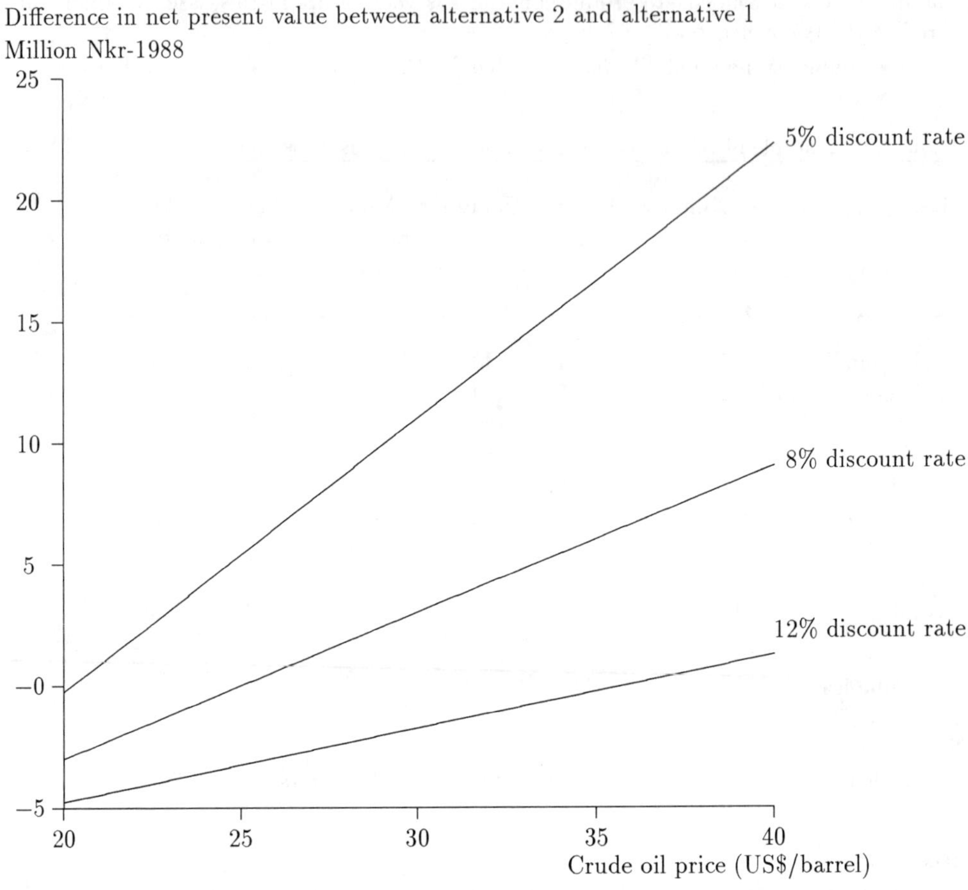

Figure 7.4: Difference in net present value between alternative 2 and alternative 1

7.3 Availability and life cycle costing related to a process instrumentation system

This section gives an example of a reliability analysis of a process instrumentation system on a Norwegian offshore installation. The example is taken from [109].

During the design of the platform process equipment it was questioned whether certain process sensors would cause too frequent erroneous shutdowns and in this way would reduce the overall oil production availability of the installation. Based on preliminary estimates it was recommended that multiple sensors with voting logic should be installed to alleviate this problem. Before a final decision was to be taken on this point a more thorough analysis was proposed by the operator. In the following this analysis is described.

Production start-up was scheduled for 1 January 1991, and from the same date production was assumed to be on plateau phase. The plateau phase will last for 8 years, from 1991 to 1998 (inclusive). The maximum production capacity during the plateau years is $17500m^3$/day. The decay phase will last from 1999 to 2008. The maximum production capacity for the years 1999-2008 is (in m^3/day) 11779, 7572, 5721, 5048, 4375, 4038, 3365, 2692, 4375 and 808, respectively.

7.3.1 Approach

Two alternative solutions were compared with respect to net present value for the production losses caused by failures of the process sensors:

Alternative 1: No voting logic (simplex system)

Alternative 2: With voting logic, for example a 2-out-of-3 system

The difference in this net present value was then compared with differences in net present values associated with investment costs and operating costs.

Calculation of difference in net present value related to production

For alternative 2 it was concluded that the failure frequency would be negligible compared to the failure frequency for alternative 1. A detailed estimation of the failure frequency for alternative 2 was therefore not carried out. Thus the net present value for alternative 2 can be set to 0.

The net present value for alternative 1 was computed using the following method:

1. Calculate the frequency f of erroneous shutdowns per year. This frequency is assumed not to depend on the year under consideration.

2. Calculate expected lost production r_i for an erroneous shutdown in year i, $i = 1, 2, \ldots, N = 18$.

3. Calculate the expected loss in Nkr (Norwegian krone) for year i: $l_i = f \times r_i \times P$, where P equals the oil price per unit of volume. The oil price is assumed to be constant.

4. Discount the expected yearly losses to the same year of reference, here set to 1990, by calculating

$$l = \sum_{i=1}^{N} \frac{l_i}{(1+s)^i}$$

where s equals the discount rate.

5. Oil that is not produced due to shutdown is not lost. It remains in the reservoir and can be produced at a later time. It is assumed as an approximation that this oil is recovered evenly distributed over the remaining years of the decay phase. This means that there is an associated revenue to every expected value of loss l_i, $i = 1, 2, \ldots, N-1$. This expected revenue is given by $g_i = l_i/minimum\{10, N-i\}$, for each of the years over which the production is distributed. By discounting these values to the reference year, we obtain a net present value

$$g = \sum_{i=1}^{N-1} \sum_{j=max\{9,i+1\}}^{N} \frac{g_i}{(1+s)^j}$$

6. Calculate total net present value by $l - g$.

In the following it is shown how the quantities f and r_i were estimated.

The process instruments in question comprised

- 20 pressure transducers
- 10 level switches

A thorough survey of applicable failure data for these units resulted in the following failure rates related to the failure mode "function without signal" (spurious trip):

0.007 per year for pressure transducers
0.04 per year for level switches

With these basic failure rates the total failure rates for alternative 1 will be 0.5 per year.

Based on experience the average downtime associated with an erroneous shutdown was estimated at 1.7 hours. This downtime will be followed by a period of 7 hours of gradual production increase, to avoid well damage. The increase is not linear but fast at first, and than gradually slower. The effective loss during this period corresponds to 2.8 hours of full production, so that

$$r_i = \frac{1.7 + 2.8}{24} \times \text{Maximum production capacity per day}$$

Calculation of difference in net present value related to other costs

The additional cost of choosing alternative 2 in place of alternative 1 is mainly related to investment cost and operating cost.

Investment costs arise from purchase of pipes, flanges, valves, sensors, wiring, electronics, and from installation and commissioning. A sum of 5 million Nkr was estimated for this.

Operating costs are mainly the routine maintenance and calibration costs. This cost was estimated to be 300 000 Nkr per year, which converted to net present value results in 3 million Nkr.

In total the additional cost for installing voting logic instead of a simplex system is approximately 8 million Nkr.

7.3.2 Conclusion

In the analysis a discount rate equal to 0.07 and an oil price equal to 18 US$/barrel (= 781 Nkr/m^3) were used as basic values. With these basic values a difference in net present value related to production of 3.8 million Nkr was calculated. Since the additional cost of installing alternative 2 is approximately twice as high (8 million Nkr), the decision should be not to introduce voting logic in this particular case.

A number of sensitivity analyses were performed to see how changes in different parameters would affect the results, and therefore also the conclusion. In particular, the changes obtained by varying the failure rate (0.2, 0.5, 1.5 per year), the oil price (14, 18, 22 US$/barrel) and the discount factor (0.05, 0.07, 0.10), were studied. The calculations show that only the combination of high failure rate and high oil price would justify the introduction of voting logic.

7.4 Discussion

In this section we will discuss some issues related to safety management and risk analysis. The discussion focuses on the petroleum sector and the three applications presented in this chapter represents the starting point for the discussion. The section is primarily based on [7,53]. The following issues are discussed:

1. Concept Safety Evaluation

2. Safety management in general

3. Risk acceptance criteria

7.4.1 Concept Safety Evaluation

The best time to initiate risk analysis in a project is at the feasibility stage, when several options for developing a new plant are being compared. From a safety point of view, one might feel tempted to evaluate the options to identify the minimum risk solution, and to recommend this to management. All experience shows that this is not the way to approach this kind of problem. It should be kept in mind that we are talking about a feasibility study. Hence, at this stage of the design, all options are in principle feasible as long as they can be made acceptable in safety terms. Emphasis should therefore not be on identifying the minimum risk solution, but on identifying the measures needed to make each solution acceptable. At the feasibility stage, it is the cost of these measures that normally matters the most to management.

By applying risk analysis in this way, it will be seen as a very constructive input to the design and decision-making process.

In 1981, the Norwegian Petroleum Directorate (NPD) introduced Guidelines for Concept Safety Evaluations (CSE) [106]. A CSE is a special kind of risk analysis to be executed at an early stage of the design when different platform concepts and field development plans are being considered.

The starting point for the CSE is that in case of an accident, personnel who are not directly affected by the accident shall be able to escape to a safe point and, if necessary, evacuate the platform safely. To ensure a consistent analysis of these aspects, three safety functions with corresponding requirements have been defined for offshore installations as follows (cf. Section 7.1.3):

- **Escape routes**
  People outside the immediate vicinity of an accident shall not be cut off from all escape routes to the shelter area (i.e. the living quarters).

- **Shelter area**
  The shelter area shall be a safe location until evacuation can be performed.

- **Structural integrity**
  The platform structure shall not lose its structural integrity (i.e. collapse) before safe evacuation can be performed.

The platform is analysed in relation to a specific list including the following types of accident:

- Blowout
- Fire
- Explosions
- Falling objects
- Ship/helicopter collisions
- Earthquake
- Extreme weather combinations
- Other relevant accident
- Relevant combinations of the above

No deterministic value can be assigned to the magnitude of the accident loads for the accident types listed above. There is however a relationship between a high load (consequences) and a corresponding low probability of occurrence. There is a high probability that the installation will encounter the low loads and be able to withstand these, but there is a definite upper limit of the magnitude of accident loads a platform can withstand. This is the key issue of the CSE. It has to be proved that the platform can withstand certain accident loads and that the probability of events which the installation cannot withstand is low.

The NPD guidelines accept that some accidents must be excluded in the analysis, and state that the sum of probability of accidents which is excluded should not exceed 10^{-4} for each safety function and accident type per platform year. For example, none of the safety functions listed above shall be impaired by the 10000 year wave, i.e. the wave height one may encounter in an area "every" 10000 years.

The accidents which the platform should tolerate are called Design Accident Events (DAE) with corresponding design accident loads. An example of a design accident load is the heat radiation load on an escape route or the explosion excess pressure on a bulkhead. Accidents which are excluded due to too low probability of occurrence are referred to as Residual Accident Events (RAEs).

The CSE provides a basis for establishing the design accident loads for the platform, and for documenting that the platform has an acceptable risk level. The CSE is used for analysis of the platform as completed in the operation phase.

The CSE can be carried out using basically the same analysis methods as in the total risk analysis presented in Section 7.1. The risk is measured by the frequencies of accidents that impair the safety functions. If we consider the event tree in Figure 7.1, this means that the risk associated with process leaks can be computed as explained in Section 7.1 with the rows B-J deleted. In the CSE emphasis should be placed on identifying and evaluating the events that have the potential to impair a safety function. In order to reduce the work, the full treatment of other accidents can be omitted.

It is often possible to simplify the calculations significantly by starting the analysis with identifying the accident loads at different places on the installation that can impair a safety function. The next step in the analysis is then to evaluate what accident corresponds to these accident loads. With this approach more emphasis is placed on the consequence analysis, than the standard approach. This alternative approach is particularly suited to simple systems. Using the method on large and complex systems there is a possibility of overlooking some accidents, with the result that the risk is incorrectly estimated.

The level of detail of the analysis must be adapted to the level of detail of the concept. There is usually a need for making a number of assumptions with respect to technical and operational conditions. It is essential that these assumptions are followed up in later phases.

The 1981 Guidelines were replaced in 1991 by Regulations for the Use of Risk Analysis [108], see Section 7.4.3.

7.4.2 Safety management in general

Safety in an organization constitutes:

1. **Safety Management** from a professional point of view. This involves establishing a safety philosophy for the company, with safety goals, programmes, etc.

2. **Safety Services**, including expertise in areas like risk analysis, emergency planning, industrial hygiene, etc.

3. **Project Safety**, including design of systems for fire fighting, gas detection, evacuation, etc.

4. **Site Safety**, including permit-to-work systems, gas measurements, safety introductions, emergency exercises, etc.

The development of an overall system in these areas in a company usually starts at the bottom, with site safety. As the company advances, it realises the need for a systematic approach to safety in projects, using risk analysis tools to make engineering decisions. Finally, the need to organize the total approach to safety in a structured safety management system is recognised.

A typical example of the above can be found in the way safety has progressed in oil and gas operations in the North Sea. Development of fields in the Norwegian sector started about 1970. In the next decade, up to 1980, safety was basically thought of in terms of site safety: when the platform was ready to be installed, safety officers moved in to take care of practicalities like hard hats and safety boots. During this decade, project safety was very much left to engineering contractors as a purely technical task.

In 1981, the Norwegian Petroleum Directorate introduced Guidelines for Concept Safety Evaluations (CSE) as mentioned in the previous section. This meant that not only did the operator have to conduct a CSE study, but he also had to introduce safety managers into the project organization, in order to co-ordinate the multi-discipline activities that follow from risk analyses.

Lately, however, many companies and the Norwegian Authorities have become very aware of the need for a safety management system approach, into which the other bits would fit to make up a total, co-ordinated system. The Norwegian experience seems to fit into a common pattern which can be observed in many industries and countries:

a) Lack of systematic risk analysis leads to unawareness of major hazards and the implementation of measures to control them.

b) As a result, a major accident takes place, which shakes up the whole industry and leads to regulatory action.

c) Risk analysis is introduced, with very strong emphasis of risk quantification and comparison with "acceptable risk targets". This leads to a major step forward in the way major hazards are considered in design.

d) After a period of risk analysis application, people see that although risk analysis brought things forward, it is not the only thing needed. Particularly the legal authorities may feel that the industry is playing a "numbers game" rather than concentrating on the real issues.

e) Safety management is introduced as a framework, in which risk analysis plays a very important part. Authorities are no longer primarily interested in reading the risk analysis reports, but on reviewing how the results are being used in a decision-making process and the steps taken to implement the recommendations and to ensure that the assumptions are valid.

The Norwegian offshore industry is moving strongly towards phase e) above, the NPD having learned that number crunching and playing around with the magic 10^{-4} target diverted attention from practical safety management. In the UK sector of the North Sea, the industry is currently in phase c) after the Piper Alpha disaster in 1988.

Traditionally, society has regulated hazardous industries by reference to engineering codes and standards, and by detailed regulations specifying hardware requirements. There are two obvious advantages to this approach: the industry knows what is expected and the authorities know what to look for. There are, however, a number of disadvantages. Here, we pin-point the following:

- Detailed regulatory requirements do not provide an incentive for the company to think about hazard potential. A company may think that as long as it conforms to the detailed regulations, it is not doing anything illegal.

- Detailed regulations have evolved with time and may take reasonable account of accidents which have occurred in the past, even though this is not always possible. This is sometimes referred to as the "tombstone syndrome"; risk factors can only be taken into account if they have already resulted in an accident. As technology develops at an increasing rate, regulators will have a hard time following up the regulations on this basis.

- Detailed regulations may present an obstacle to cost-effective solutions, since they tend to prescribe ready-made solutions which are perfect for the situation they were meant to cover, but inadequate for novel problems or applications.

Hence, regulators tend to move away from detailed regulations that require specific solutions to hazardous conditions. The alternative is functional requirements, which specify what the regulator wants to achieve, rather than the solution required.

This new freedom is a challenge to the industry. It must then subscribe to the principles of internal control; the company has the responsibility of identifying the hazards and seeing that they are controlled. There may be several ways leading to an acceptable solution, and the company is free to select the one which is optimal in the particular case. What the authorities want is evidence of proper safety management, which documents that the preferred solution is acceptable. The documentation required is often a risk analysis or other kind of safety study.

The authorities may specify "acceptable risk". The Norwegian Petroleum Directorate has proposed target values for impairment of platform safety functions, but not directly as an acceptance criterion; the 10^{-4} figure is only part of the proposed methodology. Furthermore, the number does not appear in a regulation (which is legally binding), but in a Guideline.

NPD keep risk figures out of the regulations. NPD would like the oil companies themselves to define acceptance criteria, and to document to NPD their rationale for the selection. That would certainly place the ball in the court where it belongs!

It is important that risk analysis does not become a means by which the industry without further consideration rules out potential safety problems on the grounds of low probability alone. The authorities should put primary emphasis on auditing the framework in which risk analysis is applied:

- How have the acceptance criteria been arrived at?
- For what decision-making purpose is the risk analysis used?
- How has the risk analysis influenced the decisions?
- Does the management agree with the conclusions of the study?
- Which recommendations will be implemented, and when?
- What kind of systematic follow-up of assumptions is planned?

When the authorities adopt this approach to regulating safety, the industry will benefit by being free to optimize solutions and by having a safety management system which will ensure not only the safety of personnel, public and the environment, but also the long term profitability of the company.

7.4.3 Risk acceptance criteria

The Norwegian Petroleum Directorate (NPD) issued in July 1990 Regulations concerning implementation and use of risk analyses in petroleum activities with guidelines [108]. These regulations have been in force since February 1991 and replace the Guidelines for safety evaluation of platform conceptual design, issued by NPD September 1981 [106]. A key point in the new Regulations is that the operator shall define risk acceptance criteria for the activities, where risk is related to loss of human life, personal injury, damage to the environment and loss of assets and financial interests. The results of risk analyses are to be compared to the acceptance criteria, in order to decide whether the calculated risk level is deemed acceptable, or whether risk-reducing measures should be implemented.

In this section the question about using safety/risk acceptance criteria (targets) in oil and gas production systems is addressed, and possible solutions proposed. We begin the discussion by considering safety related to material assets and production.

We will refer to safety related to loss of material assets and production as "production safety".

Production safety management by using acceptance criteria (targets) has recently been applied, or at least attempted, in several petroleum development projects. The proposed approaches differ somewhat but the three following approaches are typical of the general philosophy that seems to prevail.

Approach 1
Define a system production safety target and allocate subtargets to the items of the system.

Approach 2
Define a system production safety target and use the principles of the optimization process presented in Chapter 6 to obtain an acceptable solution.

Approach 3
Define a feasible concept of the system, calculate its production safety and call it the target.

All these management approaches have fundamental shortcomings when it comes to solving the actual problem at hand. These shortcomings will be demonstrated in the following by a specific example and a discussion of the general nature of planning complex production systems.

Application to a Gas Production Project [22]
The purpose of the project in this example was to produce natural gas from an offshore gas field and bring it ashore for delivery to purchasers. After some initial studies had been carried out, a decision was made to develop the field with a production platform and subsea pipelines for gas transportation. It was further agreed that a systematic treatment of production safety would be of benefit to the project. A study was consequently commissioned, with a conceptual description of a proposed design as input, to define a production safety acceptance criterion (target) for the platform.

The first obstacle the study team ran into was related to the definition of the overall system boundary. Their task was to define a target for the platform; however, it turned out that the performance of the production system as experienced by the gas purchasers would be very different from the performance of the platform when viewed in isolation. This was due to the large internal volume of the pipeline transportation system and the compressibility of gas, which enabled the system to be used as a buffer storage. The inherent overcapacity of the overall system thereby enabled production outages below certain volumes to be recovered by the system at the point of delivery. Hence, the study team recognised that the transportation system had to be considered in the definition of an overall production safety target.

But what should the target for the production safety for the total system be? This question resulted in considerable discussion, because no-one was able immediately to assess the consequences of choosing a specific figure. Neither could a corresponding requirement for the platform production safety be determined.

The following main conclusion was accordingly drawn by the study team:

> It would be impossible to know which level of production safety should be preferred as a target without knowing what it would require and what it would

yield in return to achieve all the possible levels.

As a result, the possibility of using Approach 1 and 2 was abandoned by the study team. An attempt was subsequently made with Approach 3, but it was soon realised that it would lead to little more than an adoption of a coarsely sketched concept as an optimal solution. Any subsequent action to optimize the design would require the targets to be changed, and a "moving target" would lose its intended meaning. Consequently, the whole concept of production targets was abandoned for this project.

Shortcomings of the Production Safety Target Approach

It is the author's opinion that any attempt to apply a production safety target approach to the problem of planning a complex oil/gas production system is a failure to recognise both the primary objectives of the activity as well as the basic properties of the planning problem itself. To substantiate this statement, let us first consider what the objectives of a project might be.

To begin with, we should acknowledge that to a given design of a production system, the categories of "correct" and "false" do not apply; we can only say that it is a good or a bad solution and this to varying degrees and maybe in different ways for different people. Likewise, there will be different conceptions with regard to the objectives of an enterprise; but the bottom line of any oil and gas project is still profit. Profit is the main objective and driving force of the industry. Other conditions, such as production safety, may have to be fulfilled to some extent, but these are nevertheless only means of reaching the primary end. From this line of argument, we can conclude that production safety should not be treated as an objective in its own right.

One might still ask whether production safety targets could not be used merely as a guideline to attain the objectives. It is often said about stated production safety targets that they are not intended as absolute levels, but only as a means of communicating a certain policy. Unfortunately, the relationship between the production safety target and the policy is seldom very well defined. Furthermore, a lot of good managers and engineers have a tendency to interpret a figure which is called a target as something one is supposed to attain. Specifying an absolute level without really meaning it could therefore prove a dangerous practice. It may restrain innovation and sound judgement, and result in an unnecessarily expensive design. But what about giving a production safety target as a range or a distribution? Or why not go all the way, and use a qualitative statement only: "Our target is to achieve normal production safety". As we can see, the whole thing is starting to get rather vague. Consequently, a production safety figure is not suitable as a policy guideline.

We conclude that as a general rule production safety targets should not be used at all. Instead a more cost-effective approach should be adopted, where attention is focused on finding the most economic (profitable) solution, rather than on attaining unfounded targets. The principles of the system optimization process explained in Chapter 6 should be implemented with the criterion "Find the best economical solution".

For a further discussion on using production safety targets and alternative approaches to production safety management, see [22].

Next we discuss the use of safety/risk acceptance criteria for safety related to loss of human life, personal injury and damage to the environment.

The principles for the system optimization process described in Chapter 6 should be applied with respect to safety related to human life, personal injury and damage to the environment. In this case the use of quantitative acceptance criteria can be justified. Assuming that we are not willing to express the value of human life and the environment in monetary terms, we need to introduce some safety related constraints on the optimization process. As a result, the optimization process becomes analogous to Approach 2 described above.

The personnel and environmental *risk* is of course closely related to monetary values. To illustrate this, consider for example the implementation of a risk reducing measure which reduces the statistically expected number of fatalities by d. If the total cost associated with the measure equals c, we can calculate a cost per statistical expected human life by c/d. This does not, however, mean that we give a monetary value to life, cf. the discussion in Section 7.1 about acceptability of accidents and risk.

In this section we will discuss some problems associated with the choice of type of safety acceptance criteria. For example, we will discuss the appropriateness of using FAR values as safety acceptance criteria.

In the following we refer to safety related to human life, personal injury and damage to the environment.

In theory, safety acceptance criteria should be related to the complete risk picture for human life, personal injury and damage to the environment. This means for instance that a specific F-N curve or an equivalent risk curve should be defined as an acceptance criterion. Furthermore, an acceptance criterion should be defined relative to the probability that a person is killed/injured by an accident, for representative personnel groups. In practice, however, it is difficult to define safety acceptance criteria related to the complete risk picture. Some of the most important reasons for this are:

A. Too many constraints are introduced into the system optimization process, and this may result in solutions that are not cost effective. For example, suppose that we have defined a specific F-N curve as an acceptance criterion. Then if a proposed design is below the acceptance criterion for accidents with many fatalities and few fatalities, but is not acceptable for accidents with a moderate number of fatalities, strict application of the criterion may result in an unnecessarily expensive design. The word "unnecessarily" can be justified provided that the probability of an accidental death/injury for a person is acceptably low.

B. In many applications, particularly in the early design phases, the information about the system and activity is so vague that calculation of the complete risk picture will be extremely difficult and the results very uncertain.

C. Establishing a complete risk picture can be very time-consuming and costly, and in view of the above comments it is not necessarily the right thing to do.

But what are the alternatives when abandoning the use of safety acceptance criteria based on a complete risk picture? In the author's view the following approach should be followed.

In respect of human life and health, there is a need for two types of safety acceptance criteria:

1. The first criterion should be related to the probability of an individual accidental death (injury), and is a measure of individual risk. This criterion may depend on where on the installation the person works and the type and extent of the actual work carried out by the individual.

   The FAR value is proportional to the average probability that an individual will be killed in an accident in a certain time period, that is:

$$FAR = \text{constant} \times \frac{1}{n} \sum_{i=1}^{n} P_i \tag{7.1}$$

   where n is the number of persons considered and P_i is the probability that individual i will be killed. The constant in the formula equals 100 million hours divided by the average number of exposed hours under risk for the persons considered (cf. Problem 1.8.5).

   This means that the FAR value can be an appropriate measure of individual risk, provided that suitable groups of personnel are defined. In general, it is not sufficient to estimate *one* FAR value for all the personnel on the installation, since there are often large individual differences in risk levels for the personnel.

   The statistically expected (average) number of lost lives, also referred to as the potential loss of life, is a commonly used measure of risk. This measure equals the sum of the P_is in the formula (7.1), and is proportional to the FAR value. The above comments on the FAR value's use as an individual risk measure also apply to this measure.

   With respect to accidental injuries, the statistically expected number of injuries that have occurred at work, per million working hours, is a suitable measure of individual risk, again provided that suitable personnel groups are defined. An acceptance criterion based on this measure is therefore appropriate. A successful use of this measure/criterion requires a precise and common definition of an *injury*.

   In a specific case, it can be a difficult task to determine the value of the above criteria for different personnel groups and work activities. In particular, it can be a problem to deal with variations in the individual risk level. We will, however, not look closer into this subject here.

2. The second criterion should be related to the probability of an accident with a large number of fatalities. The motivation for such a criterion is the additional strain that follows from such an event for society, industry and the company. The criterion is concerned with the part of the F-N curve with large N values. The problems

indicated under items B and C above also apply to this acceptance criterion. There is therefore a need to define a simpler criterion, which is more appropriate to an early design phase. The old NPD guidelines for safety evaluations [106] represent such a criterion. In fact, these are sets of criteria, as discussed in Section 7.4.1, one to each safety function (escape routes, shelter areas and structure) and each accident category (blowout, fire, explosion, etc.). A risk measure is defined as the frequency of accidents impairing the safety function, and a possible interpretation of the NPD guidelines is that this risk measure shall not exceed 10^{-4} per year for each safety function and each accident category.

The author's opinion is that one should avoid defining criteria for each accident category. What is important is the *total* risk of the activity. Specifying acceptance criteria for each accident category may result in an unnecessarily expensive design since the optimization process then has many unnecessary constraints. Furthermore, definition of only *one* safety function will simplify the decision process, provided that this safety function is sufficient informative as a risk measure. Such a safety function can for example be formulated in the following way:

> Personnel outside the immediate vicinity of the accident (for example outside the module where the accident was initiated) shall not be killed or injured as a result of the accident.

This safety function implies that for personnel outside the immediate vicinity of the accident, a safe escape and evacuation can be carried out. An acceptance criterion may be defined in relation to the frequency of accidents impairing this safety function.

Finally, we will briefly discuss the use of environmental risk acceptance criteria. It is also recommended that quantitative acceptance criteria be adopted as a tool for managing the environmental risk. Consider as an example the risk related to blowouts, which is one of the main risks to the environment. Since the blowout risk is proportional to the number of wells on an installation, possible acceptance criteria related to the blowout risk must be linked to this number or a derived quantity such as the volume of oil produced. An example of a criterion related to blowout risk is the following:

> The frequency of events that give a blowout with a duration of more than 25 days, shall not exceed 10^{-5} per year.

But to control the total environmental risk in an area, and that is what is important, we must look at the total blowout risk, and not consider one installation in isolation. The authorities control the total activity (number of platforms, number of wells drilled, etc.), and they are therefore in the position of controlling the total environmental risk. The operator of an installation must ensure that sufficient barriers are present to protect the environment against oil spills, but the usefulness of defining quantitative acceptance criteria for the installation is limited, since the blowout risk is primarily determined by the activity level.

We conclude that quantitative risk acceptance criteria should be related to the total blowout risk in the area, and not to one installation in isolation.

Of course, the above quantitative acceptance criteria should be regarded as guidelines rather than rigid rules. Risk cannot be measured exactly, and in some cases, overwhelming economic benefits might justify pursuing an activity with what would normally be regarded as intolerable risks. In other cases, risk reduction measures might be adopted for an activity with what would normally be regarded as negligible risks.

When making decisions related to acceptable risk problems, consideration will always be given to costs. In practice, the ALARP principle often applies: Reduce the risk to a level which is as low as reasonably practicable, taking into account benefits and costs, cf. Section 1.4.10.

7.5 Problems

1. Put in numerical values for the input parameters of the event tree in Figure 7.1, and calculate the risk contribution related to the tree.

2. Discuss the method used to calculate the FAR value associated with occupational accidents.

3. Do you have any comments on the risk evaluation made in Section 7.1.10?

4. Carry out the net present value calculation in Section 7.3 (use a spread sheet).

5. What are the main differences between a Concept Safety Evaluation and a Total Risk Analysis?

6. Discuss the following statement: "A CSE is a more important risk analysis than a Total Risk Analysis".

Appendix A

PROBABILITY THEORY

This appendix gives a brief introduction to probability theory. The reader is referred to textbooks on the subject for more detailed information.

A.1 Probabilities

A.1.1 Probability concepts

The probability of an event A, $P(A)$, is usually defined in the following way. If an experiment is performed n times and the event A occurs n_A times, then

$$P(A) = \lim_{n \to \infty} \frac{n_A}{n}$$

i.e. the probability of the event A is the limit of the fraction of the number of times event A occurs, when the number of experiments increases to infinity.

The probability $P(A)$ is a theoretical quantity that usually is unknown and has to be estimated from experience data. To illustrate the definition, let us look at an example.

Example A.1
We consider a fire detector of a certain type K. The function of the detector is to raise the alarm at a fire. Let A denote the event "the detector does not raise the alarm at a fire". To find $P(A)$, assume that tests of n detectors of type K have been carried out and the number of detectors that are not functioning, n_A, is registered. As n increases, the fraction n_A/n will be approximately constant and approach a certain value (this fact is called the strong law of large numbers). This limiting value is called the probability of A, $P(A)$. Thus if for example $n = 10000$ and we have observed $n_A = 50$, then $P(A) \approx 50/10000 = 5/1000 = 0.005$ (0.5%). Note that a probability is by definition a number between 0 and 1, but the quantity is also often expressed in percentages.

□

In some situations it will be possible to determine the probabilities by argument, for example when tossing a die. Here $P(\text{the dice shows two}) = 1/6$ since there are six possible outcomes which are equally likely to appear. In reliability and risk analyses it is not normally possible to determine probabilities by argument.

A probability, $P(A)$, may also be interpreted as a measure of belief. Consider for example the following statement:

> The probability that Norway will qualify for the next world championship in football, is 1%

This probability is a measure of belief. It can hardly be interpreted by means of frequencies. We call such probabilities subjective. A subjective interpretation of the probability concept is often used in situations where there are few or no experience data.

A.1.2 Rules for combining probabilities

Before we summarize some basic rules for probabilities, we will give an overview of some definitions from the set theory (in parentheses the probability interpretations are stated).

Definitions:

The empty set	$\emptyset$	A set with no elements (outcomes) (impossible event)
Basic set (Sample space)	S	A set comprising all the elements we are considering (a certain event)
Subsets	$A \subset B$	A is a subset of B, i.e. each element of A is also an element of B (if the event A occur, then the event B will also occur)
Equality	$A = B$	A has the same elements as B (if the event A occurs, then also the event B occurs, and vice versa)
Union	$A \bigcup B$	$A \bigcup B$ includes all the elements of A and B ($A \bigcup B$ occurs if either A or B occurs (or both), i.e. at least one of the events occur)
Intersection	$A \bigcap B$	$A \bigcap B$ includes only elements which are common for A and B ($A \bigcap B$ occurs if both A and B occur)

Definitions continued:

Disjoint sets	$A \bigcap B = \emptyset$	The sets have no common elements (A and B can not both occur)
Difference	$A - B$	$A - B$ includes all elements of A that are not elements in B ($A - B$ occurs if A occurs but B does not occur)
Complement	$\overline{A}$	$\overline{A}$ includes all elements of S that are not elements of A ($\overline{A}$ occurs if A does not occur)

Rules:

$$
\begin{aligned}
A \bigcup B &= B \bigcup A \\
A \bigcap B &= B \bigcap A \\
(A \bigcup B) \bigcup C &= A \bigcup (B \bigcup C) &&= A \bigcup B \bigcup C \\
(A \bigcap B) \bigcap C &= A \bigcap (B \bigcap C) &&= A \bigcap B \bigcap C \\
A \bigcap (B \bigcup C) &= (A \bigcap B) \bigcup (A \bigcap C) \\
A \bigcup (B \bigcap C) &= (A \bigcup B) \bigcap (A \bigcup C) \\
\overline{A \bigcup B} &= \overline{A} \bigcap \overline{B} \\
\overline{A \bigcap B} &= \overline{A} \bigcup \overline{B} \\
A \bigcup \overline{A} &= S
\end{aligned}
$$

Some of the above definitions are illustrated by use of Venn diagrams in Figure A.1 overleaf.

The following notation will be used:

$$
\begin{aligned}
\bigcup_{i=1}^{n} A_i &= A_1 \bigcup A_2 \bigcup \cdots \bigcup A_n \\
\bigcap_{i=1}^{n} A_i &= A_1 \bigcap A_2 \bigcap \cdots \bigcap A_n
\end{aligned}
$$

The modern probability theory is not based on any particular interpretation of probability. The starting point is a set of rules (axioms) that has to be satisfied. Let $A, A_1, A_2, \ldots$ denote events in the sample space S. For the fire detector example above the sample space comprises the events "the detector raises the alarm at a fire" and "the detector does not raise the alarm at a fire".

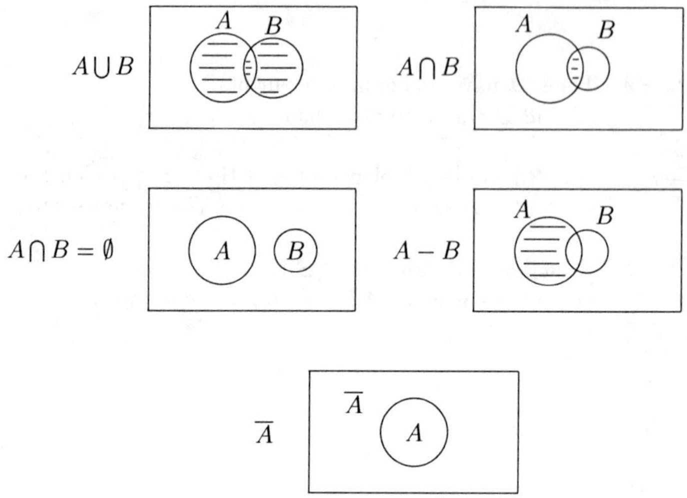

Figure A.1: Venn diagrams

The following probability axioms are assumed to hold:

1. $0 \leq P(A)$
2. $P(S) = 1$
3. $P(A_1 \bigcup A_2 \bigcup \cdots) = P(A_1) + P(A_2) + \cdots,$
   if $A_i \bigcap A_j = \emptyset$ for all i and j, $i \neq j$

To simplify some of the mathematical expressions below, we introduce the notation:

$$\begin{aligned} w_1 &= \textstyle\sum_{i=1}^{n} P(A_i) \\ w_2 &= \textstyle\sum_{i<j} P(A_i \bigcap A_j) \\ w_r &= \textstyle\sum_{i_1<i_2<\cdots<i_r} P(\bigcap_{j=1}^{r} A_{i_j}) \end{aligned}$$

Based on these axioms it is possible to deduce a set of probability rules:

$$P(\overline{A}) = 1 - P(A) \tag{A.1}$$

$$P(A_1 \bigcup A_2) = P(A_1) + P(A_2) - P(A_1 \bigcap A_2) = w_1 - w_2 \tag{A.2}$$

$$P(\bigcup_{i=1}^{n} A_i) = w_1 - w_2 + w_3 - \cdots + (-1)^{n+1} w_n \tag{A.3}$$

$$P(\bigcup_{i=1}^{n} A_i) \begin{array}{ll} \leq & w_1 \\ \geq & w_1 - w_2 \\ \leq & w_1 - w_2 + w_3 \\ \geq & w_1 - w_2 + w_3 - w_4 \\ \text{etc.} & \end{array} \tag{A.4}$$

$$A_1 \subset A_2 \quad \Rightarrow \quad P(A_1) \leq P(A_2) \tag{A.5}$$

The rule (A.4) is very often used in reliability calculations, see Chapter 3. We shall therefore prove this rule. We need, however, some more theory before we can carry out the proof.

A.1.3 Conditional probabilities

The conditional probability of the event B given the event A is denoted $P(B|A)$. As an example, consider two components and let A denote the event "component 1 is not functioning" and let B denote the event "component 2 is not functioning". The conditional probability $P(B \mid A)$ expresses the probability that component 2 is not functioning when it is known that component 1 is not functioning.

A conditional probability is defined by

$$P(B|A) = \frac{P(B \bigcap A)}{P(A)} \tag{A.6}$$

Calculation rules for standard unconditional probabilities also apply to conditional probabilities. From (A.6) we see that

$$P(A \bigcap B) = P(B \mid A)\, P(A) \tag{A.7}$$

More generally we have

$$P(\bigcap_{i=1}^{n} A_i) = P(A_1)\, P(A_2 \mid A_1) \cdots P(A_n \mid \bigcap_{i=1}^{n-1} A_i) \tag{A.8}$$

Some other important rules including conditional probabilities are:

$$P(B \mid A) = \frac{P(A \mid B)\, P(B)}{P(A)} \tag{A.9}$$

If $\bigcup_{i=1}^{r} A_i = S$ and $A_i \bigcap A_j = \emptyset$, $i \neq j$, then

$$P(B) = \sum_{i=1}^{r} P(B \bigcap A_i) = \sum_{i=1}^{r} P(B \mid A_i)\, P(A_i) \tag{A.10}$$

A.1.4 Independence

The events A and B are said to be independent if one of the following equivalent equalities hold:

$$\begin{aligned} P(B \mid A) &= P(B) \\ P(A \mid B) &= P(A) \\ P(A \cap B) &= P(A)\,P(B) \end{aligned}$$

If A and B are independent, then A and $\overline{B}$ are also independent, as well as $\overline{A}$ and B, and $\overline{A}$ and $\overline{B}$.

The events $A_1, A_2, \ldots, A_n$ are independent if

$$P(\bigcap_{j=1}^{r} A_{i_j}) = \prod_{j=1}^{r} P(A_{i_j})$$

for any set of different indices $\{i_1, i_2, \ldots, i_r\}, r = 1, 2, \ldots, n$, taken from the set $\{1, 2, \ldots, n\}$.

Example A.1 continued.
Assume that we have found that $P(A) = 0.005$, where A denotes the event "the detector does not raise the alarm at a fire". To reduce the probability of no alarm at a fire, we install two detectors. The problem is now to compute the probability of the following events:

B = "No detectors are functioning at a fire"
C = "At least one of the detectors is functioning at a fire"

To find these probabilities, let A_i, $i = 1, 2$, denote the event "detector i does not function at a fire". Then $B = A_1 \cap A_2$ and $C = \overline{A}_1 \cup \overline{A}_2$. We know that $P(A_1) = P(A_2) = 0.005$, but this information is not sufficient for calculating $P(B)$ and $P(C)$. Under the assumption that the two detectors are randomly sampled it is reasonable to assume that A_1 and A_2 are independent. Using the independence we find:

$$\begin{aligned} P(B) &= P(A_1 \cap A_2) = P(A_1)P(A_2) = 0.005^2 = 0.25 \times 10^{-4} \\ P(C) &= P(\overline{A}_1 \cup \overline{A}_2) = 1 - P(A_1)P(A_2) = 0.999975 \end{aligned}$$

Alternatively, we could have found $P(C)$ by

$$\begin{aligned} P(C) &= P(\overline{A}_1) + P(\overline{A}_2) - P(\overline{A}_1)P(\overline{A}_2) \\ &= 0.995 + 0.995 - 0.995^2 = 0.999975 \end{aligned}$$

Given that at least one of the detectors does not function, what is then the probability that detector 1 is not functioning? Intuitively, it is clear that this conditional probability will be approximately 50%. To show this formally, note that this probability can be expressed as $P(A_1 | A_1 \cup A_2)$. Use of computing rules gives:

$$\begin{aligned} P(A_1 \mid A_1 \cup A_2) &= \frac{P[A_1 \bigcap (A_1 \bigcup A_2)]}{P(A_1 \bigcup A_2)} = \frac{P(A_1)}{P(A_1)+P(A_2)-P(A_1)P(A_2)} \\ &= \frac{0.005}{0.005+0.005-0.005^2} \approx \frac{1}{2} \end{aligned}$$

A.2 Stochastic variables

When systemizing data we often focus on one or more aspects of the results. As an example, let us return to the detector example. Assume that we are considering k detectors. We are primarily interested in the *number* of detectors that are not functioning, i.e. not raising the alarm. Let X denote this number. The value of X is uniquely given when the outcome of the "experiment" is registered. If, for example $k = 2$ and it is observed that detector 1 is functioning but not detector 2, then $X = 1$. Thus X is a function from the sample space to the real numbers. We call such variables stochastic (or random) variables.

Let in general X denote a stochastic variable and assume that X is discrete, i.e. it can only take a finite number of values or a countable infinite number of values. Let $P(X = x)$ denote the probability of the event "$X = x$", where x is one of the values X can take. We call the function $f(x) = P(X = x)$ for the distribution of X. The *mean* or the *expected* value of X, EX, is defined as

$$EX = \sum_x x\,P(X = x)$$

From the definition we see that EX can be interpreted as the centre of mass of the distribution. Consider again the fire detector example. It follows from the strong law of large numbers that EX is approximately equal to the average number of detectors that are not functioning among the k, if we look at a large number of identical collections of k detectors. Hence the mean can also be interpreted as an average value.

The variance of X, $VarX$, is a measure of the spread or variability of the values of X around EX, and is defined by

$$VarX = \sum_x (x - EX)^2\,P(X = x)$$

The standard deviation of X is given by $\sqrt{VarX}$.

To describe the length of life of components and systems, continuous models are usually adopted. These models are characterized by a probability density $f(x)$ such that

$$P(a < X \leq b) = \int_a^b f(x)\,dx$$

i.e.

$$P(a < X \leq b) \approx f(x)\,(b - a)$$

if $b - a$ is small. The mean and variance are defined by

$$\begin{aligned} EX &= \textstyle\int_{-\infty}^{\infty} x\,f(x)\,dx \\ VarX &= \textstyle\int_{-\infty}^{\infty} (x - EX)^2\,f(x)\,dx \end{aligned}$$

Continuous models are treated more thoroughly in Appendix B when studying lifetime distributions.

Let $X_1, X_2, \ldots, X_n$ denote n arbitrary stochastic variables. We say that these variables are independent if

$$P(\bigcap_{i=1}^{n} X_i \leq x_i) = \prod_{i=1}^{n} P(X_i \leq x_i)$$

for all choice of $x_1, x_2, \ldots, x_n$.

Below we state some important rules for the mean and variance (a and b are constants).

$$E(aX+b) = aEX + b$$
$$EX \leq EY \text{ if } X \leq Y$$
$$E(X_1 + X_2 + \cdots + X_n) = EX_1 + EX_2 + \cdots + EX_n$$
$$Var(aX+b) = a^2 \, VarX$$

$$Eh(X) = \begin{cases} \sum_x h(x)P(X=x) & \text{if } X \text{ is discrete} \\ \int_{-\infty}^{\infty} h(x)f(x)\,dx & \text{if } X \text{ is continuous} \end{cases}$$

If the X_is are independent, then

$$Var(X_1 + X_2 + \cdots + X_n) = VarX_1 + VarX_2 + \cdots + VarX_n$$

In the general case

$$Var(X_1 + X_2 + \cdots + X_n) = \sum_{i=1}^{n} VarX_i + 2\sum_{j<l} cov(X_j, X_l)$$

where $cov(X_j, X_l) = E(X_j - EX_j)(X_l - EX_l)$ is the covariance of X_j and X_l.

One of the most useful concepts in probability theory is that of conditional probability and expectation. Let X and Y be two discrete stochastic variables. Then the conditional probability distribution of X given that $Y = y$, is given by

$$f(x|y) = P(X=x|Y=y) = \frac{P(X=x, Y=y)}{P(Y=y)}$$

for all values such that $P(Y=y) > 0$. The conditional expectation of X given $Y = y$ is defined by

$$E(X|Y=y) = \sum_x x\, f(x|y)$$

Similarly, we can define a conditional probability distribution of X and a conditional expectation for continuous stochastic variables:

$$\begin{aligned} f(x|y) &= f(x,y)/g(y) \\ E(X|Y=y) &= \int_{-\infty}^{\infty} x\, f(x|y)dx \end{aligned}$$

where $f(x,y)$ is the density function for the stochastic variables X and Y, given by

$$P(a < X \leq b, c < Y \leq d) = \int_a^b \int_c^d f(x,y)\,dy\,dx$$

and $g(y)$ is the probability density of Y. Let $E(X|Y)$ denote the function of the stochastic variable Y whose value at $Y = y$ is $E(X|Y = y)$. Note that $E(X|Y)$ is itself a stochastic variable. Then it can be shown that

$$EX = EE(X|Y) \tag{A.11}$$

If Y is a discrete stochastic variable, then this equation states that

$$EX = \sum_y E(X|Y = y)P(Y = y)$$

while if Y is a continuous stochastic variable with density $g(y)$, then it states that

$$EX = \int_{-\infty}^{\infty} E(X|Y = y)g(y)\,dy$$

Finally in this section we shall define a stochastic process:

> A *stochastic process* $X(t)$, $t \in T$, is a collection of stochastic variables. That is, for each $t \in T$, $X(t)$ is a stochastic variable. The index t is often interpreted as time and, as a result, we refer to $X(t)$ as the state of the process at time t. The set T is called the *index set* of the process. In this book T is usually $[0, \infty)$.

A.3 Some proofs

A.3.1 Proof of formula (A.4)

The inequalities (A.4) can be shown by using an induction argument. We are primarily interested in the two first inequalities of (A.4) and we will therefore restrict attention to the proof of these. First we look at the inequality

$$P(\bigcup_{i=1}^{n} A_i) \leq w_1 \tag{A.12}$$

This inequality clearly holds if $n = 1$ and 2. Assume now that it holds for $n - 1$. Then it follows that

$$\begin{aligned} P(\cup_{i=1}^{n} A_i) &= P(\cup_{i=1}^{n-1} A_i \cup A_n) \leq P(\cup_{i=1}^{n-1} A_i) + P(A_n) \\ &\leq \textstyle\sum_{i=1}^{n-1} P(A_i) + P(A_n) = w_1 \end{aligned}$$

which proves (A.12). Let us now consider the inequality

$$P(\bigcup_{i=1}^{n} A_i) \geq w_1 - w_2 \tag{A.13}$$

The reasoning is similar to the proof of (A.12). The inequality (A.13) holds for $n = 2$ since then the right hand equals the left hand. Assume now that (A.12) is valid for $n - 1$. It follows that

$$\begin{aligned} P(\cup_{i=1}^{n} A_i) &= P(\cup_{i=1}^{n-1} A_i \cup A_n) = P(\cup_{i=1}^{n-1} A_i) + P(A_n) - P(\cup_{i=1}^{n-1} A_i \cap A_n) \\ &\geq \sum_{i=1}^{n-1} P(A_i) - \sum_{i<j\leq n-1} P(A_i \cap A_j) + P(A_n) - \sum_{i=1}^{n-1} P(A_i \cap A_n) \\ &= w_1 - w_2 \end{aligned}$$

which proves (A.13).

Alternatively, we could have proved (A.12) and (A.13) by a more direct argument. Let $I(A)$ denote the indicator function for an arbitrary event A, defined by

$$I(A) = \begin{cases} 1 & \text{if A occurs} \\ 0 & \text{if A does not occur} \end{cases}$$

Note that $P(A) = EI(A)$. Then

$$P(\bigcup_{i=1}^{n} A_i) = EI(\bigcup_{i=1}^{n} A_i) \leq E\sum_{i=1}^{n} I(A_i) = \sum_{i=1}^{n} EI(A_i) = w_1$$

which proves (A.11). We have used that

$$I(\bigcup_{i=1}^{n} A_i) \leq \sum_{i=1}^{n} I(A_i)$$

This inequality clearly holds, for if the left hand equals one, this means that at least one of the A_is occur, which in turn implies that the right hand must be at least one.

Similarly, we can show (A.13) by using the inequality

$$I(\bigcup_{i=1}^{n} A_i) \geq \sum_{i=1}^{n} I(A_i) - \sum_{i<j} I(A_i \cap A_j)$$

A.3.2 Probability calculations in event trees

Consider an event tree as described in Chapter 2. Assume that the initiating event has occurred. Let K denote a specific terminating event (consequence) in the tree and let H_i denote the ith event sequence leading to K. The Hs are disjoint and together they give K. It follows from axiom 3 that

$$P(K) = P(\bigcup_i H_i) = \sum_i P(H_i)$$

To calculate $P(H_i)$ the rule (A.8) is used, with

$$H_i = \bigcap_j A_j \quad \text{where the } As \text{ are the event in the actual sequence}$$

To compute the frequency of K, F_K, which equals the expected number of times K occurs in the relevant period of time, we must multiply $P(K)$ with the frequency of the initiating event. To show this, let N denote the number of times the initiating event occurs and let F denote the frequency of the initiating event. Furthermore let K_i denote the terminating event K the ith time the initiating event occurs. It follows that if it is given that $N = n$, then

$$F_K = E\sum_{i=1}^{n} I(K_i) = \sum_{i=1}^{n} EI(K) = n\,P(K)$$

which means that we unconditionally have

$$F_K = EN\,P(K) = F\,P(K)$$

We have used the rule (A.11).

A.3.3 Proof of an error bound for the approximations (4.2) and (4.3)

Consider the approximation formulae (4.2) and (4.3), with their associated error term δ. We will show that δ is in fact an error bound for (4.2) and (4.3). That is:

$$F_1 - \delta \;\leq\; P(\phi(\mathbf{X}) < \phi_k) \leq F_1 + \delta \qquad \text{(A.14)}$$
$$F_2 - \delta \;\leq\; E[1 - \phi(\mathbf{X})/\phi_M] \leq F_2 + \delta \qquad \text{(A.15)}$$

where

$$F_1 = \sum_{i,r:\,\phi(r_i,\mathbf{M})<\phi_j} p_{ir}$$
$$F_2 = \sum_{i,r} \{1 - \phi(r_i,\mathbf{M})/\phi_M\}\, p_{ir}$$

First we will show (A.15). It is not difficult to see that

$$E\{1-\phi(\mathbf{X})/\phi_M\} \leq E\{\sum_{i=1}^{n}[1-\phi(X_i,\mathbf{M})/\phi_M]+I(\text{two or more component failures})\} \leq F_2+\delta,$$

hence it suffices to show the first inequality of (A.15). Now, observe that

$$\begin{aligned}
1 - \phi(\mathbf{X})/\phi_M \;&\geq\; (1 - \phi(\mathbf{X})/\phi_M)\; I(\phi(\mathbf{X}) < \phi_M) \\
&\geq\; (1 - \phi(\mathbf{X})/\phi_M)\; I(\bigcup_{i=1}^{n}\{\phi(X_i,\mathbf{M}) < \phi_M\}) \\
&\geq\; (1 - \phi(\mathbf{X})/\phi_M)[\sum_{i=1}^{n} I(\phi(X_i,\mathbf{M}) < \phi_M) \\
&\quad - \sum_{i,j:\,i<j} I(\phi(X_i,\mathbf{M}) < \phi_M \bigcap \phi(X_j,\mathbf{M}) < \phi_M)],
\end{aligned}$$

the second inequality holds since $\phi(X_i, \mathbf{M}) < \phi_M$ implies that $\phi(\mathbf{X}) < \phi_M$ by the definition of a monotone structure function, while the last inequality is an application of an inclusion-exclusion type formula which is easy to prove by induction.

By taking expectations and noting that $1 \geq 1 - \phi(\mathbf{X})/\phi_M \geq 1 - \phi(X_i, \mathbf{M})/\phi_M$, it follows that

$$\begin{aligned} E\{1 - \phi(\mathbf{X})/\phi_M\} &\geq E\sum_{i=1}^{n}[1 - \phi(X_i, \mathbf{M})/\phi_M]\, I(\phi(X_i, \mathbf{M}) < \phi_M) - \\ &\quad E\sum_{i,j:i<j} I(\phi(X_i, \mathbf{M}) < \phi_M \bigcap \phi(X_j, \mathbf{M}) < \phi_M) \\ &\geq E\sum_{i=1}^{n}[1 - \phi(X_i, \mathbf{M})/\phi_M] - E\sum_{i,j:i<j} I(X_i < M_i)\, I(X_j < M_j) \\ &\geq F_2 - \delta \end{aligned}$$

which proves the first inequality of (A.15).

Using similar arguments with $I(\phi(\mathbf{X}) < \phi_k)$ in place of $1 - \phi(\mathbf{X})/\phi_M$, and F_1 in place of F_2, it is seen that (A.14) holds, i.e. δ is an error bound for (4.2) and (4.3).

A.4 Problems

1. Show the formulae (A.1), (A.2), (A.3), (A.5), (A.9) and (A.10).

2. A system comprises two components K_1 and K_2. The probability that component K_1 is functioning at a specific point in time, equals 0.95, whereas the corresponding probability for component K_2 is 0.90. The probability that both components are functioning is 0.87. What is the probability that no components are functioning?

3. We consider two machines A and B. Let X and Y denote the number of failures that occur on machines A and B during a week, respectively. The probability distribution of X and Y is given by:

i	0	1	2
$P(X = i)$	0.95	0.04	0.01
$P(Y = i)$	0.90	0.08	0.02

We assume that X and Y are independent.

a) Find EX, EY and $E(X + Y)$. Give an interpretation of these quantities.

b) Find $P(X = i \bigcap Y = j)$ for $i, j = 0, 1, 2$. Calculate the probability that more failures occur on machine A than on machine B.

c) Given that in total two failures have occurred on the two machines, what is then the probability that no failures have occurred on machine A?

4. A company employs k persons. Let X_i be defined as 1 if person i is killed in an accident the next year, and as 0 otherwise, $i = 1, 2, \ldots, k$. Furthermore, let Y denote the total number of accidental deaths in the company in the same period.

   a) Explain why

   $$Y = \sum_{i=1}^{k} X_i$$

   and $EX_i = P(X_i = 1)$.

   b) Use *a*) above to show that the expected portion of accidental deaths in the company the next year equals the average probability that a specific person will be killed, i.e. show that

   $$E\frac{Y}{k} = \frac{1}{k}\sum_{i=1}^{k} P(X_i = 1)$$

5. Let C be a discrete stochastic variable representing the loss as a result of undesirable events in relation to a specific activity, cf. Section 1.4.6. As a measure of risk related to the activity we define

   $$Ef(C) = \sum_{c} f(c)\,P(C = c)$$

   where $f(c)$ is a given function. Discuss this measure. Explain the relation between this measure and the *statistically expected loss*. Generalize the measure to include continuous distributions.

6. The number of accidents occurring in a factory in a week is a stochastic variable with mean μ and variance σ^2. The number of individuals injured in single accidents are independently distributed, each with mean ν and variance τ^2. Find the mean and variance of the number of individuals injured in a week, assuming that the number of individuals injured per accident is independent of the number of accidents.

Appendix B

STOCHASTIC FAILURE MODELS

This appendix gives an overview of the most important stochastic failure models that are used in reliability analyses. The appendix comprises two main sections. First we consider units that are not repaired at failure. Then we look into the case that repair or replacement is carried out at failure. Often repairable units are analysed as if they were non-repairable, with the result that the results and conclusions become meaningless. It has been an objective of this appendix to make clear the difference in modelling for the two situations.

Refer to [6,24,78,84,94,95] for a detailed analysis of the various failure models and for presentation of other types of models and distributions.

B.1 Non-repairable units

B.1.1 Basic concepts

We consider the lifetime, T, of a unit of a certain type from its installation until it fails, i.e. to the point when it is not able to perform its intended function. We put $t = 0$ initially. It is natural to assume that the lifetime T is a stochastic variable since we cannot in advance say how long it will function.

Let $F(t)$ denote the lifetime distribution of T, i.e.

$$F(t) = P(T \leq t)$$

In view of the frequency interpretation of the probability concept, we interpret $F(t)$ as the portion of units that will fail within t units of time, when a large number of units of this type is installed at time $t = 0$.

Often we are more interested in $P(T > t)$, and we therefore introduce a special notation for this probability:

$$R(t) = P(T > t) = 1 - F(t)$$

We call $R(t)$ the survivor function. In the following we assume that T has a continuous distribution, such that

$$F(t) = \int_0^t f(s)\, ds$$

The function $f(t)$ is called the probability density of T. The relation between $F(t)$ and $f(t)$ is illustrated in Figure B.1 below.

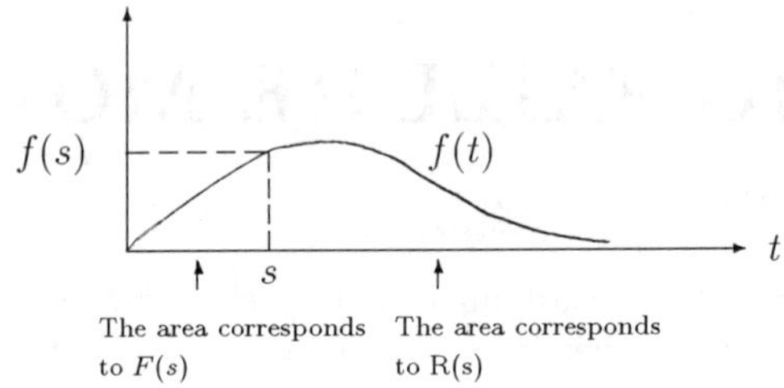

Figure B.1: Relation between $F(t)$ and $f(t)$

Another important parameter in reliability modelling is the failure rate, $z(t)$, defined by

$$z(t) = \frac{f(t)}{R(t)} \tag{B.1}$$

To see the physical interpretation of the failure rate, consider a small time interval $(t, t+h)$ and assume that the unit has survived t. Then we find that

$$\begin{aligned} \tfrac{1}{h}P(T \leq t+h \mid T > t) &= \tfrac{1}{h}\tfrac{P(t<T\leq t+h)}{P(T>t)} \\ = \tfrac{F(t+h)-F(t)}{h}\tfrac{1}{R(t)} &\rightarrow \tfrac{f(t)}{R(t)} = z(t) \quad \text{when } h \rightarrow 0\,, \end{aligned}$$

Thus

$$P(T \leq t+h \mid T > t) \approx z(t)\, h$$

for small values of h. We see that the failure rate expresses the proneness of the unit to fail at time (age) t. A high failure rate means that there is a high probability that the unit will fail "soon", whereas a small failure rate means that there is a small probability that the unit will fail "in a short time". If a large number of identical units are put into operation at time 0, $z(t)h$ will be approximately equal to the fraction of units that fail in the interval $(t, t+h)$ among the units that function at time (age) t.

A typical shape of the failure rate is the so-called bath-tub curve which is shown in Figure B.2.

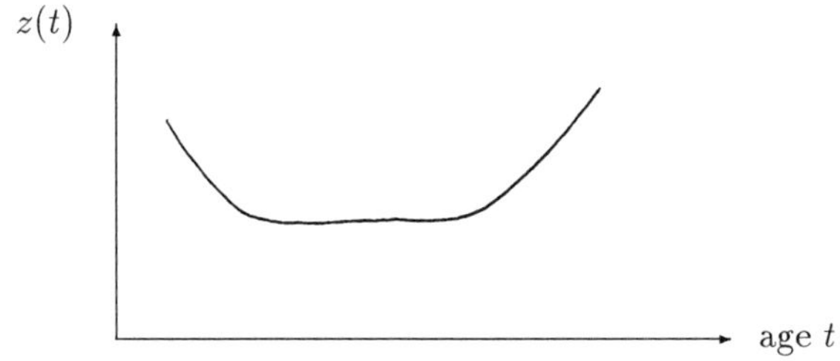

Figure B.2: Bath-tub shape of the failure rate

The decreasing failure rate in the first period is due to problems such as manufacturing errors or inappropriate design. This period is known as the infant mortality, burn-in or de-bugging phase. Often the units are tested before they are sent to the users so that most units with such defects are removed. For units that survive the infant mortality phase, the failure rate is approximately constant. The failures that occur in this phase (the useful life phase) are "purely by chance". This is one of the main reason why we so often in practice use the *exponential* lifetime distribution, which is characterized by a constant failure rate. The increasing failure rate phase represents the wear-out or fatigue phase. With an appropriate maintenance strategy the units might be replaced before problems arise to due wear-out and fatigue.

From (B.1) it follows that

$$z(t) = -\frac{d}{dt} ln\ R(t)$$

such that

$$R(t) = e^{-\int_0^t z(s)\,ds}$$

The integral in the above expression is often called the hazard or the cumulative failure rate, and it will be denoted by $Z(t)$. A distribution with a non-decreasing failure rate is called an IFR (Increasing Failure Rate) distribution. Analogously, a distribution with a non-increasing failure rate is called a DFR (Decreasing Failure Rate) distribution. Note that an IFR distribution has a convex hazard, whereas a DFR distribution has a concave hazard, see Figure B.3. A constant failure rate corresponds to a linear hazard as shown in Figure B.4. The failure rate is given by the angle of inclination of the line.

The mean (expected) lifetime, ET, is defined by

$$ET = \int_0^\infty t\,f(t)\,dt = \int_0^\infty R(t)\,dt$$

Thus the mean lifetime is given by the area below the survivor function $R(t)$. The quantity ET is often referred to as the Mean Time To Failure (MTTF).

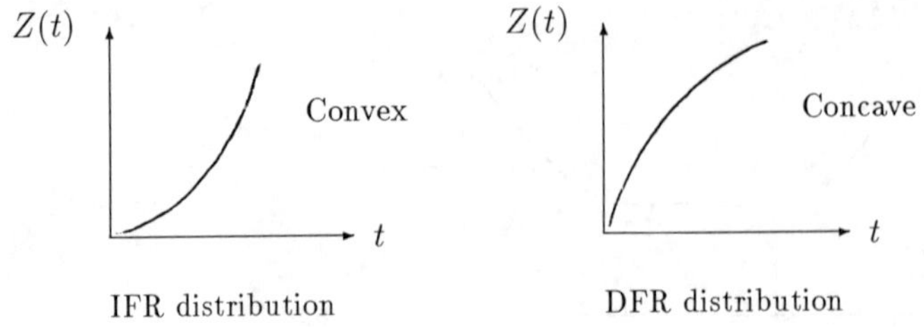

Figure B.3: Hazard for IFR and DFR distributions

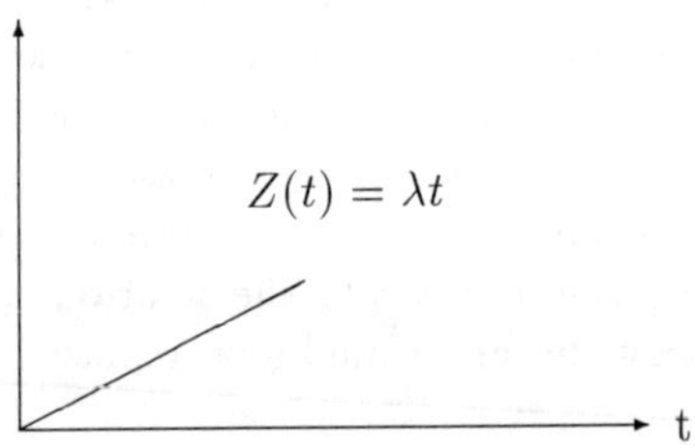

Figure B.4: Hazard for the exponential distribution

Some remarks

In the above model we consider the unit on a "high level". We only register whether the unit is functioning or not. We do not look into the causes of failure and mechanisms leading to failure. The unit is regarded as a "black box", and the lifetime distribution is determined from experience and tests.

The model can easily be generalized to the case that the unit has more than one failure mode. We then define a failure rate $z_i(t)$ related to failure mode i :

$$z_i(t) = \lim_{h \to 0} \frac{1}{h} \, P(T \leq t + h, \text{ failure mode } i \text{ occurs } \mid T > t)$$

Hence the total failure rate $z(t)$ of the unit can be expressed as

$$z(t) = \sum_i z_i(t)$$

If the mechanisms leading to failure are to be taken into account, there is a need for more physical and basic failure models. One such model is the Stress-Strength Interference model (SSI). In this model it is assumed that the unit at a fixed point in time has a certain strength S. At the same time the unit is exposed to a stress (load) L. The unit is functioning as long as $S > L$. The unit fails if $S < L$. The reliability of the unit, i.e. the probability that the unit is functioning at the same point in time, equals $P(S > L)$.

The model SSI is a static model where time variations with respect to strength and stress are not taken into account. Refer to [78] for a more thorough presentation of the SSI model and to other strength stress models, including models that allow for time dependencies.

B.1.2 Some common lifetime distributions

Below the most common lifetime distributions are described.

The exponential distribution
The lifetime T is said to be exponentially distributed with parameter λ (> 0) if

$$F(t) = 1 - e^{-\lambda t}\,, \quad t \geq 0$$

Thus the exponential distribution is characterized by a constant failure rate. A unit having an exponential failure time distribution has a tendency to failure that does not depend on the age of the unit. Assume that the unit has survived u hours. The probability that the unit then will survive an additional v hours is given by:

$$\begin{aligned} P(T > u + v \mid T > u) &= \frac{P(T>u+v \bigcap T>u)}{P(T>u)} \\ &= \frac{P(T>u+v)}{P(T>u)} = \frac{e^{-\lambda(u+v)}}{e^{-\lambda u}} \\ &= e^{-\lambda v} = P(T > v) \end{aligned}$$

Thus the probability of survival of the additional v hours is not dependent on how long the unit has functioned. The exponential distribution is the only distribution with this property. The lack of memory simplifies the mathematical modelling.

The fact that the failure rate is constant for large values of t may seem unrealistic for practical applications. We must, however, remember that we are usually interested in studying the lifetime in a limited period of time. The failure rate assumed outside this period will then not be critical. In addition, the probability that the unit will really last so long will be small, since the mass of the probability density function is positioned around "small" values of t, see Figure B.5.

The exponential distribution usually gives a good description of the lifetime of electrical and electronic units. In some cases it has also been useful for modelling units comprising a large number of mechanical components, for example pumps, when the unit has been in operation for a relatively long period of time and maintenance has led to different ages of the various components of the unit.

The mean and variance in the exponential distribution are given by:

$$ET = \frac{1}{\lambda} \text{ and } VarT = \frac{1}{\lambda^2}$$

If the lifetime distribution is IFR, but not necessarily exponential, then

$$R(t) \geq e^{-t/ET}, \quad \text{for } t < ET$$

Thus the exponential distribution represents a lower limit for the survivor function $R(t)$ when $t < ET$. Assuming an exponential distribution when the true distribution is another IFR distribution, gives an underestimated value of $R(t)$ for these values of t.

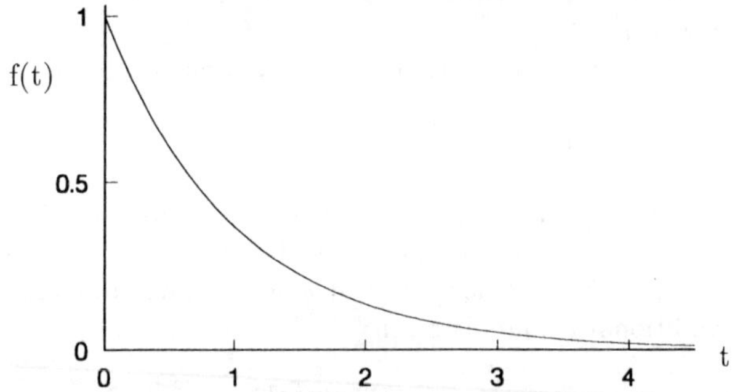

Figure B.5: Density function of the exponential distribution, $\lambda = 1$

Weibull distribution

A distribution that is often used when the failure rate is increasing or decreasing, is the Weibull distribution. A lifetime T is said to be Weibull distributed with parameters $\lambda (> 0)$ and $\beta\ (> 0)$ if the distribution is given by

$$F(t) = 1 - e^{-(\lambda t)^\beta}, \quad t \geq 0$$

We call λ and β the scale and form parameter, respectively. If we choose $\beta = 1$, then the failure rate becomes a constant. Hence the exponential distribution is a special case of the Weibull distribution. When $\beta > 1$, the failure rate is increasing, and when $\beta < 1$, it is decreasing. In Figure B.6 the density function and the failure rate are shown graphically for some values of the parameters. Note that

$$R(\frac{1}{\lambda}) = e^{-1} = 0.3679 \text{ for all } \beta > 0$$

The quantity $1/\lambda$ is often called the *characteristic lifetime.*

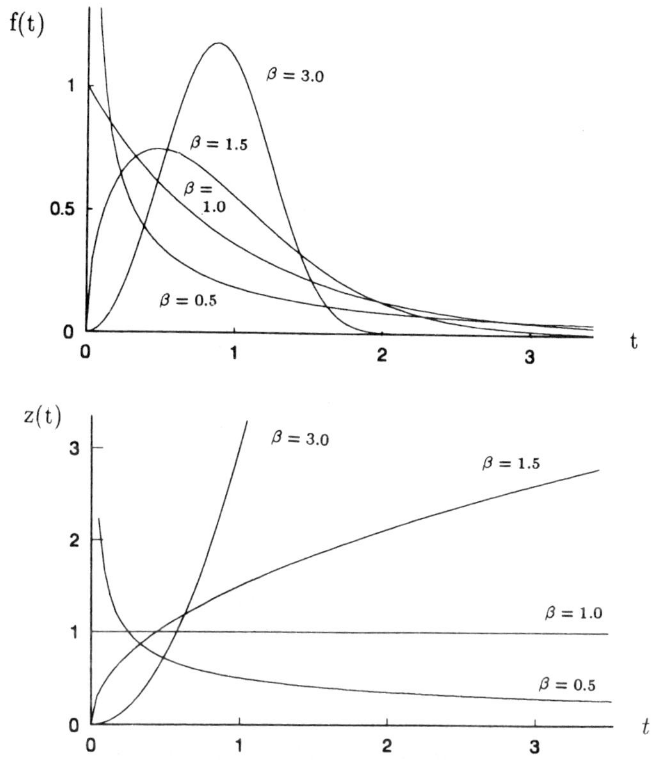

Figure B.6: Density function and failure rate of the Weibull distribution, $\lambda = 1$

The mean (expected) lifetime of the Weibull distribution is given by:

$$ET = MTTF = \int_0^\infty R(t)\,dt = \frac{1}{\lambda}\Gamma(1+\frac{1}{\beta})$$

where $\Gamma(\cdot)$ is the gamma function, see Section B.4.

The variance of T becomes:

$$VarT = \frac{1}{\lambda^2}[\Gamma(1+\frac{2}{\beta}) - \Gamma^2(1+\frac{1}{\beta})]$$

The Weibull distribution has for example been used in reliability analyses of items such as vacuum tubes, ball bearings and electrical insulation. The Weibull distribution is widely used in maintenance analyses for optimization of test and replacement intervals.

The gamma distribution

If $T_1, T_2, \ldots, T_n$ are independent and exponentially distributed stochastic variables with parameter λ, then $T = T_1 + T_2 + \cdots + T_n$ is gamma distributed with parameters λ and n, i.e.

$$f(t) = \frac{\lambda}{\Gamma(n)} (\lambda t)^{n-1} e^{-\lambda t}, \quad t \geq 0 \tag{B.2}$$

Assume for example that n units of a certain type have exponentially distributed lifetimes $T_1, T_2, \ldots, T_n$ with failure rate λ and that the units are put into operation one by one as a unit fails. Then the total lifetime equals the sum of the Ts.

The parameter n in (B.2) does not need to be restricted to the positive integers. If it is a positive integer, we can write the survivor function in the following form:

$$R(t) = \sum_{i=0}^{n-1} \frac{(\lambda t)^i}{i!} e^{-\lambda t}$$

where $i! = 1 \cdot 2 \cdot 3 \cdot \ldots \cdot (i-1) \cdot i$.

The mean and variance of the gamma distribution are given by

$$\begin{aligned} ET &= MTTF = \tfrac{n}{\lambda} \\ VarT &= \tfrac{n}{\lambda^2} \end{aligned}$$

In Figure B.7 the density function and the failure rate is illustrated graphically for some values of the parameters.

Other distributions

The **normal distribution** is probably the most important and widely used distribution in the entire field of statistics and probability. Although having some important applications in reliability evaluation, it is of less significance in this field than many other distributions. The normal distribution is sometimes used to model wear-out failure, but it is also used as a lifetime distribution for batteries and condensators.

The **lognormal distribution** has quite a limited application as a failure time distribution. It is, however, frequently used when modelling, for example, pipe burst, and it has shown to be appropriate as a repair time distribution.

Histograms, as for example illustrated in Figure 2.2, are used in many situations, particularly for repair times. A histogram column may represent repair time and the probability of a specific failure mode.

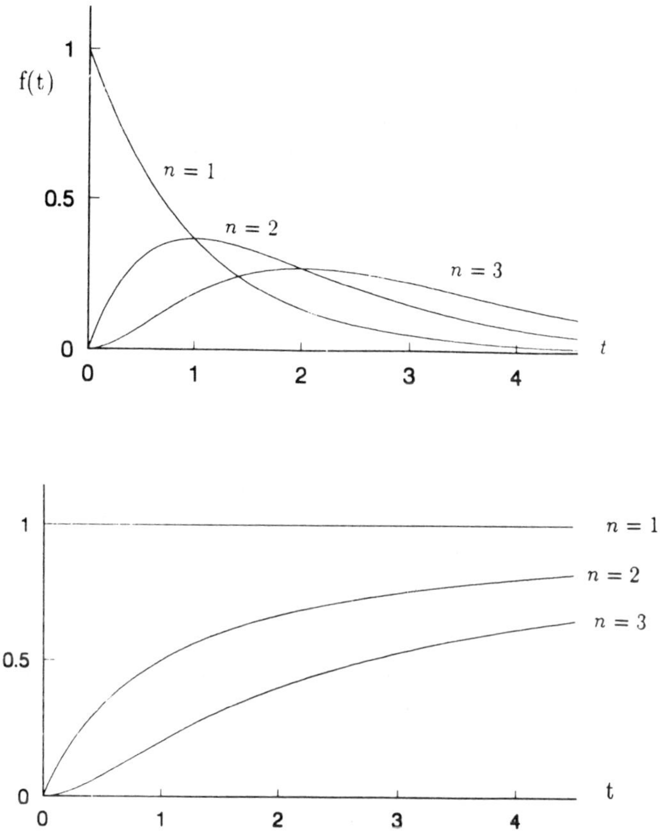

Figure B.7: Density function and failure rate for the gamma distribution, $\lambda = 1$

B.2 Repairable units

Assume now that a unit is repaired at failure, or possibly replaced by a new unit of the same type. We disregard the repair/replacement times. We then obtain a sequence of lifetimes $T_1, T_2, \ldots$. If these are independent and identically distributed, we have a model that is characterized by a lifetime distribution as described in Section B.1. In a model with repair/replacement we will focus not only on the distribution of the T_is but also on for example, the distribution of

$$N(t) = \text{ the number of failures in } [0, t]$$

when disregarding the time the unit is not operating. If the T_is are independent and identically distributed, we call $N(t)$ a *renewal process.* For most distributions of the T_is extensive numerical computations are necessary to find the distribution of $N(t)$. A renewal process with exponential distributed is an exception. In this case the distribution of $N(t)$ is given by a Poisson distribution with parameter λt:

$$P(N(t) = i) = \frac{(\lambda t)^i}{i!} e^{-\lambda t}, \quad i = 0, 1, 2, \ldots$$

where λ represents the failure rate of T_i. The process $N(t)$ is known as the Poisson process with intensity λ. Note that for a Poisson process we have

$$EN(t) = \lambda t$$

Hence we can interpret λ as the mean number of failures per unit of time.

Obviously, the assumption that the unit is as good as new after repair can be unrealistic in many cases. We might for example think of a situation where the unit is deteriorating and that possible repairs bring the unit to a state or condition which is approximately as good as it was immediately before the failure occurred. Such a repair is often called a *minimal repair.* In this case a *non-homogeneous Poisson process* with intensity $\lambda(t)$ describes the failure process. In this process $N(t)$ is Poisson distributed with parameter $\int_0^t \lambda(s)ds$. This parameter equals the mean of $N(t)$; hence for small values of h we have

$$E[N(t+h) - N(t)]/h \approx \lambda(t) \tag{B.3}$$

We see that the intensity $\lambda(t)$ may be interpreted as the expected number of failures per unit of time at t. For the non-homogeneous Poisson process the intensity does not depend on the history of the process up to time t.

The intensity $\lambda(t)$ must not be confused with the failure rate $z(t)$ for a non-repairable unit; $z(t)h$ is approximately equal to the probability that the *first* failure occurs in $(t, t+h)$, whereas $\lambda(t)\,h$ is approximately equal to the probability that a failure, not necessarily the first, occurs in $(t, t+h)$.

As for the failure rate $z(t)$ the bath-tub shape is typical of the intensity $\lambda(t)$, see Figure B.8.

The decreasing intensity in the first period is due to various types of initial weaknesses. Next follows a period when the randomization in the failure pattern of the components of the unit, as a result of replacements and maintenance, makes the intensity approximately constant. The failures in this phase are mainly "random failures". The intensity then increases due to wear-out and fatigue. Often the units are tested before they are sent to the users, so that most initial weaknesses are eliminated before the units are put into operation. As a consequence of an appropriate maintenance strategy, the units might be replaced before problems occur due to wear and fatigue. Hence the operating period is characterized by the part of the bath-tub curve that has an approximately constant intensity. This is one of the reasons why the Poisson process is so often used to model repairable systems. With a constant intensity the distinction between repairable and non-repairable units is not important.

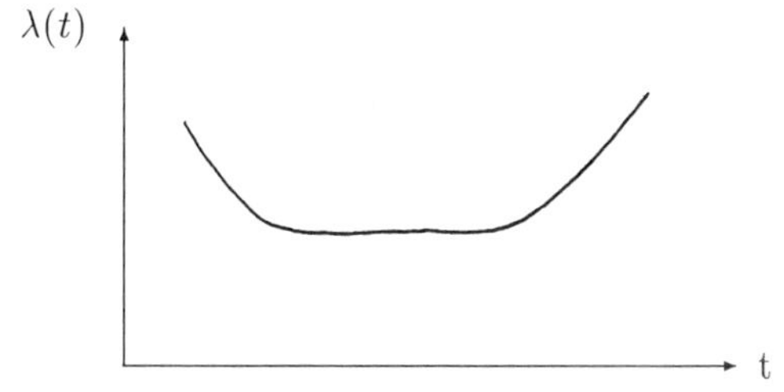

Figure B.8: Bath-tub shape of the failure intensity

Various parametric forms of the intensity $\lambda(t)$ have been proposed. The following forms have shown to be suitable in many situations:

$$\begin{aligned}\lambda(t) &= \lambda\beta(\lambda t)^{\beta-1} \quad \text{(Power law model)}\\ \lambda(t) &= \lambda e^{\beta t} \quad \text{(Log linear model)}\end{aligned}$$

If it is not realistic to assume that the repairs are "minimal", we can still apply the above model if we define the intensity $\lambda(t)$ by (B.3). Then we will have, as before,

$$EN(t) = \int_0^t \lambda(s)\, ds,$$

but now $N(t)$ is no longer Poisson distributed.

The model can easily be extended to the case that the unit has more than one failure mode. We then define a process $N_i(t)$ with failure intensity $\lambda_i(t)$ related to each failure mode. See the Appendix D, Problem D.7.3.

The processes discussed in this section are put into a more advanced form in Appendix E.

B.3 Binomial distribution

The binomial distribution is used in situations where a series of independent trials is performed, and where each trial results in either success or failure. If the probability of success is denoted p and there are k trials, then the number of successes, which we denote X, is binomial distributed with parameters k and p, i.e.

$$P(X = x) = \binom{k}{x} p^x (1-p)^{k-x}$$

where the binomial coefficient

$$\binom{k}{x} = k!/x!(k-x)!$$

For a binomial distribution it can be shown that

$$EX = kp \quad \text{and} \quad VarX = kp(1-p)$$

In the fire detector example in Appendix A, X is binomially distributed with parameters k and $p = 0.995$. Note that if X is binomially distributed with parameters k and p, then $k - X$, which represents the number of failures, is binomially distributed with parameters k and $1 - p$.

B.4 Gamma function

The gamma function, $\Gamma(x)$, is defined by

$$\Gamma(x) = \int_0^\infty t^{x-1}\, e^{-t}\, dt, \quad x > 0$$

In particular,

$$\Gamma(n+1) = n!\,, \quad n = 0, 1, 2, \ldots$$

To find the gamma function for other values, the formula

$$\Gamma(x+1) = x\,\Gamma(x)$$

is used repeatedly, together with Table B.1.

Table B.1: The gamma function for values between 1 and 2

x	$\Gamma(x)$	x	$\Gamma(x)$	x	$\Gamma(x)$	x	$\Gamma(x)$
1.00	1.000	1.25	0.906	1.50	0.886	1.75	0.919
1.01	0.994	1.26	0.904	1.51	0.887	1.76	0.921
1.02	0.989	1.27	0.903	1.52	0.887	1.77	0.924
1.03	0.984	1.28	0.901	1.53	0.888	1.78	0.926
1.04	0.978	1.29	0.899	1.54	0.888	1.79	0.929
1.05	0.974	1.30	0.897	1.55	0.889	1.80	0.931
1.06	0.969	1.31	0.896	1.56	0.890	1.81	0.934
1.07	0.964	1.32	0.895	1.57	0.890	1.82	0.937
1.08	0.960	1.33	0.893	1.58	0.891	1.83	0.940
1.09	0.955	1.34	0.892	1.59	0.892	1.84	0.943
1.10	0.951	1.35	0.891	1.60	0.893	1.85	0.946
1.11	0.947	1.36	0.890	1.61	0.894	1.86	0.949
1.12	0.944	1.37	0.889	1.62	0.896	1.87	0.952
1.13	0.940	1.38	0.889	1.63	0.897	1.88	0.955
1.14	0.936	1.39	0.888	1.64	0.899	1.89	0.958
1.15	0.933	1.40	0.887	1.65	0.900	1.90	0.962
1.16	0.930	1.41	0.887	1.66	0.902	1.91	0.965
1.17	0.927	1.42	0.886	1.67	0.903	1.92	0.969
1.18	0.924	1.43	0.886	1.68	0.905	1.93	0.972
1.19	0.921	1.44	0.886	1.69	0.907	1.94	0.976
1.20	0.918	1.45	0.886	1.70	0.909	1.95	0.980
1.21	0.916	1.46	0.886	1.71	0.911	1.96	0.984
1.22	0.913	1.47	0.886	1.72	0.913	1.97	0.988
1.23	0.911	1.48	0.886	1.73	0.915	1.98	0.992
1.24	0.909	1.49	0.886	1.74	0.917	1.99	0.996
						2.00	1.000

B.5 Problems

1. A unit has a lifetime T which is exponentially distributed with $MTTF = 1000$ hours.

   a) Find the probability that the lifetime exceeds 1000 hours.

   b) Given that the unit has survived 1000 hours, what is the probability that it survives 1200 hours? Comment on the result.

   c) Let T_1 and T_2 denote the lifetime of two units of this type. Assume that T_1 and T_2 are independent.
   Find the distribution of $T_1 + T_2$? Calculate $P(T_1 + T_2 \leq 1000)$.

2. A unit has a lifetime T which is Weibull distributed with shape parameter $\beta = 2$ and scale parameter $\lambda = 10^{-3}$.

   a) The same as Problem 1a).

   b) The same as Problem 1b).

   c) Calculate the mean lifetime.

3. The lifetime of a certain type of unit is assumed to have a constant failure rate $\lambda = 5 \times 10^{-4}$ (per hours). If the component fails, it is immediately replaced by a new unit of the same type. Thus there is always one unit in operation. We assume that the lifetimes are independent.

   a) Explain to a person who is not an expert in reliability analyses what is meant by "the component has a failure rate equal to $\lambda = 5 \times 10^{-4}$".

   b) Find the mean number of failures in a period of two years.

   c) Find the probability of having no failures in a period of 730 hours. Find also the probability of at least three failures in such a period of time.

4. A machine in a production system is put into operation at time $t = 0$. At failure the machine is immediately repaired. We disregard the repair times.
   A failure trend analysis is to be carried out for the machine. It is of particular interest to obtain knowledge of how the failure pattern is affected by wear and fatigue as the machine becomes older. A so-called expert in reliability analysis states that to correctly model the wear-out period of the machines, a Weibull distribution should be assumed for the times between failures, since the failure rate in this distribution is an increasing function of time for certain values of the parameters. What are your comments on this statement?

5. We consider 5 independent units of a certain type A. The probability that one specific unit will function at a given point in time, equals 0.90. Let X denote the number of units among the 5 that are functioning.

   Determine the distribution of X. Compute $P(X = x)$, $x = 0, 1, 2, 3, 4, 5$. Find EX, i.e. the mean number of units that is functioning.

Appendix C

STATISTICAL ANALYSIS OF RELIABILITY DATA

In this appendix we give an overview of some of the most important statistical methods that are used to analyse reliability and lifetime data. First we present a method, Hazard plotting, to identify the underlying distribution of the lifetimes. In a real-life situation the determination of lifetime distribution will normally be based on a combination of methods like Hazard plotting, experience and expert judgements. Another useful method for identification of lifetime distribution, is the TTT (Total Time on Test) plot. This method, however, will not be presented here. Refer to [30] for a thorough examination of this method.

Next we look at various techniques to estimates the parameters of the models. Then we discuss statistical analysis of non-homogeneous Poisson processes, which are used to model repairable systems. Finally, an overview of some relevant data bases are given.

Statistical analysis of reliability and lifetime data is the topic of many textbooks and papers. Some relevant references are [3,6,30,65,84,93,95].

C.1 Identification of lifetime distribution, Hazard plotting

We consider the model described in Section B.1.1. To be able to identify the lifetime distribution F in a specific situation, experience data must be collected and analysed. These data can be obtained from experiments, operational experience and/or data banks, cf. Section C.4.

We assume that lifetime data are available in the following form:

$$(T_1, \delta_1), (T_2, \delta_2), \ldots, (T_n, \delta_n)$$

where T_i represents the ith lifetime and δ_i is an indicator variable defined by

$$\delta_i = \begin{cases} 1 & \text{if } T_i \text{ equals a failure time} \\ 0 & \text{if } T_i \text{ equals a censored lifetime} \end{cases}$$

A lifetime T_i is said to be censored (more precisely, censored on the right) if the failure time is unknown, but it is known that it is greater than or equal to T_i. We assume that the T_is are ordered such that $T_1 \leq T_2 \leq \ldots \leq T_n$. For the sake of simplicity in the following we will write T_i in place of (T_i, δ_i).

Lifetimes can be censored due to a number of different reasons, for example, the units under study have not failed when the testing/observations are finished, the test equipment breaks down, or the units fail due to reasons other than those of interest. In the literature, four types of censoring are usually discussed:

I The units are observed to a predetermined age t_0. The data information consists of the v $(v \leq n)$ observed lifetimes $T_1 \leq T_2 \leq \cdots \leq T_v$, and the fact that $n - v$ units have survived t_0.

II The units are observed until the vth failure $(v \leq n)$. We assume that all the units are activated at the same time. The data information here consists of the observed lifetimes $T_1 \leq T_2 \leq \cdots \leq T_v$, and the fact that $n - v$ units have survived T_v.

III The units are observed until t_0 and the sth failure time, whichever occurs first. We assume that all units are activated at the same time. The data information consists of the v observed lifetimes $T_1 \leq T_2 \leq \cdots \leq T_v$, and the fact that $n - v$ units have survived minimum$\{t_0, T_s\}$.

IV Unit i is observed until age S_i, $i = 1, 2, \ldots, n$. It is assumed that the S_is are independent stochastic variables, and also independent of the failure times. S_i might for example be a constant. The data information consists of the v observed lifetimes $T_{i_1} \leq T_{i_2} \leq \cdots \leq T_{i_v}$, and the fact that unit i_j has survived $S_{i_j}, j = v+1, v+2, \ldots, n$.

Figure C.1 illustrates how the failure history of a unit generates lifetimes T_i. For each unit we determine a starting point of the failure history, points of failure and completed repairs/replacements, and a stop point of the failure history (Figure C.1a). From these histories, it is then possible to determine the times to the first failures (Figure C.1b). If repaired (replaced) units can be considered as "good as new", these times should also be included (Figure C.1c).

Hazard plotting

Hazard plotting or Nelson plotting is a graphic method to identify the underlying lifetime distribution. The method is based on the estimation of the hazard $Z(t)$ using the so-called Nelson estimator, $\hat{Z}(t)$, given by

$$\hat{Z}(t) = \sum_{j:\delta_j=1, T_j \leq t} \frac{1}{n-j+1}$$

Thus for each observed failure time T_j, a contribution $1/(n-j+1)$ is added to $\hat{Z}(\cdot)$.

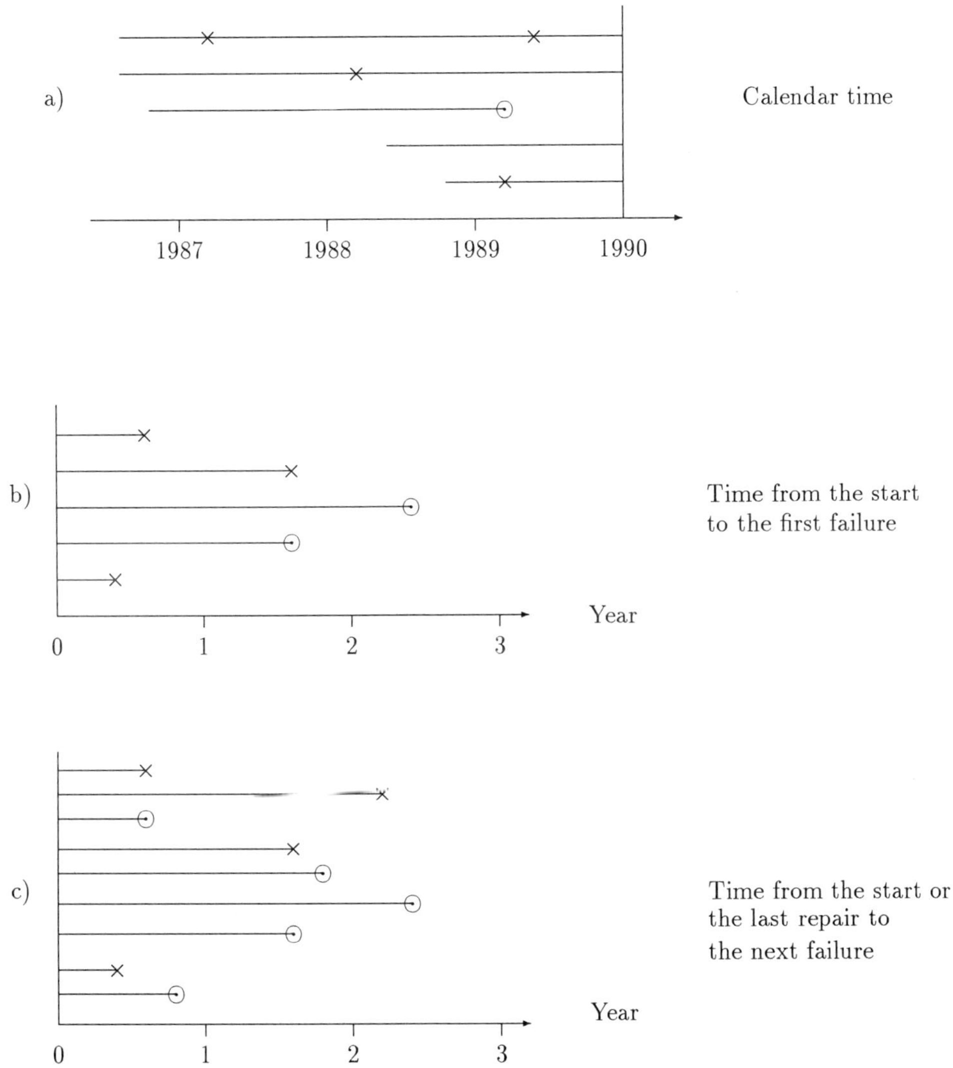

Figure C.1: Failure history of five units with different starting times,
In the figure × denotes failure,
whereas ⊙ denotes censoring.

To justify the definition of the Nelson estimator, consider a small time interval $[t, t+h)$ and assume that $j-1$ units have failed or been censored before time t. Then the expected number of failures in this interval is approximately equal to

$$(n-j+1)\, z(t)\, h \approx (n-j+1) \int_t^{t+h} z(s)\, ds \tag{C.1}$$

since there are $n-j+1$ units under test/observation just before time t. But a natural estimator for this expected number is 1 if a failure occurs in the interval, and 0 otherwise. We can disregard the possibility that two or more failures will occur in the interval since we have assumed that h is small. Hence $1/(n-j+1)$ is a natural estimator for the integral in (C.1) if a failure occurs in this interval, and 0 otherwise. The expression for the Nelson estimator follows.

The Nelson estimator and generalizations of this estimator, have been thoroughly studied in the literature, see e.g. [3,84,103]. It can for example be shown that the estimator is (approximately) unbiased, i.e.:

$$E\hat{Z}(t) \approx Z(t)$$

and under certain conditions the estimator will be consistent, i.e.

$$P(|\ \hat{Z}(t) - Z(t)\ | < \epsilon) \to 1 \text{ for all } \epsilon > 0, \text{ when } n \to \infty$$

Hence with an increasing number of observations, the estimator will "approach" the hazard.

A Hazard plot is constructed by plotting $(T_i, \hat{Z}(T_i))$, possibly $(\hat{Z}(T_i), T_i)$. With this plot the form of the hazard is studied. If the points in the plots fall roughly in a straight line, this indicates that the hazard is linear, i.e. the failure rate is constant. A convex (concave) form of the plot indicates an IFR (DFR) distribution.

Example C.1

The following lifetimes are given, where the underlined times represent failure times: $\underline{32}, \underline{40}, \underline{58}, \underline{65}, \underline{70}, 75, 75, 75, 75, \underline{100}, \underline{110}, 125$. The computation of the Nelson estimator is carried out in Table C.1 and the Hazard plot is shown in Figure C.2. We see from Figure C.2 that the plot has a convex form, indicating that the failure rate is increasing. An exponential model does not seem to be appropriate.

□

In order to compare the data with other distributions, it is common practice to use a *Hazard plotting paper*. On this type of plotting paper the axes are transformed, such that the connection between $Z(t)$ and t is linear. For the Weibull distribution we have

$$log\, Z(t) = \beta\, log\, \lambda + \beta\, log\, t$$

Table C.1: Computation of the Nelson estimator for Example C.1

j	Failure time	$\hat{Z}$
1	32	$\frac{1}{12} = 0.083$
2	40	$0.083+\frac{1}{11} = 0.174$
3	58	$0.174+\frac{1}{10} = 0.274$
4	65	$0.274+\frac{1}{9} = 0.385$
5	70	$0.385+\frac{1}{8} = 0.510$
10	100	$0.510+\frac{1}{3} = 0.844$
11	110	$0.844+\frac{1}{2} = 1.344$

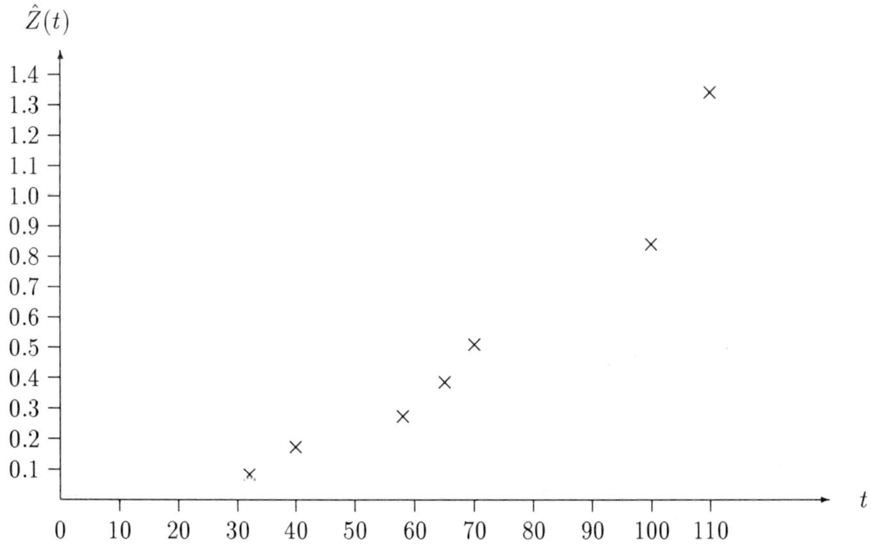

Figure C.2: Hazard plot for the data in Table C.1.

By drawing the Hazard plot on a double logarithmic paper, the points will fall roughly on a straight line if the data can be described by a Weibull distribution.

Hazard plotting paper can also be used to estimate the parameters of the distribution, as illustrated in following example.

Example C.1 continued
The Hazard plot in Figure C.2 showed a poor fit to the exponential distribution. The same points are plotted on *Weibull hazard plotting paper* in Figure C.3. The figure shows that the fit to the Weibull distribution is good. We therefore assume that the data are from a Weibull distribution. The slope of the estimated straight line through the points provides an estimate of the shape parameter β. This slope can be measured directly from the fitted line, or alternatively by drawing a parallel line through the point Q. The point of intersection between this line and the line "Shape parameter" gives the estimate. The point Q is placed on the paper such that the distance from the point to the shape parameter line equals one unit length on this line. We find that $\hat{\beta} = 2.0$. Since $1/\lambda$ is the value that corresponds to $Z(t) = 1$ (100%), $\hat{\lambda} = 1/100$ will be a natural estimate for λ. The hazard at time $t = 40$, for example, is estimated from Figure C.3:

$$\hat{Z}(40) = 0.15$$

The probability that the unit does not survive to this point in time, $F(40)$, can be estimated by

$$\hat{F}(40) = 1 - \hat{R}(40) = 1 - e^{-\hat{Z}(40)} = 1 - e^{-0.15} = 0.14$$

or by reading the value on the "Probability %" scale of Figure C.3 (we then get approximately the same values in this case).

C.2 Estimation of parameters in lifetime distributions

We assume that the lifetime distribution belongs to a known parametric class of distributions, for example the exponential or the Weibull distribution. The problem is to estimate the parameters of the distribution. We have already looked at some graphic methods based on Hazard plotting. These estimators are usually quite rough. In this appendix we shall present two more "advanced techniques", maximum likelihood estimation and Bayes estimation.

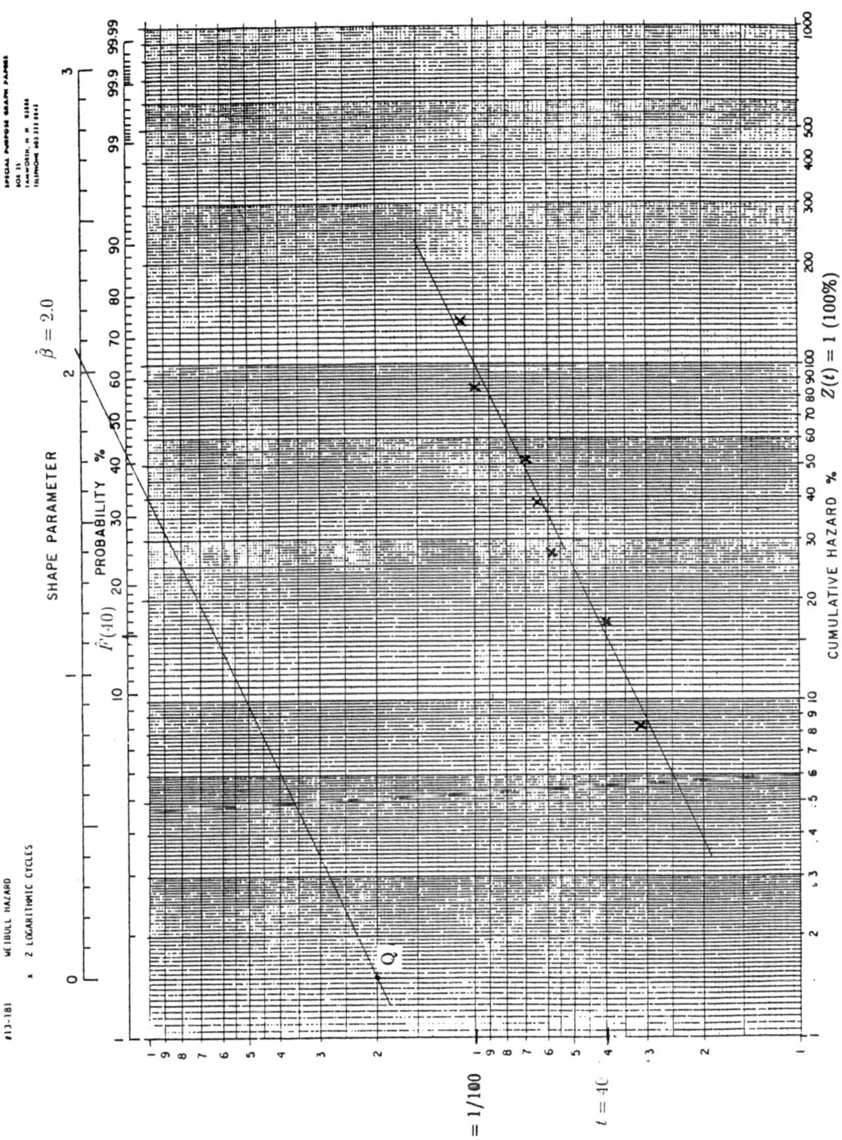

Figure C.3: Weibull hazard plotting paper for Example C.1

Maximum likelihood estimation

Let $\theta = (\theta_1, \theta_2, \ldots, \theta_p)$ be the parameter in the distribution and let $f(t,\theta)$ and $R(t,\theta)$ denote the probability density and survival function, respectively. For the exponential distribution $\theta = \lambda$, and for the Weibull distribution $\theta = (\beta, \lambda)$.

In this model we can write the probability density of the data in the following form:

$$\prod_{i \in F} f(T_i, \theta) \prod_{i \in C} R(T_i, \theta)$$

where F and C denote the set of failure times and censored times, respectively. As a function of the parameter θ, the density is called the likelihood function and is denoted $L(\theta)$. The likelihood function is a measure of the likelihood of the observed result as a function of the unknown parameter.

The maximum likelihood estimator (MLE) of θ, θ^*, maximizes $L(\theta)$. In other words, MLE is the value of θ that makes the observed result most likely. It can be shown that in most cases the MLE will be a good estimator; it will be unbiased and consistent. In practice the MLE is determined by differentiating the likelihood function and setting the derivative equal to zero.

For the exponential model we find that

$$\theta^* = \frac{r}{T}$$

where

$$\begin{array}{rcccl} r & = & \sum_i \delta_i & = & \text{Number of observed failure times} \\ T & = & \sum_i T_i & = & \text{Total test time (observation time)} \end{array}$$

In general it is not possible to find an explicit expression for the MLE. Numerical methods must then be used.

As a measure of data variation, a confidence interval (region) for θ is often presented in addition to the estimate of the parameter.

An interval (θ_L, θ_H) is said to be a $(1-\alpha)100\%$ confidence interval if there is a probability of $1-\alpha$ that the interval contains θ, that is

$$P(\theta_L < \theta < \theta_H) = 1 - \alpha$$

The level $1-\alpha$ is a measure of our confidence that the interval contains θ. The most common values of α are 0.10 and 0.05. Notice that θ_L and θ_H are stochastic variables. When the confidence interval is calculated, the resulting interval either contains the true value of θ or it does not, but in the long run if the experiment is repeated many times, the confidence interval will include θ $(1-\alpha)100\%$ of the times. The level of confidence $1-\alpha$ therefore expresses a property of the method that we are using to determine the interval.

In the exponential model a (approximate) $(1-\alpha)100\%$ confidence interval is given by:

$$(\lambda_L, \lambda_H) = \left(\frac{z_{1-\alpha/2,2r}}{2T}, \frac{z_{\alpha/2,2(r+1)}}{2T}\right)$$

where $z_{\alpha,v}$ equals the α quantile in the chi-square distribution with v degrees of freedom. For censoring of type II the number of degrees of freedom can be set to $2r$ for both quantiles. A table of quantiles in the chi-square distribution is given at the end of this appendix, Table C.2.

Example C.2

A type of pressure gauge has an accumulated operating time of 1.48×10^6 hours on a oil producing platform. A total of 19 failures have been registered. This gives an estimated failure rate equal to

$$\frac{19}{1.48 \times 10^6 \text{ hours}} = 12.8 \times 10^{-6} \text{ hours}^{-1} = 0.11 \text{ years}^{-1}$$

A 90% confidence interval is given by (time units hours)

$$\lambda_L = \frac{24.91}{2 \times 1.48 \times 10^6} = 8.4 \times 10^{-6}$$

$$\lambda_H = \frac{55.76}{2 \times 1.48 \times 10^6} = 18.8 \times 10^{-6}$$

Here 24.91 is found by linear interpolation in Table C.2.

Bayes estimation

The above method requires a large or moderately large number of failure times to give reasonable estimates of the parameters. In real life there are often few failure times available, in some cases only one or two registered failures, or no failures at all. Estimation in such situations requires that engineering judgements and other expert judgements are applied. A method that can be used in such situations is Bayes estimation. This method is based on two types of information:

- A priori information
- Observed lifetime data

The a priori information includes all available information "without looking at the data". It might include information from manufacturers, other reliability studies, operational experience, etc. The a priori information is expressed by the so-called *a priori* distribution for the parameter θ. When this a priori distribution is established, it is combined with the observed lifetime data by using standard probability calculus to find the so-called *posterior* distribution. If $p(\theta)$ expresses the a priori probability density, then the posterior density $p(\theta \mid Data)$, is given by

$$p(\theta \mid Data) = c\, L(\theta)\, p(\theta)$$

where c is a constant such that the integral over $p(\theta \mid Data)$ becomes 1. The posterior density expresses our uncertainty with respect to the true value of θ when the data are observed, and includes all the available information about θ. Based on the posterior distribution we can establish estimators and *credibility intervals.* This will be illustrated below for the exponential distribution.

Assume in the rest of this section that the underlying lifetime distribution is exponential with failure rate λ. If the a priori density $p(\lambda)$ takes the form

$$p(\lambda) = b^a \lambda^{a-1} e^{-b\lambda} / \Gamma(a)$$

i.e. p is a gamma density with parameters a and b, then we find that the posterior density is also a gamma density, with parameters $a+r$ and $b+T$. A natural estimator for λ is the mean of the posterior distribution

$$\hat{\lambda} = \frac{a+r}{b+T} \tag{C.2}$$

A $(1-\alpha)100\%$ credibility interval for λ, (λ_L, λ_H), is determined by the posterior probability

$$P(\text{“}\lambda_L < \lambda < \lambda_H\text{”} \mid Data) = 1 - \alpha$$

With a gamma a priori density it can be shown that the interval

$$(\frac{z_{1-\alpha/2, 2(a+r)}}{2(b+T)}, \frac{z_{\alpha/2, 2(a+r)}}{2(b+T)})$$

is an approximate $(1-\alpha)100\%$ credibility interval. Note that a credibility interval is interpreted given that the data are observed, in contrast to the classic confidence interval which is interpreted before the data are observed.

The most critical part of the Bayes estimation is to establish the a priori distribution. The choice of a gamma a priori density gives simple mathematics. In addition, it can be shown that if the a priori distribution is non-informative, i.e.

$$p(\lambda) = \begin{cases} \frac{1}{M} & \text{for } 0 \le \lambda \le M \quad \text{(M large)} \\ 0 & \text{otherwise,} \end{cases}$$

then the resulting posterior distribution is an approximate gamma distribution. We can think of a situation where we started with a non-informative a priori distribution and this distribution was updated to a gamma distribution. The parameter $a-1$ in the gamma distribution can be interpreted as the number of observations in an earlier experiment (real or fictitious) and b as the corresponding test time.

An alternative method for determining a and b is presented in [95]. In this method quantities $\lambda_1, \lambda_2, q_1$ and q_2 are specified such that

$$\begin{aligned} P(\text{“}\lambda < \lambda_1\text{”}) &= q_1 \\ P(\text{“}\lambda < \lambda_2\text{”}) &= q_2 \end{aligned}$$

see the example below.

Example C.3
The problem is to estimate the failure rate of a valve of a specific type. According to expert judgement, a 90% a priori credibility interval for the failure rate equals (0.1,0.4) per year. We assume a gamma a priori distribution and will determine the parameters a and b by specifying $\lambda_1, \lambda_2, q_1$ and q_2 as explained above. Let $\lambda_1 = 0.1, \lambda_2 = 0.4, q_1 = 0.05$ and $q_2 = 0.95$. From [95] it follows that $a \approx 6.5$ and $b \approx 20$.

With these values of a and b, the mean of the a priori distribution becomes $6.5/20 = 0.325$.
The observed data are $r = 1$ and $T = 5$ year. The posterior distribution is a gamma distribution with parameters $6.5 + 1 = 7.5$ and $20 + 5 = 25$. The estimator (C.2) equals

$$\hat{\lambda} = \frac{7.5}{25} = 0.30 \text{ years}^{-1}$$

A 90% credibility interval equals

$$(7.26/50, 25/50) = (0.145, 0.500)$$

C.3 Statistical analysis of non-homogeneous Poisson processes

A non-homogeneous Poisson process as defined in Appendix B, is characterized by a failure intensity $\lambda(t)$ and a cumulative intensity

$$\Lambda(t) = \int_0^t \lambda(s)\, ds \tag{C.3}$$

To determine the shape of $\lambda(t)$ and $\Lambda(t)$ assume that we have observations from k independent Poisson processes $N_1(t), N_2(t), \ldots, N_k(t)$. Here $N_i(t)$ represents the number of observed failures in the interval $[0, t]$ for unit i. The process i is assumed to be observed in the interval $(a_i, b_i]$. Let $U_1 \leq U_2 \leq \cdots \leq U_N$ denote the observed failure times when we consider all processes together. A good estimator for $\Lambda(t)$ is the Nelson estimator $\hat{\Lambda}(t)$, given by

$$\hat{\Lambda}(t) = \sum_{i:U_i \leq t} \frac{1}{Y(U_i)}$$

where $Y(t)$ equals the number of processes that are under observation just before time t. If for example $a_i = 0$ and $b_i = b$ for all i, then $Y(t) = k$ for all t in the interval $[0, b]$, such that

$$\hat{\Lambda}(t) = \frac{\sum_{i=1}^k N_i(t)}{k} = \frac{\text{Total number of failures in } [0, t]}{k}$$

The analysis is now analogous to the analysis of $\hat{Z}(t)$ in Section C.1. For example we can find out whether the Power Law model $\Lambda(t) = (\lambda t)^\beta$ fits the data by plotting $(\hat{\Lambda}(U_i), U_i)$ on

Hazard plotting paper for the Weibull distribution. An approximate straight line indicates such a form. The parameters λ and β can be estimated in the same way as in Section C.1.

Maximum likelihood estimation and Bayes estimation can be carried out as in Section C.2 with the likelihood function

$$L(\theta) = \prod_{i=1}^{N} \lambda(U_i)\, e^{-\int_0^b \lambda(t,\theta)Y(t)\,dt}$$

where b is the largest b_i value. We find for example that if the cumulative intensity is in a log-linear form $(\lambda t)^\beta$ (the Power Law model) and $a_i = 0$ and $b_i = b$ for all i, then the MLE for the parameters are given by

$$\lambda^* = \frac{N}{b^{\beta^*}}$$

$$\beta^* = \frac{N}{\sum_{i=1}^{N} ln\,(b/U_i)}$$

The above methods can also be used where $N(t)$ is not a non-homogeneous Poisson process; what is required is that $N(t)$ has a mean function on the form (C.3), cf. the comment at end of Section B.2.

Example C.4 [17]

Assume that $k = 3$ systems are observed in $[0, 15]$, and assume the following failure time data:

System 1: 2.4, 6.2, 7.7, 11.4
System 2: 5.0, 9.8, 12.0, 13.7
System 3: 4.0, 8.7, 10.9, 12.5, 13.5, 14.5

In Figure C.4 we have plotted $(U_i, \hat{\Lambda}(U_i))$. The plot indicates that the failure intensity is slowly increasing. The systems seems to deteriorate over time. By plotting the data on Hazard plotting paper it is seen that the Power Law model fits the data quite well.

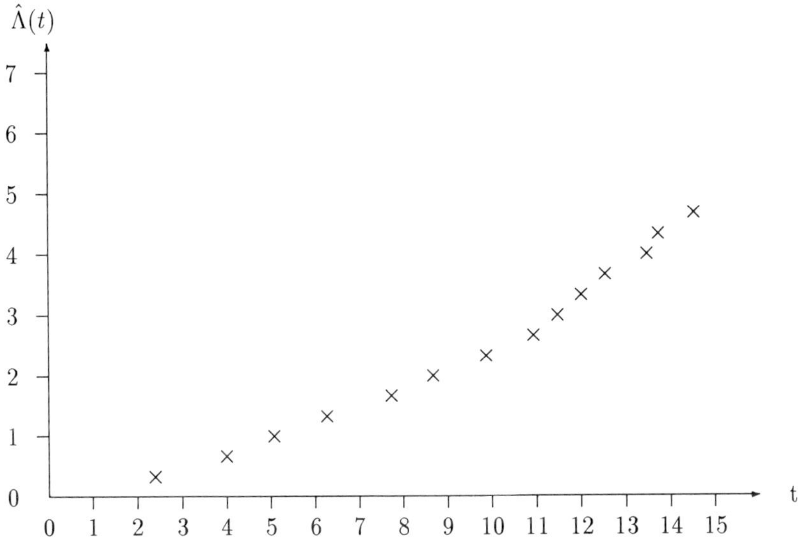

Figure C.4: The Nelson estimator for the data of Example C.4

Table C.2: Table over quantiles in the chi-square distribution with v degrees of freedom
$z_{\alpha,v} = \alpha$ quantile with v degrees of freedom, i.e.
$P(Z > z_{\alpha,v}) = \alpha$, where Z is chi-square distributed with v degrees of freedom

	α							
v	0.995	0.990	0.975	0.950	0.050	0.025	0.010	0.005
1	0.00	0.00	0.00	0.00	3.84	5.02	6.63	7.88
2	0.01	0.02	0.05	0.10	5.99	7.38	9.21	10.60
3	0.07	0.11	0.22	0.35	7.81	9.35	11.34	12.84
4	0.21	0.30	0.48	0.71	9.49	11.14	13.28	14.86
5	0.41	0.55	0.83	1.15	11.07	12.83	15.09	16.75
6	0.68	0.87	1.24	1.64	12.59	14.45	16.81	18.55
7	0.99	1.24	1.69	2.17	14.07	16.01	18.48	20.28
8	1.34	1.65	2.18	2.73	15.51	17.53	20.09	21.95
9	1.73	2.09	2.70	3.33	16.92	19.02	21.67	23.59
10	2.16	2.56	3.25	3.94	18.31	20.48	23.21	25.19
11	2.60	3.05	3.82	4.57	19.68	21.92	24.72	26.76
12	3.07	3.57	4.40	5.23	21.03	23.34	26.22	28.30
13	3.57	4.11	5.01	5.89	22.36	24.74	27.69	29.82
14	4.07	4.66	5.63	6.57	23.68	26.12	29.14	31.32
15	4.60	5.23	6.26	7.26	25.00	27.49	30.58	32.80
16	5.14	5.81	6.91	7.96	26.30	28.85	32.00	34.27
17	5.70	6.41	7.56	8.67	27.59	30.19	33.41	35.72
18	6.26	7.01	8.23	9.39	28.87	31.53	34.81	37.16
19	6.84	7.63	8.91	10.12	30.14	32.85	36.19	38.58
20	7.43	8.26	9.59	10.85	31.41	34.17	37.57	40.00
21	8.03	8.90	10.28	11.59	32.67	35.48	38.93	41.40
22	8.64	9.54	10.98	12.34	33.92	36.78	40.29	42.80
23	9.26	10.20	11.69	13.09	35.17	38.08	41.64	44.18
24	9.89	10.86	12.40	13.85	36.42	39.36	42.98	45.56
25	10.52	11.52	13.12	14.61	37.65	40.65	44.31	46.93
26	11.16	12.20	13.84	15.38	38.89	41.92	45.64	48.29
27	11.81	12.88	14.57	16.15	40.11	43.19	46.96	49.64
28	12.46	13.56	15.31	16.93	41.34	44.46	48.28	50.99
29	13.12	14.26	16.05	17.71	42.56	45.72	49.59	52.34
30	13.79	14.95	16.79	18.49	43.77	46.98	50.89	53.67
40	20.71	22.16	24.43	26.51	55.76	59.34	63.69	66.77
50	27.99	29.71	32.36	34.76	67.50	71.42	76.15	79.49
60	35.53	37.48	40.48	43.19	79.08	83.30	88.37	91.95
70	43.28	45.44	48.76	51.74	90.53	95.02	100.42	104.21
80	51.17	53.54	57.15	60.39	101.88	106.63	112.32	116.31
90	59.20	61.75	65.65	69.13	113.14	118.13	124.11	128.29
100	67.33	70.06	74.22	77.93	124.34	129.56	135.80	140.16

Table C.3: Table over the standard normal distribution with mean 0 and variance 1: $\Phi(x) = P(X \leq x)$, $x \geq 0$. To compute $\Phi(x)$ for $x < 0$, we can use that $\Phi(x) = 1 - \Phi(-x)$.

x	0.00	0.01	0.02	0.03	0.04	0.05	0.06	0.07	0.08	0.09
0.0	0.500	0.504	0.508	0.512	0.516	0.520	0.524	0.528	0.532	0.536
0.1	0.540	0.544	0.548	0.552	0.556	0.560	0.564	0.567	0.571	0.575
0.2	0.579	0.583	0.587	0.591	0.595	0.599	0.603	0.606	0.610	0.614
0.3	0.618	0.622	0.626	0.629	0.633	0.637	0.641	0.644	0.648	0.652
0.4	0.655	0.659	0.663	0.666	0.670	0.674	0.677	0.681	0.684	0.688
0.5	0.691	0.695	0.698	0.702	0.705	0.709	0.712	0.716	0.719	0.722
0.6	0.726	0.729	0.732	0.736	0.739	0.742	0.745	0.749	0.752	0.755
0.7	0.758	0.761	0.764	0.767	0.770	0.773	0.776	0.779	0.782	0.785
0.8	0.788	0.791	0.794	0.797	0.800	0.802	0.805	0.808	0.811	0.813
0.9	0.816	0.819	0.821	0.824	0.826	0.829	0.831	0.834	0.836	0.839
1.0	0.841	0.844	0.846	0.849	0.851	0.853	0.855	0.858	0.860	0.862
1.1	0.864	0.867	0.869	0.871	0.873	0.875	0.877	0.879	0.881	0.883
1.2	0.885	0.887	0.889	0.891	0.893	0.894	0.896	0.898	0.900	0.901
1.3	0.903	0.905	0.907	0.908	0.910	0.911	0.913	0.915	0.916	0.918
1.4	0.919	0.921	0.922	0.924	0.925	0.926	0.928	0.929	0.931	0.932
1.5	0.933	0.934	0.936	0.937	0.938	0.939	0.941	0.942	0.943	0.944
1.6	0.945	0.946	0.947	0.948	0.949	0.951	0.952	0.953	0.954	0.954
1.7	0.955	0.956	0.957	0.958	0.959	0.960	0.961	0.962	0.962	0.963
1.8	0.964	0.965	0.966	0.966	0.967	0.968	0.969	0.969	0.970	0.971
1.9	0.971	0.972	0.973	0.973	0.974	0.974	0.975	0.976	0.976	0.977
2.0	0.977	0.978	0.978	0.979	0.979	0.980	0.980	0.981	0.981	0.982
2.1	0.982	0.983	0.983	0.983	0.984	0.984	0.985	0.985	0.985	0.986
2.2	0.986	0.986	0.987	0.987	0.987	0.988	0.988	0.988	0.989	0.989
2.3	0.989	0.990	0.990	0.990	0.990	0.991	0.991	0.991	0.991	0.992
2.4	0.992	0.992	0.992	0.992	0.993	0.993	0.993	0.993	0.993	0.994
2.5	0.9940	0.9940	0.9942	0.9943	0.9945	0.9946	0.9948	0.9949	0.9951	0.9952
2.6	0.9953	0.9955	0.9956	0.9957	0.9959	0.9960	0.9961	0.9962	0.9963	0.9964
2.7	0.9965	0.9966	0.9967	0.9968	0.9969	0.9970	0.9971	0.9972	0.9973	0.9974
2.8	0.9974	0.9975	0.9976	0.9977	0.9977	0.9979	0.9979	0.9980	0.9980	0.9981
2.9	0.9981	0.9982	0.9982	0.9983	0.9984	0.9984	0.9985	0.9985	0.9986	0.9986
3.0	0.9987	0.9987	0.9988	0.9988	0.9988	0.9989	0.9989	0.9989	0.9990	0.9990
3.1	0.9990	0.9991	0.9991	0.9991	0.9992	0.9992	0.9992	0.9992	0.9993	0.9993
3.2	0.9993	0.9993	0.9994	0.9994	0.9994	0.9994	0.9994	0.9995	0.9995	0.9995
3.3	0.9995	0.9995	0.9996	0.9996	0.9996	0.9996	0.9996	0.9996	0.9996	0.9997
3.4	0.9997	0.9997	0.9997	0.9997	0.9997	0.9997	0.9997	0.9997	0.9998	0.9998
3.5	0.9998	0.9998	0.9998	0.9998	0.9998	0.9998	0.9998	0.9998	0.9998	0.9998
3.6	0.9998	0.9998	0.9999	0.9999	0.9999	0.9999	0.9999	0.9999	0.9999	0.9999
3.7	0.9999	0.9999	0.9999	0.9999	0.9999	0.9999	0.9999	0.9999	0.9999	0.9999

C.4 Data sources

To collect reliability and lifetime data, different approaches can be used:

1. *Determine the component reliabilities experimentally in laboratories.* To reduce the time involved in such experiments, *accelerated* testing is often introduced, i.e. the components are exposed to higher loads/stresses than in a normal operational situation. To obtain relevant results for the normal operational situation, the same deteriorating mechanisms must be relevant in accelerated testing as in the normal operating phase.

2. *Collect operational data.* Operational data usually give the best information about the component reliabilities. To obtain good reliability estimates, the data must be relevant with respect to type and operational and environmental conditions.

3. *Ask the component supplier about data.* The component supplier usually has detailed information about the performance of the components, but in some cases the supplier restricts the availability of reliability data and failure statistics. The reason for this might be that he is afraid that the data will be misused and that his product might have a negative reception.

4. *Use data from reliability data banks and data books.* In a reliability analysis we will often make use of data from several data sources of this type. The following list gives some relevant reliability data banks and books:

    - Systems Reliability Services (SRS) [134]
    - Non-electronic Reliability Notebook [115]
    - Military Handbook 217B [97]
    - IEEE Standard 500 [70]
    - Reactor Safety Study Wash-1400 [135]
    - European Reliability Data Bank, EuReData
    - Electronic Reliability Data [71]
    - Non-electronic Parts Reliability Data [113]
    - Offshore Reliability Data, OREDA [110]. This book/data base comprises reliability data of equipment that is used in offshore petroleum activity in the North Sea
    - Reliability Technology [59]
    - Reliability and Maintainability in Perspective [120]
    - Handbook of Human Reliability Analysis [130]

For a survey of data sources for offshore applications, see [137].

Collection, systematization and retention of reliability data is an important area of reliability technology. The quality of reliability and risk analyses is to large extent dependent on the quality of the reliability data. It is, however, outside the scope of this book to discuss in detail how collection, systematization and retention of reliability data should be carried out. See [64,73] and the above references under the list of relevant data books and banks.

C.5 Problems

1. What is the purpose of Hazard plotting? Explain how you would proceed to check whether an exponential model is appropriate.

2. The following lifetime data are given: 0.9, 1.4, 2.4, 3.0, 4.5, 6.5, 7.7 and 9.0. There is no censoring.

   a) Compute the Nelson estimator and construct a Nelson plot on standard paper and on Weibull plotting paper.

   b) What is your conclusion when looking at the plots? Estimate the parameters of the fitted distribution. Estimate also $MTTF$ and $R(4)$.

3. As Problem 2 above, but with the following data: $\underline{0.9}, \underline{1.4}, \underline{2.4}, \underline{3.0}, \underline{4.5}, \underline{6.5}, \underline{7.7}, \underline{9.0}, 9.0, 9.0, 9.0, 9.0$. The underlined lifetimes represent failure times, whereas the remaining lifetimes are censored lifetimes.

4. A unit of a specific type has an accumulated operating time of 2×10^6 hours on an installation. It has registered 10 failures. Under the assumption of a constant failure rate, estimate the failure rate for this type of unit. Compute also an approximate 90% confidence interval for the failure rate.

5. A machine is installed at time $t = 0$. At failure the machine is immediately repaired. Assume that the following times between failures have been registered: 1, 2, 4, 7, 15, 35 and 100. The first failure occurred at time $t = 1$, the next at time $t = 3$, and so forth. We disregard the repair times. In the standard model for analysing failure time data, the starting point is the times between failures; it is assumed that they are independent and identically distributed stochastic variables. Do you think such a model is appropriate for the example presented in this problem?

6. Perform the analysis described in Example C.4. Try to fit the data to the Power Law model $(\Lambda(t) = (\lambda t)^{\beta})$.

Appendix D

MARKOV MODELS

D.1 Introduction

This appendix gives a short introduction to Markov models for analysing reliability and risk. The presentation of the theory is to a large extent based on [91]. For a more detailed presentation of Markov models refer to [32,64,65,79,116].

We consider a system which can be in one out of several possible states, from the perfect functioning state to the complete failure state. Markov models are used to describe and analyse the movement of the system among the various states. This movement can be described by a state diagram, also called a Markov diagram. Examples of such diagrams are presented in the following sections.

Markov models are particularly suitable for analysing smaller systems with complicated maintenance policies and possible dependencies between components. The weakness of the method is that the complexity of the model soon becomes unwieldy, even with a computer. Markov models are therefore preferably used for analysing sub-systems of a larger system. It is also limiting that all lifetimes and repair times in a Markov model must be assumed to be exponential.

In a Markov model we can calculate:

- The probability that the system is in a specific state at a given point in time
- The steady state distribution, i.e. the probability that the system is in a specific state after a very long time of operation
- The average time (portion of time) the system stays in the various states
- The average number of times the system visits various states during a specific period. This quantity can be used to estimate the need for spares and maintenance personnel
- The average time it takes for the system to reach a specific state, possibly the probability distribution of this time

A Markov analysis is primarily a quantitative analysis technique, but the construction of the state diagram will often give increased system knowledge.

In the following sections we give some examples of Markov models. In Section D.6 we give a general description of a Markov process. First we consider some simple examples with two states only.

D.2 One component. No repair

Consider a component with lifetime T. Let state 1 represent the state in which the component is functioning and state 0 that the component has failed. Then the state of the component, as a function of time, can be represented by the curve shown in Figure D.1, where $X(t)$ denote the state of the component at time t, i.e.

$$X(t) = \begin{cases} 1 & \text{if } t < T \\ 0 & \text{if } t \geq T \end{cases}$$

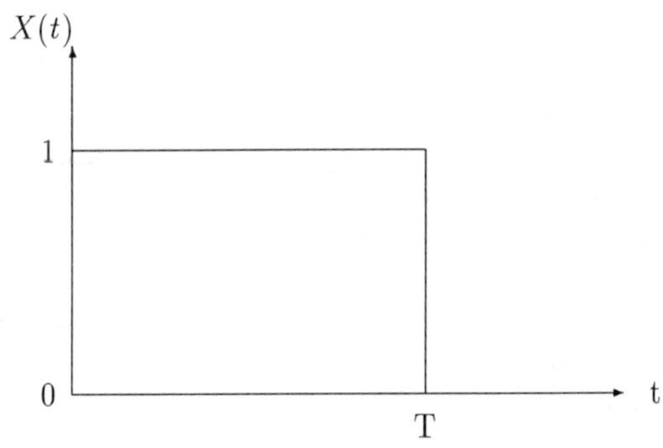

Figure D.1

Assume that the component has a constant failure rate λ. This means that if the component has survived t units of time, the probability that it fails before time $t + h$ is approximately equal to λh when h is small. We illustrate the process by the state diagram in Figure D.2.

The state diagram is interpreted as follows: As long as $X(t) = 1$, there is a possibility of a transition from state 1 to state 0, with rate λ.

The state process of the component, $X(t)$, as a function of time t, is an example of a simple Markov process. Such a process is characterized by no memory, i.e. if the process is in a specific state j at time t, the future states of the process will depend only on the state j at time t and not on "the rest of the history" of the component before time t.

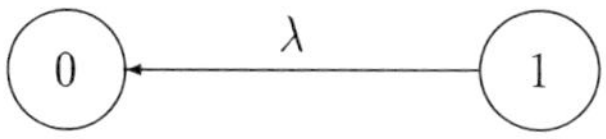

Figure D.2

Since the failure rate is equal to λ, we know that

$$P(T > t) = e^{-\lambda t}$$

i.e. T is exponentially distributed. The following shows how Markov techniques can be used to establish this result. Let

$$P_j(t) = P(X(t) = j), \quad \text{for } j = 0, 1$$

Then it follows by using the rule of total probability (A.10) that

$$\begin{aligned} P_1(t+h) = P(X(t+h) = 1) &= P(X(t+h) = 1 \mid X(t) = 1)\, P(X(t) = 1) \\ &+ P(X(t+h) = 1 \mid X(t) = 0)\, P(X(t) = 0) \end{aligned} \quad \text{(D.1)}$$

Here,

$$P(X(t) = 1 \mid X(t) = 0) = 0$$

since the component cannot function at time $t + h$ if it is in the failure state at time t. Furthermore,

$$\begin{aligned} P(X(t+h) = 1 \mid X(t) = 1) &= 1 - P(X(t+h) = 0 \mid X(t) = 1) \\ = 1 - P(T \le t+h \mid T > t) &= 1 - \lambda h + o(h) \end{aligned}$$

where $o(h)/h \to 0$ when $h \to 0$. Hence (D.1) can be written

$$P_1(t+h) = (1 - \lambda h)\, P_1(t) + o(h)$$

which can also be formulated in the following way

$$\frac{P_1(t+h) - P_1(t)}{h} = -\lambda\, P_1(t) + \frac{o(h)}{h}$$

By letting $h \to 0$, we obtain the following differential equation:

$$P_1'(t) = -\lambda\, P_1(t)$$

with the constraint $P_1(0) = 1$. The solution of this differential equation gives

$$P_1(t) = e^{-\lambda t}$$

D.3 One repairable component

Let the situation be as above, but assume now that the component is repaired at failure. The state of the component as a function of time, may then be represented by a curve of the form illustrated in Figure 3.6. Thus in this case we have

$$X(t) = \begin{cases} 1 & \text{if the component is functioning at time } t \\ 0 & \text{if the component is under repair at time } t \end{cases}$$

As above we assume that the component has a constant failure rate λ as long as it is in failure state 1, i.e.

$$P(X(t+h) = 0 \mid X(t) = 1) \;=\; \lambda h + o(h)$$

Hence the probability that the process makes a transition from state 1 to state 0 in the period from t to $t+h$, is approximately equal to λh.

It remains to model the repair time. In a Markov model this is done analogously to the modelling of the lifetime. This means that the repair times must be assumed to be exponentially distributed. More precisely, we introduce a parameter μ such that

$$P(X(t+h) = 1 \mid X(t) = 0) \;=\; \mu h + o(h)$$

Thus the repair time is exponentially distributed with parameter μ, i.e. the expected repair time equals $1/\mu$.

The state diagram for this process is given in Figure D.3.

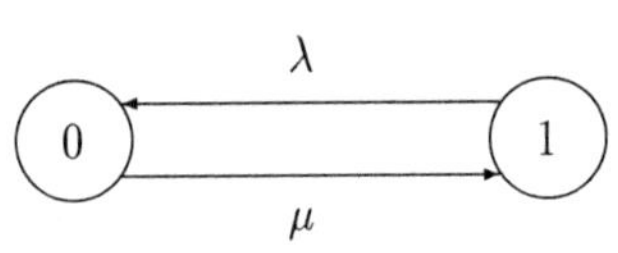

Figure D.3

Next we will study the problem of computing the probabilities $P_0(t)$ and $P_1(t)$. We return to (D.1), which now gives:

$$P_1(t+h) = (1-\lambda h)P_1(t) + \mu h P_0(t) + o(h)$$

Rewriting this expression we obtain

$$\frac{P_1(t+h) - P_1(t)}{h} = -\lambda P_1(t) + \mu P_0(t) + \frac{o(h)}{h} \tag{D.2}$$

It follows that

$$P_1'(t) = -\lambda\, P_1(t) + \mu P_0(t)$$

By using $P_0(t) = 1 - P_1(t)$ we can write this equation in the following form:

$$P_1'(t) + (\lambda + \mu)\, P_1(t) = \mu$$

which has the general solution

$$P_1(t) = \mu/(\lambda + \mu) \; + \; C\, e^{-(\lambda+\mu)t}$$

where C is a constant that has to be determined from a constraint on $P_1(t)$. If we start at time 0 with a functioning component, then $P_1(0) = 1$, which gives $C = \lambda/(\lambda + \mu)$. As a result we obtain the following probability that the component is functioning at time t:

$$P_1(t) = \mu/(\lambda + \mu) \; + \; \lambda/(\lambda + \mu)\, e^{-(\lambda+\mu)t} \qquad \text{(D.3)}$$

Hence formula (3.11) has been shown.

If the process $X(t)$ has run for a long time, i.e. t is large, then we see that $P_1(t)$ is approximately equal to $\mu/(\lambda + \mu) = MTTF/(MTTF + MTTR)$, with the notation of Section 3.4.1. Thus if we call the limiting values of $P_i(t)$ for P_i as $t \to \infty$, we can write

$$P_1 = \mu/(\lambda + \mu) \quad P_0 = 1 - P_1 = \lambda/(\lambda + \mu) \qquad \text{(D.4)}$$

This distribution is called the steady state distribution of the process. In many situations it is sufficient to compute P_1 and P_0, since we know that the process has run for a long time. It that case we could have found P_1 and P_0 directly without establishing the differential equation. By letting $t \to \infty$ in equation (D.2) it follows that

$$-\lambda P_1 + \mu P_0 = 0 \qquad \text{(D.5)}$$

which gives (D.4) using that $P_0 + P_1 = 1$.

The equation (D.5) may also be written

$$\lambda P_1 = \mu P_0$$

The two sides in this formula can be interpreted as the "expected (average) number of departures from state 1 per units of time" and the "expected (average) number of arrivals at state 1 per unit of time", respectively. This is explained as follows.

The expected number of transitions from state 1 to state 0 in the interval $(t, t+h)$ is approximately equal to the probability that a transition from state 1 to 0 occurs in this interval, which again is approximately equal to the probability that $X(t) = 1$ (i.e. P_1) multiplied by the probability of a transition from state 1 to 0 in $(t, t+h)$, in other words $P_1\lambda h$. It follows that the expected number of departures per unit of time equals $P_1\lambda$.

Similarly, the probability of a transition to state 1 in $(t, t+h)$ is approximately given by $P_0\mu h$, i.e. the expected number of arrivals per unit of time equals $P_0\mu$.

In a steady state the expected number of arrivals should be equal to the expected number of departures., i.e. (D.5) should hold.

D.4 Two repairable components

In this section we will generalize the model of the previous section by considering several components. The starting point is a system comprising two components. For a system with more than two components we can use a similar approach as the one described below for two components.

We assume that component i ($i = 1, 2$) has a failure rate and a repair rate given by λ_i and μ_i, respectively. Furthermore, we assume in the first part of this section that the components fail and are repaired independently of each other.

For each component we can now construct a state diagram as in Figure D.3. If we combine the states of the components, we obtain, however, a state description with four states as shown in Table D.1.

Table D.1: System states

State	Comp. 2	Comp. 1	Interpretation
0	0	0	1 and 2 down
1	0	1	1 up, 2 down
2	1	0	1 down, 2 up
3	1	1	1 and 2 up

The state diagram for the system is shown in Figure D.4. We see from the figure that the system makes a transition from state 3 to state 1 as a result of the failure of component 2. Similarly, the system returns to state 3 as a result of the completed repair of component 2. The transition rate from state 3 to state 1 equals λ_2, which corresponds to the failure rate of component 2, whereas the transition rate from state 1 to state 3 equals μ_2, i.e. the repair rate of component 2.

If the system state at time t is 3, then it will be at time $t + h$:

1 with probability approximately equal to $\lambda_2 h$
2 with probability approximately equal to $\lambda_1 h$
3 with probability approximately equal to $1 - (\lambda_1 + \lambda_2)h$

and similarly for the other states.

From the assumption of independence we can conclude that

$$P(X_1(t) = x_1, X_2(t) = x_2) = P(X_1(t) = x_1)P(X_2(t) = x_2)$$

where $X_i(t)$ represents the state process of component i.

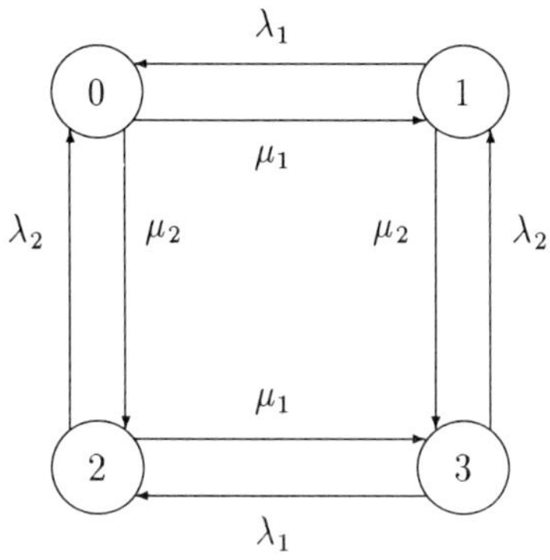

Figure D.4

The probabilities $P_j(t) = P(X(t) = j), j = 0, 1, 2, 3$, and the limits as t approaches infinity, can be calculated by using the results of Section D.3. We will therefore not go further into this. Instead we will show how the state diagram in Figure D.4 can be modified to take into account dependency between components.

The assumption of independence implies that the components have a constant failure rate regardless of the states of the other components. In some cases the load will increase on for example component 1 if component 2 fails. The failure rate for a transition from state 1 to state 0 will then be greater than for the transition from state 3 to 1. Such an extension of the model can therefore easily be carried out.

The assumption of independence also implies that the components are repaired as they fail. In some situations it might be appropriate to repair the components simultaneously when the system reaches state 0. In that case instead of transitions from state 0 to state 1, and from state 0 to state 2, direct transitions from state 0 to state 3 are obtained.

Since two independent components cannot fail at the same time, there are no transitions from state 3 to state 0 when assuming independent components. If the system is exposed to common mode failures, a transition directly from state 3 to state 1 will be possible. Common mode failures can therefore easily be modelled by including such a transition in the state diagram.

As an example, we will compute the steady state probabilities for a system with components that fail independently of each other with the same failure rate λ. The components may also have a common mode failure, the failure rate is ρ. Hence the total failure rate for each component equals $\lambda + \rho$. Note that we assume that the common mode failure may also occur when one of the components is down. We assume also that repair can only be carried out when both components are in the failure state, and that the components then are repaired simultaneously with a rate of μ. The state diagram of the process is shown in Figure D.5

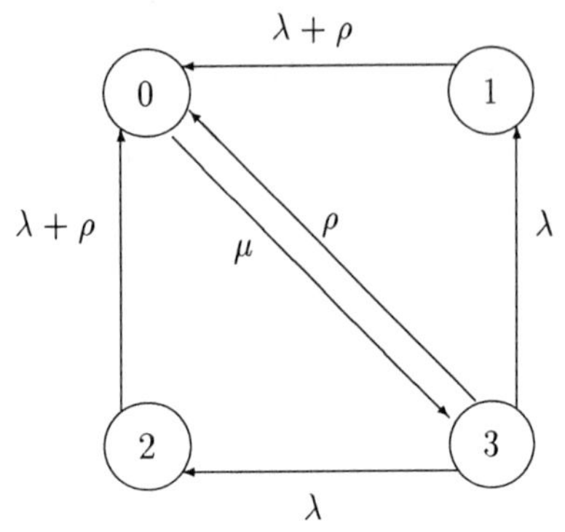

Figure D.5

The steady state distribution is found as explained in Section D.3. For each state the expected number of departures per unit of time is set equal to the expected number of arrivals per unit of time:

$$\begin{array}{lrcl} & \underline{\text{Departures}} & & \underline{\text{Arrivals}} \\ \text{State 3:} & (2\lambda+\rho)P_3 & = & \mu P_0 \\ \text{State 2:} & (\lambda+\rho)P_2 & = & \lambda P_3 \\ \text{State 1:} & (\lambda+\rho)P_1 & = & \lambda P_3 \\ \text{State 0:} & \mu P_0 & = & \rho P_3 + (\lambda+\rho)(P_1+P_2) \end{array}$$

This gives the following steady state distribution:

$$\begin{aligned} P_3 &= (\lambda+\rho)\mu/S \\ P_2 &= \lambda\mu/S \\ P_1 &= P_2 \\ P_0 &= (\lambda+\rho)(2\lambda+\rho)/S \end{aligned}$$

where

$$S = (\lambda+\rho)(2\lambda+\rho) + \mu(\lambda+\rho) + 2\lambda\mu$$

If the components are in parallel, the states 3, 2 and 1 will be "up-states", whereas state 0 will be a "down-state". Hence the probability that the system is up at an arbitrary point in time equals:

$$P_3 + P_2 + P_1 = 1 - P_0$$

This probability may also be interpreted as the average portion of time the system is up.

D.5 Birth and death processes

We consider a system that can be in one out of $n+1$ states: $0, 1, 2, \ldots, n$. Let q_{ij} denote the transition rate from state i to state j. We assume that

$$\begin{aligned} q_{i,i+1} &= \lambda_i \quad i = 0, 1, \ldots, n-1 \\ q_{i,i-1} &= \mu_i \quad i = 1, 2, \ldots, n \\ q_{ij} &= 0 \quad \text{for } \mid i-j \mid > 1 \end{aligned}$$

The process is called a birth and death process, with birth rate λ_i and death rate μ_i. We will focus on the problem of computing the steady state distribution for this process. By setting the expected number of departures per unit of time equal to the expected number of arrivals per unit of time, we obtain the following equations:

$$\begin{aligned} -\lambda_0 P_0 + \mu_1 P_1 &= 0 \\ \lambda_{i-1} P_{i-1} - (\lambda_i + \mu_i) P_i + \mu_{i+1} P_{i+1} &= 0 \quad , \quad i = 1, \ldots, n-1 \\ \lambda_{n-1} P_{n-1} - \mu_n P_n &= 0 \end{aligned}$$

This gives

$$\begin{aligned} P_i &= \frac{\prod_{j=0}^{i-1} \lambda_j}{\prod_{j=1}^{i} \mu_j} P_0 \equiv r_i P_0, \quad i = 1, 2, \ldots, n \qquad \text{(D.6)} \\ P_0 &= \frac{1}{1 + \sum r_i} \end{aligned}$$

see Problem D.7.6. In Section 3.4.3 some formulae were established, which are based on the above steady state probabilities, formulae (3.16) and (3.17). Below the proof of these formulae are presented.

Formula (3.16)
In addition to the assumptions made in Section 3.4.3 it is assumed that the components have repair rates equal to $1/MTTR$. It then follows that the process given by the number of components that are down (not functioning), is a birth and death process with

$$\lambda_i = \begin{cases} (n-1)\lambda & i = 0 \\ (n-i)\lambda & i = 1, 2, \ldots, n \end{cases}$$

and

$$\mu_i = u_i / MTTR \quad i = 1, 2, \ldots, n$$

The steady state distribution for these birth and death rates is given by the formula (3.16).

Formula (3.17)
It is sufficient to show that S_i^* has the form given by (3.17), since S_i is expressed by (3.16). We assume as above that the components have repair rates that are constant, equal to $1/MTTR$. It follows that the process defined by the number of components that are down, is a birth and death process, with

$$\begin{aligned} \lambda_i &= (n-1-i)\lambda, \quad i = 0, 1, \ldots, n-1 \\ \mu_i &= v_i/MTTR, \quad i = 1, 2, \ldots, n-1 \end{aligned}$$

The expression for S_i^* follows.

D.6 General formulation of a Markov process

So far we have considered examples of Markov models. In this section we will describe a general Markov process and state some results that apply for such processes. For rigorous proof of these results, refer to textbooks in the field.

Let $X(t)$ denote the state of the system at time t. We assume that $X(t)$ can be in one out of $n+1$ states, which we denote $0, 1, \ldots, n$. Let

$$P_i(t) = P(X(t) = i)$$

The stochastic process $X(t), t \geq 0$, is a Markov process if it satisfies the following property (the Markovian property):

> If the process is in a specific state i at time t, then the future states of the system will depend on the state of the system at time t and not of the states the system has been in before time t. Mathematically, this can be expressed in the following way:
>
> $$P(X(s) = j \mid X(t) = i, X(u) = x(u), 0 \leq u < t) = P(X(s) = j \mid X(t) = i), s > t$$

The conditional probabilities $P(X(s) = j \mid X(t) = i)$ are called the transition probabilities in the Markov process. Here we will limit ourselves to Markov processes with *stationary* transition probabilities, i.e. the transition probabilities $P(X(t+v) = j \mid X(t) = i)$ do not depend on t. All processes considered in this appendix have been Markov processes with stationary transition probabilities. This class of processes also includes the homogeneous Poisson process.

We define transition rates, $q_{ij}, i \neq j$, by

$$q_{ij} = \lim_{h \to 0} \frac{P(X(t+h) = j \mid X(t) = i)}{h}$$

Let $\mathbf{Q}$ denote the matrix with elements $q_{ij}, i \neq j$, and diagonal elements q_{jj} given by

$$q_{jj} = - \sum_{k=0,k\neq j}^{n} q_{jk}$$

We call $\mathbf{Q}$ for the intensity matrix of the process.

Let $\mathbf{P}(\mathrm{t})$ denote the matrix with elements

$$P_{ij}(t) = P(X(s+t) = j \mid X(s) = i)$$

and $\mathbf{P}'(t)$ the matrix with elements $P'_{ij}(t)$. Then it can be shown that:

$$\mathbf{P}'(t) = \mathbf{P}(t)\mathbf{Q} = \mathbf{Q}\mathbf{P}(t) \tag{D.7}$$

We will sketch the proof of these equations. Using the Markovian property, it is not difficult to see that the Chapman-Kolmogorov relation holds:

$$P_{ik}(t+s) = \sum_{j=0}^{n} P_{ij}(t)\, P_{jk}(s)$$

which can be written in matrix notation as

$$\mathbf{P}(t+s) = \mathbf{P}(t)\,\mathbf{P}(s) \tag{D.8}$$

Let $h > 0$ and let $\mathbf{I}$ denote the identity matrix whose elements are one on the diagonal and 0 otherwise. Then

$$\lim_{h\to 0} \frac{\mathbf{P}(h) - \mathbf{I}}{h} = \mathbf{Q}$$

Using this formula and referring to (D.8) we have

$$\frac{\mathbf{P}(t+h) - \mathbf{P}(t)}{h} = \frac{\mathbf{P}(t)[\mathbf{P}(h) - \mathbf{I}]}{h} = \frac{\mathbf{P}(h) - \mathbf{I}}{h}\mathbf{P}(t)$$

Letting $h \to 0$ leads to the differential equations (D.7). To be precise, we should write the right-hand derivative. However with a little more care we could derive the same relation involving the left-hand derivative, see Problem D.7.7. The equations (D.7) can be solved using various methods, see e.g. [79].

For a process that has run for a long time, where it can make transitions between all the states, the probability that it will be in a specific state will not depend very much on the initial state of the process. The process converges towards one where the probability of being in state j becomes

$$P_j = \lim_{t\to\infty} P_{ij}(t)$$

These asymptotic probabilities are called the steady state probabilities of the process. By letting $t \to \infty$ in (D.7) it is seen that these steady state probabilities satisfy the following equation:

$$\mathbf{pQ} = \mathbf{0} \tag{D.9}$$

where $\mathbf{p} = (P_0, P_1, \ldots, P_n)$ and $\mathbf{0} = (0, 0, \ldots, 0)$. To find the P_is we make use of n of the $n+1$ equations in (D.9), and the fact that the sum of the steady state probabilities equals 1.

For a Markov process with transition rates q_{ij} the sojourn time in each state j is exponentially distributed with parameter $-q_{jj}$. At the moment of a transition from state j, the process will jump to state i with a probability equal to $-q_{ji}/q_{jj}$.

D.7 Problems

1. Consider the Markov process given by Figure D.5. Assume that $\lambda = 1$, $\rho = 0.2$ and $\mu = 100$. Compute the steady state distribution of the process in this case.

2. We will look closer at the Markov process given by Figure D.4, but we assume now that there is some dependency between the two components: if component 2 fails, the failure rate of component 1 increases to $2\,\lambda_1$. Similarly the failure rate of component 2 increases to $2\,\lambda_2$ if component 1 fails.

   a) Find the intensity matrix for this process.

   b) Compute the steady state distribution of the process.

3. A unit has two failure modes, which are denoted 1 and 2. At failures the unit is repaired. Thus the unit can be in one out of three states:

   3: The functioning state

   2: Under repair due to failure 2

   1: Under repair due to failure 1

   The transition rate from state 3 to state i, $i = 1, 2$, equals λ_i. Let $MTTR_i$ denote the mean repair time for failure mode i, $i = 1, 2$.

   a) Specify what is required to ensure that the state process shall be a Markov process with stationary transition probabilities. Assume in the rest of this problem that the state process is such a process.

   b) Construct the state diagram with its associated transition matrix for this process.

   c) Show that the steady state distribution equals:

   $$\begin{aligned} P_3 &= 1/(1 + q_1 + q_2) \\ P_2 &= q_2 P_3 \\ P_1 &= q_1 P_3 \end{aligned}$$

   where

   $$q_i = \lambda_i \, MTTR_i$$

   Comment on the result.

d) Generalize the model by including an arbitrary number of failure modes.

4. We consider the process given by the state diagram in Figure D.4. Let m_i denote the mean time until the process visits state 0, given that it starts in state i at time $t = 0$. Then the following equations hold:

$$
\begin{aligned}
m_3 &= \frac{1}{\lambda_1+\lambda_2} + \frac{\lambda_2}{\lambda_1+\lambda_2} m_1 + \frac{\lambda_1}{\lambda_1+\lambda_2} m_2 \\
m_2 &= \frac{1}{\lambda_2+\mu_1} + \frac{\mu_1}{\lambda_2+\mu_1} m_3 \\
m_1 &= \frac{1}{\lambda_1+\mu_2} + \frac{\mu_2}{\lambda_1+\mu_2} m_3
\end{aligned}
$$

   a) Show that these equations hold. Use the results stated in the last paragraph of Section D.6.

   b) Find the m_is by solving the equations. Give numerical values to λ_i and μ_i.

5. Again we look at the state diagram of Figure D.4. The problem is now to find the distribution of time until both components are down, i.e. the process is in state 0, given that both components are functioning at time 0.

   Explain how we can find this distribution by putting the departure intensities for state 0 equal to 0, i.e. the process is *absorbed* in state 0. Generalize the method.

6. Prove formula (D.6).

7. Show that formula (D.7) also holds for the left-hand derivative.

Appendix E

MORE ON STOCHASTIC PROCESSES

In this appendix we present some definitions and results from the theory of renewal processes, renewal reward processes, regenerative processes and counting processes, based on [3,24,35,60,116]. Some simple proofs are included. The definitions and results are used to prove some results of Chapter 3 and 5.

The purpose of this appendix has not been to give an all-inclusive presentation of the stochastic process theory. Only definitions and results needed for establishing the results of Chapters 3 and 5 are covered.

E.1 Renewal processes

Let $T_j, j = 1, 2, \ldots$, be a sequence of non-negative independent random variables with a common distribution F. To avoid trivialities, we assume that $P(T_j = 0) < 1$. From the non-negativity of T_j, it follows that ET_j exists, though it may be infinite, and we denote

$$ET = ET_j = \int_0^\infty P(T_j > t)\, dt$$

Let

$$S_0 = 0 \quad S_j = \sum_{i=1}^{j} T_i, \quad j \geq 1$$

and define

$$N(t) = sup\{j : S_j \leq t\}$$

The process $N(t), t \geq 0$, is a *Renewal Process*. We say that a renewal occurs at t if $S_j = t$ for some j. The random variable $N(t)$ represents the number of renewals in $[0, t]$. Since the *inter-arrival times* T_j are independent and identically distributed, it follows that after each renewal the process starts over again.

Let $M(t) = EN(t)$, $0 \leq t < \infty$. The function $M(t)$ is called the renewal function. It can be shown that $M(t)$ is finite for all t.

Result E.1
With probability one,

$$\frac{N(t)}{t} \to \frac{1}{ET} \quad \text{as} \quad t \to \infty$$

Proof
By definition of $N(t)$, it follows that

$$S_{N(t)} \leq t \leq S_{N(t)+1}$$

Hence,

$$\frac{S_{N(t)}}{N(t)} \leq \frac{t}{N(t)} \leq \frac{S_{N(t)+1}}{N(t)}$$

Now the strong law of large numbers states that with probability one, $S_j/j \to ET$ as $j \to \infty$. Since $N(t) \to \infty$ as $t \to \infty$, we obtain

$$\frac{S_{N(t)}}{N(t)} \to ET \quad \text{as} \quad t \to \infty \qquad \text{(with probability one)}$$

By the same argument, we also see that with probability one,

$$\frac{S_{N(t)+1}}{N(t)} = \frac{S_{N(t)+1}}{N(t)+1} \frac{N(t)+1}{N(t)} \to ET \cdot 1 = ET \quad \text{as} \quad t \to \infty$$

The result follows.

Result E.2 - The Elementary Renewal Theorem

$$\frac{M(t)}{t} \to \frac{1}{ET} \quad \text{as} \quad t \to \infty$$

Result E.3 - The Key Renewal Theorem
Assume that the distribution of T_j is not periodic, i.e. a discrete variable with values $\{c, 2c, 3c, \ldots\}$ for some c. If $G(t)$ is a bounded, non-negative and non-increasing function, such that $\int_0^\infty G(t)\,dt < \infty$, then

$$\lim_{t\to\infty} \int_0^t G(t-s)\,dM(s) = \frac{1}{ET} \int_0^\infty G(s)\,ds$$

Result E.4 - Blackwell's Theorem
For a renewal process with a non-periodic distribution of T_j we have

$$\lim_{t\to\infty} [M(t) - M(t-s)] = \frac{s}{ET}$$

Result E.5 - The Central Limit Theorem for Renewal Processes
Assume that $\mu = ET < \infty$ and $\sigma^2 = VarT < \infty$. Then if $t \to \infty$, $N(t)$ has an asymptotically normal distribution with mean t/μ and variance $t\sigma^2/\mu^3$. Thus

$$\lim_{t\to\infty} P(\frac{N(t) - t/\mu}{\sqrt{t\sigma^2/\mu^3}} < x) = \frac{1}{\sqrt{2\pi}} \int_{-\infty}^{x} e^{-\frac{1}{2}u^2}\, du$$

If the distribution of T_j is non-periodic, then

$$\lim_{t\to\infty} [M(t) - \frac{t}{\mu}] = \frac{\sigma^2 - \mu^2}{2\mu^2}$$
$$\lim_{t\to\infty} \frac{VarN(t)}{t} = \frac{\sigma^2}{\mu^3}$$

E.2 Renewal reward processes

Suppose that a "reward" ("cost") Y_j is associated with the ith renewal. The random variable Y_j may depend on T_j, but we suppose that the pairs (T_j, Y_j), $j = 1, 2, \ldots$, are independent and identically distributed. Let EY denote the mean of the Y_js. Let $Y(t)$ denote the total reward earned by time t. We see that if the reward is earned at the time of the renewal,

$$Y(t) = \sum_{j=1}^{N(t)} Y_j$$

If all returns are non-negative, then

$$\sum_{j=1}^{N(t)} Y_j \le Y(t) \le \sum_{j=1}^{N(t)+1} Y_j$$

The limiting value of the average return is given by the following result.

Result E.6
If $E|Y|$ and ET are finite, then

(i) with probability 1,

$$\frac{Y(t)}{t} \to \frac{EY}{ET} \quad \text{as } t \to \infty$$

(ii)

$$\frac{EY(t)}{t} \to \frac{EY}{ET} \quad \text{as } t \to \infty$$

Proof. (i) Assume that the returns are all non-negative. Then

$$\frac{\sum_{j=1}^{N(t)} Y_j}{t} \leq \frac{Y(t)}{t} \leq \frac{\sum_{j=1}^{N(t)+1} Y_j}{t}$$

Writing

$$\frac{\sum_{j=1}^{N(t)} Y_j}{t} = \frac{\sum_{j=1}^{N(t)} Y_j}{N(t)} \frac{N(t)}{t}$$

and using the strong law of large numbers and Result E.1, it follows that $\sum_{j=1}^{N(t)} Y_j/t$ converges to EY/ET. Similarly, it is shown that $\sum_{j=1}^{N(t)+1} Y_j/t$ converges to EY/ET. The result follows.

A similar argument holds when the returns are non-positive, and the general case follows by breaking up the returns into their positive and negative parts and applying the above argument separately to each.
(ii) See [116].

Many results from renewal theory can be generalized to renewal reward processes. For example Blackwell's Theorem holds:

$$\lim_{t\to\infty}[Y(t) - Y(t-s)] = \frac{sEY}{ET}$$

E.3 Regenerative processes

Consider a stochastic process $X(t), t \geq 0$, with state space $\{0, 1, 2, \ldots\}$, having the property that there exist time points at which the process (probabilistic) restarts itself. That is, suppose that with probability one, there exists a time T_1, such that the continuation of the process beyond T_1 is a probabilistic replica of the whole process starting at 0. Note that this property implies the existence of further times $T_2, T_3, \ldots$, having the same property as T_1. Such a stochastic process is known as a *regenerative process.*

From the above it follows that $T_1, T_2, \ldots$, forms a renewal process; and we shall say that a cycle is completed every time a renewal occurs.
Let

$$P_k(t) = P(X(t) = k)$$

for all $k \geq 0$.

Result E.7
If T_1 has an absolutely continuous component, i.e. has a density on some interval, and $ET_1 < \infty$, then

$$\lim_{t\to\infty} P_k(t) = \frac{E[\text{amount of time in state } k \text{ during one cycle}]}{E[\text{time of one cycle}]} = \frac{E\int_0^{T_1} I(X(t) = k)\,dt}{ET_1}$$

for all $k \geq 0$, where $I(\cdot)$ is the indicator function, which equals 1 if the argument is true and 0 otherwise.

Result E.8
If $\lim_{t\to\infty} P_k(t) = P_k$ exists, then

$$\lim_{t\to\infty} \frac{1}{t} E \int_0^t I(X(s) = k)\, ds \;=\; P_k$$

Remark
We see that the quantity $(1/t)E\int_0^t I(X(s) = k)\, ds$ represents the expected portion of time the process is in state k in $[0, t]$. Since

$$\frac{1}{t} E \int_0^t I(X(s) = k)\, ds = \frac{1}{t} \int_0^t EI(X(s) = k)\, ds = \frac{1}{t} \int_0^t P_k(s)\, ds$$

this quantity is also equal to the average probability that the process is in state k.

Proof of Result E.8
Given $\epsilon > 0$, there exists t_0 sufficiently large so that $|P_k(t) - P_k| < \epsilon$ for all $t > t_0$. For $t > t_0$ it follows that

$$\frac{1}{t} \int_0^t P_k(s)\, ds \;\leq\; \frac{1}{t} \int_0^{t_0} P_k(s)\, ds \;+\; \frac{1}{t} \int_{t_0}^t (P_k + \epsilon)\, ds$$

Thus since t_0 is fixed, we have

$$\limsup_{t\to\infty} \frac{1}{t} \int_0^t P_k(s)\, ds \;\leq\; P_k + \epsilon$$

Similarly, we obtain

$$\liminf_{t\to\infty} \frac{1}{t} \int_0^t P_k(s)\, ds \;\geq\; P_k - \epsilon$$

Since ϵ is arbitrary, the proof is complete.

E.4 Counting processes

A multivariate counting process $(N_1(t), N_2(t), \ldots, N_n(t))$, $t \geq 0$, is an n-dimensional stochastic process with components $N_j(t)$ which have sample paths which are non-decreasing, right-continuous step functions, 0 at time 0, and with jumps of unit size. Moreover it is assumed that the processes $N_j(t)$s are finite and have no common jumps.

The processes $N_j(t)$ can be thought of as counting the occurrence, as time t proceeds, of n different types of events, $N_j(t)$ being the number of type j events in the time interval $[0, t]$.

The development in time of the multivariate counting process is governed by its *intensity process* $(\lambda_1(t), \lambda_2(t), \ldots, \lambda_n(t))$, $t \geq 0$, which is given as follows. Consider a small interval $(t, t+h)$. Then $\lambda_j(t)h$ is the conditional probability that the process $N_j(t)$ jumps in the interval $(t, t+h)$, given all that has happened till time t. If we let $\mathcal{F}_t$ denote everything that has happened up to t, then we can write informally

$$\frac{E(N_j(t+h) - N_j(t) \,|\, \mathcal{F}_t)}{h} = \lambda_j(t) \tag{E.1}$$

The history $\mathcal{F}_t$ includes a complete specification of the path of the multivariate counting process in the time interval $[0, t]$ as well as all other events implicitly or explicitly included in the model which have happened in the interval $[0, t]$. As a consequence we have $\mathcal{F}_s \subseteq \mathcal{F}_t$, whenever $s \leq t$, reflecting the fact that as time proceeds more and more is learnt about the process.

Mathematically, $\mathcal{F}_t$, $t \geq 0$, given on a probability space $(\Omega, \mathcal{F}, P)$, is a non-decreasing family of sub-σ-fields of $\mathcal{F}$ such that $\mathcal{F}_t$, $t \geq 0$, is right-continuous and $\mathcal{F}_0$ includes all null sets. The processes $N_j(t)$ are adapted to $\mathcal{F}_t$, meaning that $N_j(t)$ is $\mathcal{F}_t$-measurable for all t. In a formal mathematical framework, the intensity concept (E.1) can be defined in the following way.

Definition E.1
Let $\lambda_j(t)$ be a progressively measurable process (for example a right-continuous or left-continuous process) adapted to $\mathcal{F}_t$ such that

$$\int_0^t \lambda_j(s)\, ds < \infty$$

If for all non-negative predictable processes $C(t)$ adapted to $\mathcal{F}_t$, the equality

$$E \int_0^\infty C(t)\, dN_j(t) = E \int_0^\infty C(t)\, \lambda_j(t)\, dt \tag{E.2}$$

holds, then we say that $N_j(t)$ admits the $\mathcal{F}_t$-intensity $\lambda_j(t)$.

A predictable process is a process whose value at time t is known just before time t. If a process is adapted and has left-continuous sample paths, then it is predictable. Moreover, any measurable (for example a left- or right-continuous) deterministic function is predictable.

Definition E.2
If for all $0 \leq s \leq t$ and $k = 0, 1, 2, \ldots$,

$$P(N_j(t) - N_j(s) = k \,|\, \mathcal{F}_s) = e^{-\int_s^t \lambda_j(u)\, du} \frac{(\int_s^t \lambda_j(u)\, du)^k}{k!}$$

where $\lambda_j(t)$ is a measurable, deterministic function with $\int_0^t \lambda_j(u)\,du < \infty$ for all $t < \infty$, we call N_j a (non-homogeneous) $\mathcal{F}_t$-Poisson process with intensity $\lambda_j(t)$.

Remark. If N_j is an (non-homogeneous) $\mathcal{F}_t$-Poisson process with intensity $\lambda_j(t)$, then $N_j(t)$ admits the $\mathcal{F}_t$-intensity $\lambda_j(t)$.

Result E.9
Assume that $N_j(t)$ admits the $\mathcal{F}_t$-intensity $\lambda_j(t)$, where $\lambda_j(t)$ is a non-negative, deterministic function, $j = 1, 2, \ldots, n$. Then the $N_j(t)$ processes are independent $\mathcal{F}_t$-Poisson processes with the intensities $\lambda_j(t)$, $j = 1, 2, \ldots, n$, respectively.

Result E.10
Assume that $N_j(t)$ admits the $\mathcal{F}_t$-intensity $\lambda_j(t)$, $j = 1, 2, \ldots, n$. Then the counting process

$$N(t) = \sum_{j=1}^{n} N_j(t)$$

admits the $\mathcal{F}_t$-intensity $\sum_{j=1}^{n} \lambda_j(t)$.

Finally in this section we define a stopping time.

Definition E.3. A $[0, \infty]$-valued random variable T is called an $\mathcal{F}_t$-*stopping time* if

$$\{T \le t\} \in \mathcal{F}_t \quad \text{for all} \quad t \ge 0$$

This means that the event "the stopping time is less than or equal to t" is determined by the history of the process up to time t.

In the stochastic models considered in this book it is tacitly understood that we have given a probability space $(\Omega, \mathcal{F}, P)$ and a history $\mathcal{F}_t$, which the stochastic processes are adapted to.

E.5 Proofs of some results in Chapters 3 and 5

E.5.1 Section 3.4.1

Consider the model presented in Section 3.4.1. The model is a special case of the model described in Sections E.1 - E.3, with regenerative point $T_1 = T_{i1} + D_{i1}$, a "reward" $Y_1 = T_{i1}$ and $k = 1$. Note that in this case we have

$$\int_0^{T_1} I(X(t) = k)\,dt = Y_1$$

so that the limits in Results E.6, E.7 and E.8 are identical. The results of Section 3.4.1 follow.

Proof of (3.12) [24]
Let $N_i^*(t)$ denote the renewal process generated by the subsequent operation periods T_{ij}, $j = 1, 2, \ldots$, and let $U(t)$ and $D(t)$ denote the accumulated operating time and downtime by time t, respectively. It is seen that the number of failures of component i in $[0, t]$, $N_i(t)$, is given by

$$N_i(t) \equiv N_i^*(U(t))$$

Using result E.1 and observing that with probability one $U(t) \to \infty$ as $t \to \infty$, we can conclude that with probability one,

$$\lim_{t\to\infty} \frac{N_i^*(U(t))}{U(t)} = \frac{1}{MTTF_i} \tag{E.3}$$

We may now derive the desired result. The proof is based on the following inequalities:

$$\sum_{i=1}^{n} \sum_{j=1}^{N_i(t)-1} D_{ij} \leq D(t) \leq \sum_{i=1}^{n} \sum_{j=1}^{N_i(t)} D_{ij}$$

The left-hand inequality results from the fact that none of the final downtime beginning before time t is included; the right-hand inequality results from the fact that the final downtime beginning before time t may actually extend beyond time t.

Since $U(t) + D(t) \doteq t$,

$$\frac{U(t)}{t} = [1 + \frac{D(t)}{U(t)}]^{-1} \geq [1 + \sum_{i=1}^{n} \frac{1}{N_i(t)} \sum_{j=1}^{N_i(t)} D_{ij} \frac{N_i^*(U(t))}{U(t)}]^{-1} \tag{E.4}$$

By the strong law of large numbers,

$$\frac{1}{N_i(t)} \sum_{j=1}^{N_i(t)} D_{ij} \to MTTR_i$$

with probability one. Using this fact, together with (E.3), it follows that the following inequality holds with probability one:

$$\liminf_{t\to\infty} \frac{U(t)}{t} \geq \frac{1}{1+\sum_{i=1}^n \frac{MTTR_i}{MTTF_i}}$$

Similarly, it is proved that

$$\limsup_{t\to\infty} \frac{U(t)}{t} \leq \frac{1}{1+\sum_{i=1}^n \frac{MTTR_i}{MTTF_i}}$$

Thus with probability one,

$$\lim_{t\to\infty} \frac{U(t)}{t} = \frac{1}{1+\sum_{i=1}^n \frac{MTTR_i}{MTTF_i}}$$

Using this result and Lebesgue dominated convergence theorem we can also conclude that

$$\lim_{t\to\infty} \frac{EU(t)}{t} = \frac{1}{1+\sum_{i=1}^n \frac{MTTR_i}{MTTF_i}}$$

Proof of (3.13) and (3.14)

To prove (3.13), observe that

$$EN(t) = E\sum_{i=1}^n \int_0^t [\phi(1_i, \mathbf{X}(s)) - \phi(0_i, \mathbf{X}(s))]dN_i(s)$$

where $N_i(t)$ represents the number of failures of component i in $[0,t]$. This equality is established noting that with probability one a system failure coincides with a failure of a component that is critical for the system, i.e. the system functions if and only if the component functions. Now using this expression and the independency between the processes $N_i(t)$ and $\phi(1_i, \mathbf{X}(t)) - \phi(0_i, \mathbf{X}(t))$, we find that

$$\begin{aligned} EN(t) &= E\sum_{i=1}^n \int_0^t [\phi(1_i, \mathbf{X}(s)) - \phi(0_i, \mathbf{X}(s))]dEN_i(s) & \text{(E.5)} \\ &= \sum_{i=1}^n \int_0^t E[\phi(1_i, \mathbf{X}(s)) - \phi(0_i, \mathbf{X}(s))]dEN_i(s) & \text{(E.6)} \end{aligned}$$

For a formal proof of (E.5), see e.g. [19]. The equality (E.6) follows from Fubini's Theorem. Note that if λ_i is the failure rate of component i, then by (E.2),

$$EN_i(t) = E\int_0^t X_i(s)\lambda_i\, ds = \int_0^t \lambda_i A_i(s)\, ds$$

Now by dividing by t and using The Elementary Renewal Theorem, the result follows noting that $E[\phi(1_i, \mathbf{X}(t)) - \phi(0_i, \mathbf{X}(t))] \to [h(1_i, \mathbf{A}) - h(0_i, \mathbf{A})]$ as $t \to \infty$. The formula (3.14) is seen to hold, writing

$$E[N(u+s) - N(u)] = \sum_{i=1}^n \int_u^{u+s} E[\phi(1_i, \mathbf{X}(t)) - \phi(0_i, \mathbf{X}(t))]dEN_i(t)$$

and using Blackwell's Theorem for renewal reward processes.

Assuming that the process ϕ has regenerative points, then it follows from the theory of renewal reward processes (Result E.6) that with probability one, $\lim_{t\to\infty} N(t)/t$ exists and equals the limit of (3.13). Thus, these results hold if for example all lifetime distributions are exponential.

If the counting process $N_i(t)$ has an intensity process $\lambda_i(t)$, then it can be shown that

$$E[N(u+s) - N(u)] = \sum_{i=1}^{n} \int_{u}^{u+s} E\{[\phi(1_i, \mathbf{X}(t)) - \phi(0_i, \mathbf{X}(t))]\lambda_i(t)\}\, dt \tag{E.7}$$

This result is not restricted to independent components, the intensity may depend on the history of all components up to time t. The formula (E.7) follows from (E.2) by introducing a predictable (left-continuous) modification of $\phi(1_i, \mathbf{X}(t)) - \phi(0_i, \mathbf{X}(t))$.

E.5.2 Section 3.4.2

We will show that the equalities of (3.15) hold. The model considered in Section 3.4.2 is a special case of the model described in Sections E.2 and E.3 above, with regenerative point $L_i + R_i + W_i$, a "reward" $Y_1 = L_i$ and $k = 1$. As in the previous section, the limits in Results E.6, E.7 and E.8 are identical. The first two equalities of (3.15) follow, observing that the conditions of Result E.7 are satisfied.

It remains to show the third equality of (3.15). Consider the non-homogeneous Poisson process $N_i(t)$, $t \geq 0$, representing the number of failures when no replacement is performed. Now suppose we flip a coin at the time of each failure time of N_i. We designate the outcome at time t by $Z(t)$, which takes the value one (head) with probability $\alpha_i(t)$ and the value zero (tail) with probability $\beta_i(t)$, independent of the history up to time t. Let $N_i^{(\alpha)}(t)$ be the modified process which counts all the failure times with associated "head", i.e.

$$N_i^{(\alpha)}(t) = \sum_{j=1}^{N_i(t)} Z(U_j)$$

where $U_1, U_2, \ldots$, are the successive failure (arrival) times of the process $N_i(t)$. The modified process $N_i^{(\alpha)}(t)$ is a non-homogeneous Poisson process with intensity function $\alpha_i(t)\lambda_i(t)$. A rigorous proof of this is given below. It is, however, easy to give a heuristic proof: given the history of the process up to time t, the expected number of jumps of the process in the time interval $[t, t+h)$ is approximately $\lambda_i(t)h\alpha_i(t)$, observing that a jump of the process $N_i^{(\alpha)}(t)$ corresponds to a jump of the process $N_i(t)$ and the coin shows heads.

Similarly, we introduce a non-homogeneous Poisson process $N_i^{(\beta)}(t)$, which counts all the failure times with associated "tail". Clearly,

$$N_i^{(\alpha)}(t) + N_i^{(\beta)}(t) = N_i(t)$$

The processes $N_i^{(\alpha)}(t)$ and $N_i^{(\beta)}(t)$ are independent, see proof below. It is not difficult to see that for $t < T_i$,

$$P(L_i > t) = P(N_i^{(\alpha)}(t) = 0) = e^{-\int_0^t \alpha_i(s)\lambda_i(s)\,ds}$$

and

$$P(L_i = T_i) = P(N_i^{(\alpha)}(T_i) = 0) = e^{-\int_0^{T_i} \alpha_i(s)\lambda_i(s)\,ds}$$

Furthermore

$$EL_i = E\int_0^{T_i} I(L_i > t)\,dt = \int_0^{T_i} EI(L_i > t)\,dt = \int_0^{T_i} P(L_i > t)\,dt$$

It is easy to see that the expected downtime associated with the replacement equals

$$EW_i = v_i\,P(L_i < T_i) + u_i\,P(L_i = T_i)$$

Hence it remains to look at the downtime associated with minimal repairs. We must show that

$$ER_i = \int_0^{T_i} r_i(t)P(L_i > t)\,\beta_i(t)\,\lambda_i(t)\,dt \tag{E.8}$$

We will first give a "simple" proof of (E.8), assuming that the length of a minimal repair at time t is deterministic and given by $r_i(t)$.

First proof of (E.8)

Let $V_1, V_2, \ldots,$ denote the successive failure (arrival) times of the process $N_i^{(\beta)}(t)$. It is not difficult to see that V_j has a distribution with density $P(N_i^{(\beta)}(t) = j-1)\,\beta_i(t)\,\lambda_i(t)$, observing that the event "jth failure occurs in $(t, t+h)$" means that there are $j-1$ failures in $[0,t]$ and one failure occurs in $(t, t+h)$. It follows that the expected downtime associated with minimal repairs equals

$$\begin{aligned} ER_i &= E\textstyle\sum_{j=1}^{\infty} r_i(V_j)\,I(V_j < T_i)I(V_j < L_i) \\ &= \textstyle\sum_{j=1}^{\infty} E\,r_i(V_j)I(V_j < T_i)\,I(V_j < L_i) \\ &= \textstyle\sum_{j=1}^{\infty} \int_0^{T_i} r_i(t)\,P(t < L_i)P(N_i^{(\beta)}(t) = j-1)\,\beta_i(t)\,\lambda_i(t)dt \\ &= \textstyle\int_0^{T_i} r_i(t)P(L_i > t)\,\beta_i(t)\,\lambda_i(t)\,dt \end{aligned}$$

which proves the result.

Second proof of (E.8)

Let $\Psi_i(t)$ be a random variable representing the length of a minimal repair at time t. We assume that $\Psi_i(t)$, $t \geq 0$, is a left-continuous process. The mean length of a minimal repair at time t is $r_i(t)$, given no replacement in $[0, t]$. Now, writing

$$ER_i = E\int_0^\infty \Psi_i(t)\, I(t \leq T_i)\, I(t \leq L_i)\, dN_i^{(\beta)}(t)$$

and using (E.2), it follows that

$$\begin{aligned} ER_i &= E\textstyle\int_0^{T_i} \Psi_i(t)\, I(t \leq L_i)\, \beta_i(t)\lambda_i(t)\, dt \\ &= \textstyle\int_0^{T_i} E\Psi_i(t)\, I(t \leq L_i)\, \beta_i(t)\lambda_i(t)\, dt \\ &= \textstyle\int_0^{T_i} r_i(t)\, P(L_i > t)\, \beta_i(t)\lambda_i(t)\, dt \end{aligned}$$

This proves the result.

Results for the processes $N_i^{(\alpha)}(t)$ and $N_i^{(\beta)}(t)$

It follows from Result E.9 that the two processes are independent non-homogeneous Poisson processes if we can show that

$$E\int_0^\infty C(t)\, dN_i^{(\alpha)}(t) = E\int_0^\infty C(t)\alpha_i(t)\lambda_i(t)\, dt \tag{E.9}$$

and

$$E\int_0^\infty C(t)\, dN_i^{(\beta)}(t) = E\int_0^\infty C(t)\beta_i(t)\lambda_i(t)\, dt \tag{E.10}$$

for all non-negative predictable processes. It sufficies to show that (E.9) and (E.10) hold for $C(t) = I(A)I(a < t \leq b)$, where a and b are constants such that $0 \leq a \leq b$ and $A \in \mathcal{F}_a$ [35]. Consider first formula (E.9). Writing $N_i^{(\alpha)}(t) = \int_0^t Z(s)dN_i(s)$ and observing that $I(A)$ and $Z(t)$, $a < t \leq b$, are independent, it follows that

$$\begin{aligned} E\textstyle\int_0^\infty C(t)\, dN_i^{(\alpha)}(t) &= E\textstyle\int_0^\infty C(t)\, Z(t)\, dN_i(t) &= \\ E\textstyle\int_0^\infty C(t)\, Z(t)\lambda_i(t)dt &= \textstyle\int_0^\infty EC(t)\, EZ(t)\lambda_i(t)dt &= \\ E\textstyle\int_0^\infty C(t)\, \alpha_i(t)\lambda_i(t)dt & & \end{aligned}$$

This completes the proof of (E.9). The proof of (E.10) is analogous. The third equality of (3.15) follows.

E.5.3 Section 5.4

Consider the multistate monotone system analysed in Section 5.4. We shall show that the intensity in this case is given by

$$\lambda(t) = \sum_{i=1}^{n} z_i(t)X_i(t)(1 - \phi(0_i, \mathbf{X}(t)))$$

To this end we use (E.2). It follows that

$$\begin{aligned}
E\int_0^\infty C(t)dN(t) &= E\sum_{i=1}^n C(R_i)I(R_i = \eta) \\
&= \sum_{i=1}^n E\int_0^\infty C(t)\phi(\mathbf{X}(t^-))(1-\phi(0_i,\mathbf{X}(t^-)))dN_i(t) \\
&= \sum_{i=1}^n E\int_0^\infty C(t)\phi(\mathbf{X}(t^-))(1-\phi(0_i,\mathbf{X}(t^-)))z_i(t)X_i(t)dt \\
&= E\int_0^\infty C(t)\lambda(t)\phi(\mathbf{X}(t))dt;
\end{aligned}$$

we have used that $\phi(\mathbf{X}(t^-))(1-\phi(0_i,\mathbf{X}(t^-)))$ is left-continuous and hence predictable, and that $N_i(t)$ has intensity $z_i(t)X_i(t)$. This proves the desired result.

Appendix F

AN ALGORITHM FOR CALCULATING SYSTEM RELIABILITY

In this appendix we present an algorithm for exact reliability calculation of multistate systems as described in Chapter 4. The algorithm, which is based on Doulliez and Jamoulle's decomposition method [44], is reported in [18]. In [18] a computer program for implementing the algorithm is also given. A binary version is reported in [15]. The binary version is implemented in the computer program CARA [123]. The presentation of the algorithm is based on [15,18,90].

Consider the model introduced in Sections 4.1 and 4.2. The problem is to calculate

$$g = P(\phi(\mathbf{X}) < \phi_k) \quad \text{and} \quad h = P(\phi(\mathbf{X}) \geq \phi_k)$$

where ϕ_k is a fixed level. The component states X_i are assumed to be independent. Two versions of the algorithm will be included: algorithm $1a$ requires the minimal cut vectors (at the level ϕ_k) to be known; algorithm $1b$ requires the minimal paths to be known. Otherwise, the two algorithms are basically identical.

The basic idea of the algorithm is to compute g and h as the sum of terms which are each of the form

$$P(\mathbf{c} \leq \mathbf{X} \leq \mathbf{d}) = \prod_{i=1}^{n} P(c_i \leq X_i \leq d_i) = \prod_{i=1}^{n} \sum_{j:\, c_i \leq x_{ij} \leq d_i} p_{ij} \qquad \text{(F.1)}$$

where $\mathbf{c}$ and $\mathbf{d}$ are state vectors such that $\mathbf{c} \leq \mathbf{d}$. In the following we shall let $[\mathbf{c}, \mathbf{d}]$ denote the set of state vectors $\mathbf{x}$ such that $\mathbf{c} \leq \mathbf{x} \leq \mathbf{d}$, and call the set an interval of vectors. Formula (F.1) is thus a general formula for computing the probability that $\mathbf{X}$ is contained in an interval $[\mathbf{c}, \mathbf{d}]$. This formula is used repeatedly in the algorithm.

For the sake of notational convenience, we shall in the following write $(i_1, i_2, \ldots, i_n)$ instead of $(x_{1i_1}, x_{2i_2}, \ldots, x_{ni_n})$. All vectors considered are state vectors, i.e. the ith element of the vector is an element of S_i.

Algorithm 1a

We say that a vector $\mathbf{x}$ is *acceptable* if $\phi(\mathbf{x}) < \phi_k$, and *non-acceptable* if $\phi(\mathbf{x}) \geq \phi_k$. To start the algorithm, define the vectors $\mathbf{b}^0 = (0, 0, \ldots, 0)$ and $\mathbf{b} = (M_1, M_2, \ldots, M_n)$. (These vectors will be changed throughout the algorithm.) Put $g = h = 0$. Consider the interval of vectors $[\mathbf{b}^0, \mathbf{b}]$. The set $[\mathbf{b}^0, \mathbf{b}]$ is decomposed into three disjoint sets:

- A set A of acceptable states
- A set B of non-acceptable states
- A set C of unspecified states

such that $[\mathbf{b}^0, \mathbf{b}] = A \bigcup B \bigcup C$. This is done by first constructing "critical" vectors $\mathbf{v}^0$ and $\mathbf{v}$ in $[\mathbf{b}^0, \mathbf{b}]$ such that a vector $\mathbf{x}$ in $[\mathbf{b}^0, \mathbf{b}]$ is acceptable if $\mathbf{b}^0 \leq \mathbf{x} \leq \mathbf{v}^0$ and non-acceptable if $x_i > v_i$ for at least one i. The vectors that do not satisfy any of these two criteria are left unspecified at this stage of the algorithm and put into the set C for further treatment later. The acceptable and non-acceptable states are put into A and B, respectively.

In order to reduce computing time, the critical vectors $\mathbf{v}^0$ and $\mathbf{v}$ should be chosen so that the set C of unspecified vectors becomes as small as possible. This may be achieved by choosing $\mathbf{v}^0$ as large as possible and $\mathbf{v}$ as small as possible, in some sense. In the algorithm this is done as follows. First, pick out the "largest" minimal cut vector $\mathbf{z}^s$ among all the minimal cut vectors $\mathbf{z}^j$ satisfying $\mathbf{z}^j \geq \mathbf{b}^0$. Technically this is done by picking out a minimal cut vector $\mathbf{z}^j \geq \mathbf{b}^0$ such that $\mathbf{z}^j$ maximizes the sum

$$H(\mathbf{z}^j) = \sum_{i=1}^{n} [min\{z_i^j, b_i\} - b_i^0]$$

Now the following vector $\mathbf{v}^0$ is chosen:

$$v_i^0 = \min\{z_i^s, b_i\}, \quad i = 1, 2, \ldots, n$$

From the set $A = [\mathbf{b}^0, \mathbf{v}^0]$ of acceptable states, we get a contribution to the g value:

$$P(\mathbf{X} \in A) = P(\mathbf{b}^0 \leq \mathbf{X} \leq \mathbf{v}^0) = \prod_{i=1}^{n} P(b_i^0 \leq X_i \leq v_i^0) = \prod_{i=1}^{n} \sum_{j:\, b_i^0 \leq x_{ij} \leq v_i^0} p_{ij}$$

Next, we must define a vector $\mathbf{v}$ in $[\mathbf{b}^0, \mathbf{b}]$ such that any vector $\mathbf{x}$ in $[\mathbf{b}^0, \mathbf{b}]$ with $x_i > v_i$ for at least one i is non-acceptable. In the algorithm this is done by first choosing a vector $\mathbf{z}$ such that

$$z_i = \max\{z_i^j : \mathbf{z}^j \geq \mathbf{b}^0\}$$

or, in other words, $\mathbf{z}$ is the vector with the ith element equal to the maximum of the ith elements of all the minimal cut vectors $\mathbf{z}^j$ with $\mathbf{z}^j \geq \mathbf{b}^0$. Then we define $\mathbf{v}$ by $v_i = \min\{z_i, b_i\}$.

The set B of non-acceptable states and the set C of unspecified states are usually not of the interval form $[\mathbf{c}, \mathbf{d}]$. It is possible, however, in a simple way, to write B and C as disjoint unions of such intervals:

For $j = 1, 2, \ldots, n$, define B_j as the set of vectors $\mathbf{x}$ in B such that $x_i \leq v_i$ for $i = 1, 2, \ldots, j-1$ and $x_j > v_j$. Thus B_j is the set of vectors in B for which the jth element is the first one to strictly exceed v_j. The B_js are clearly disjoint.

Now we can compute the probabilities $P(\mathbf{X} \in B_j)$ by the formula (F.1), and the sum of these probabilities over j is added to the old value of h.

Similarly, the set C is written as a union of disjoint intervals C_j: let C_j be the set of vectors $\mathbf{x}$ in C such that $x_i \leq v_i^0$ for $i = 1, 2, \ldots, j-1$ and $x_j > v_j^0$. The sets C_j generated at a certain stage are now to be put on a "waiting list" of sets of unspecified states that have not yet been further decomposed.

The algorithm continues by choosing a set C_j from the "waiting list" and defining the new $\mathbf{b}^0$ and $\mathbf{b}$ to be the limiting points of the interval form of C_j. Then the steps described above are run through once more, to obtain a decomposition of the set into new sets A, B and C. This gives new contributions to be added to g and h, and new sets of unspecified sets to be put on the "waiting list". In this way the algorithm continues until the whole state space is exhausted, i.e. until there are no more sets of unspecified states (or, possibly, until g and 1-h are close enough).

To summarize, algorithm 1a runs as follows:

1. Set $\mathbf{b}^0 = (0, 0, \ldots, 0)$ and $\mathbf{b} = (M_1, M_2, \ldots, M_n)$, and $g = h = 0$.

2. Given the limiting state points $\mathbf{b}^0$ and $\mathbf{b}$, compute the "critical" vectors $\mathbf{v}^0$ and $\mathbf{v}$ by the method given above. The sets A, $\{B_j\}$ and $\{C_j\}$ can then be defined. Note that each such set is given as an interval of states. (It is therefore sufficient to record the lower and upper limits of each set.)
   Let
   $$g \leftarrow g + P(\mathbf{X} \in A)$$
   and
   $$h \leftarrow h + \sum_j P(\mathbf{X} \in B_j)$$
   where $P(\mathbf{X} \in A)$ and $P(\mathbf{X} \in B_j)$ are computed by formula (F.1). The sets $\{C_j\}$ are put on a list $\{C_{(k)}\}$ of sets of unspecified sets.

3. Choose a set $C_{(k)}$ that has not been yet been considered. If no such non-empty set exists, go to Step 4. Otherwise, define $\mathbf{b}^0$ and $\mathbf{b}$ to be limit points of $C_{(k)}$ and go to Step 2.

4. g and h equal the sought values.

Each time Step 2 above has been performed, the current values of g and $1-h$ are lower and upper bounds for g, respectively. It is therefore possible to stop the process as soon as these bounds are close enough.

It should be clear from the description of the algorithm that if an exact value of g is sought, then h need not be considered at all, and therefore the decompositions of the B sets into disjoint intervals B_j are unnecessary.

Algorithm 1b

For this version of the algorithm we say that a vector $\mathbf{x}$ is acceptable if $\phi(\mathbf{x}) \geq \phi_k$, and non-acceptable if $\phi(\mathbf{x}) < \phi_k$.

The algorithm $1b$ starts with $\mathbf{b} = (0, 0, \ldots, 0)$ and $\mathbf{b}^0 = (M_1, M_2, \ldots, M_n)$. In algorithm $1b$ we define the vector $\mathbf{v}^0$ by

$$v_i^0 = \max\{y_i^s, b_i\}, \quad i = 1, 2, \ldots, n$$

where $\mathbf{y}^s$ is a minimal path vector maximizing

$$H(\mathbf{y}^j) = \sum_{i=1}^{n} [b_i^0 - \max\{y_i^j, b_i\}], \quad \mathbf{y}^j \leq \mathbf{b}^0$$

Here $\mathbf{y}^j$ are the minimal path vectors of the system. The set A is given by $A = [\mathbf{v}^0, \mathbf{b}^0]$. The vector $\mathbf{v}$ is defined by

$$v_i = \max\{y_i, b_i\}$$

where

$$y_i = \min\{y_i^j : \mathbf{y}^j \leq \mathbf{b}^0\}$$

The sets B, B_j and C_j are defined as in algorithm $1a$, but with "$<$" and $\geq$ instead of "$>$" and $\leq$, respectively.

Otherwise, this algorithm is similar to algorithm $1a$.

Remark

Doulliez and Jamoulle [44] considered a flow network system similar to those studied in Chapter 4. This system is a special case of a multistate system, so the above algorithm applies. It is, however, possible to carry out the calculations without knowing the minimal cut (path) vectors. Another procedure for determining the "critical" vectors $\mathbf{v}^0$ and $\mathbf{v}$ must then be used. Below Doulliez and Jamoulle's solution is sketched. In the algorithm a vector $\mathbf{x}$ is acceptable if $\phi(\mathbf{x}) \geq \phi_k$, and non-acceptable if $\phi(\mathbf{x}) < \phi_k$.

Send a flow equal to ϕ_k through the network, from the source node to the demand node. Let f_i denote the flow of component (arc) i. Define the vector $\mathbf{v}^0$ by letting v_i^0 be the smallest state (capacity value) of S_i that is equal to or greater than f_i. The set A is given by $A = [\mathbf{v}^0, \mathbf{b}^0]$.

The vector $\mathbf{v}$ is determined in the following way: given the limiting states (capacities) b_j^0, $j \neq i$, and the existing flow ϕ_k in the network, find the maximal flow F_i that can still go from the origin to the end of arc i without using arc i. The value F_i can be obtained by Ford-Fulkerson's method [49] and corresponds to the largest feasible decrease in flow for arc i. If $F_i < f_i$, then arc i is in a minimal cut separating the source node and the demand node and v_i is the smallest capacity value which is greater than or equal to $f_i - F_i$. If $F_i \geq f_i$, then $v_i = 0$ and the associated B set is empty.

Appendix G

A MINIMIZATION TECHNIQUE

In this appendix we study in general terms how the minimization of a function $B^T = M^T/S^T$, where T is an element of a some set, can be carried out by first minimizing δ functions $C_\delta^T = M^T - \delta S^T$. The technique is very suitable where it is easy to find solutions to δ problems; in replacement applications this is often the case. The analysis in this appendix is taken from [20].

Let $\mathcal{T}$ be a set and M and S functions defined on $\mathcal{T}$ such that

a) $-\infty < M^T < \infty$ for all $T \in \mathcal{T}$,

b) $0 \leq S^T < \infty$ for all $T \in \mathcal{T}$,

c) $S^T = 0 \Rightarrow M^T > 0$,

d) there exists a $T \in \mathcal{T}$ such that $S^T > 0$.

Define the function B from $\mathcal{T}$ to $(-\infty, \infty]$ by

$$B^T = \frac{M^T}{S^T}$$

and let

$$\delta^* = \inf_{T \in \mathcal{T}} B^T$$

Clearly $-\infty \leq \delta^* < \infty$. If there exists a $T \in \mathcal{T}$ such that $B^T = \delta^*$, we say that T is optimal.

Let the δ functions C_δ^T, $\delta \in (-\infty, \infty)$ from $\mathcal{T}$ to $(-\infty, \infty)$ be defined by

$$C_\delta^T = M^T - \delta S^T$$

A result that has been used in a number of applications is the following [25,116,132]:

Result G.1. If $T_\delta \in \mathcal{T}$ minimizes C_δ^T and $C_\delta^{T_\delta} = 0$, then T_δ is optimal and $\delta = B^{T_\delta} = \delta^*$.

The proof of this result is left to the reader. It should be noted that if there exists a T which is optimal, then T minimizes C_{δ^*} and $C^T_{\delta^*} = 0$.

We now make the following assumption:

e) There exists a $T_\delta \in \mathcal{T}$ such that T_δ minimizes C^T_δ for each $\delta \in (-\infty, \infty)$.

We shall deduce some consequences of this assumption. Define

$$M(\delta) = M^{T_\delta} \qquad S(\delta) = S^{T_\delta}$$
$$B(\delta) = B^{T_\delta} \qquad C(\delta) = C^{T_\delta}$$

and observe that

$$\begin{aligned} B(\delta) < \delta &\Leftrightarrow C(\delta) < 0 \\ B(\delta) = \delta &\Leftrightarrow C(\delta) = 0 \\ B(\delta) > \delta &\Leftrightarrow C(\delta) > 0 \end{aligned}$$

We now prove the following results.

Result G.2

i) For each δ such that $\delta > \delta^*$ we have $C(\delta) < 0$ or $B(\delta) < \delta$

ii) For each δ such that $\delta < \delta^*$ we have $C(\delta) > 0$ or $B(\delta) > \delta$

iii) $C(\delta)$ is non-increasing and concave

Remark. The concavity of $C(\delta)$ implies that $C(\delta)$ is continuous.

Proof

i) Let T be such that $\delta^* \le B^T < \delta$. Using e) we get

$$C(\delta) \le C^T_\delta < M^T - B^T S^T = 0$$

ii) The result follows by noting that $B(\delta) \ge \delta^* > \delta$

iii) Let $\delta_1 \le \delta_2$. Using e) we obtain

$$C(\delta_2) \le M(\delta_1) - \delta_2 S(\delta_1) \le M(\delta_1) - \delta_1 S(\delta_1) = C(\delta_1)$$

hence $C(\delta)$ is non-increasing. To show the concavity, let $\alpha, \delta_1, \delta_2$ be given numbers, $0 \le \alpha \le 1$, and define $\delta = \alpha\delta_1 + (1-\alpha)\delta_2$. Then

$$\begin{aligned} C(\delta) &= M(\delta) - (\alpha\delta_1 + (1-\alpha)\delta_2)S(\delta) \\ &= \alpha[M(\delta) - \delta_1 S(\delta)] + (1-\alpha)[M(\delta) - \delta_2 S(\delta)] \\ &\ge \alpha C(\delta_1) + (1-\alpha)C(\delta_2) \end{aligned}$$

i.e. $C(\delta)$ is concave.

Result G.3. If $\delta^* > -\infty$, then $C(\delta^*) = 0$ or equivalently $B(\delta^*) = \delta^*$, i.e. T_{δ^*} is optimal.

Proof. The result follows by using *Result G.2 i),ii)* and the fact that $C(\delta)$ is continuous.

Result G.4.

i) Let $\delta^* < \delta \leq \delta_1 < \delta_2$. Then we have

$$S(\delta) < -\frac{C(\delta_2)}{\delta_2 - \delta_1}$$

ii) $M(0) \leq M^T$ for all $T \in \mathcal{T}$

Proof.

i) By e), the fact that $\delta \leq \delta_1$ and *Result G.2 i)* we get

$$\begin{aligned} C(\delta_2) &\leq M(\delta) - \delta_2 S(\delta) = M(\delta) - \delta_1 S(\delta) - (\delta_2 - \delta_1) S(\delta) \\ &\leq C(\delta) - (\delta_2 - \delta_1) S(\delta) < -(\delta_2 - \delta_1) S(\delta) \end{aligned}$$

and i) follows.

ii) The result is an immediate consequence of assumption e) (let $\delta = 0$).

Result G.5. If $\delta^* = -\infty$, then

$$M(\delta) < 0 \quad \text{for all} \quad \delta \leq 0 \quad \text{and} \quad S(\delta) \to 0 \quad \text{as} \quad \delta \to -\infty$$

Proof. The statement follows from *Result G.2 i)* and *G.4 i)*.

Now we present an algorithm which always produces a sequence converging to δ^*.

Result G.6. Choose any δ_1 such that $S(\delta_1) > 0$, and set iteratively $\delta_{n+1} = B(\delta_n)$, $n = 1, 2, \ldots$. Then

$$\lim_{n \to \infty} \delta_n = \delta^*$$

Proof. By induction, $\delta^* \leq \delta_{n+1} = B(\delta_n) \leq \delta_n < \infty$, $n \geq 2$. Hence there exists a $\delta' \geq \delta^*$ such that $\lim \delta_n = \delta'$. We must show that $\delta' = \delta^*$. Assume first that $\delta^* > -\infty$. It then suffices to prove that $\delta' \leq B(\delta')$ or equivalently $C(\delta') \geq 0$ since then $\delta' = \delta^*$ by *Result G.2 i)*. Now writing

$$C(\delta_n) = M(\delta_n) - \delta_n S(\delta_n) = \delta_{n+1} S(\delta_n) - \delta_n S(\delta_n) = (\delta_{n+1} - \delta_n) S(\delta_n)$$

and noting that $S(\delta_n)$ is bounded (*Result G.4 i)*) and $\delta_{n+1} - \delta_n \to 0$, we see that $C(\delta_n) \to 0$ as $n \to \infty$. Since $C(\delta') \geq C(\delta_n)$ (*Result G.2 iii)*) we can conclude that $C(\delta') \geq 0$. This proves the convergence if $\delta^* > -\infty$. Assume now that $\delta^* = -\infty$. We must show that $\delta' = -\infty$. Suppose $\delta' > -\infty$. Then by the proof above we have $C(\delta') \geq 0$. From *Result G.2* we conclude that $\delta^* = \delta'$. But this is a contradiction since $\delta^* = -\infty$. Thus $\delta' = -\infty$. The proof is complete.

Observation. It is easily seen that $\delta_{n+1} < \delta_n$ unless $\delta_n = \delta^*$ $(n \geq 2)$.

Some final remarks.

i) Suppose that assumption *e)* is replace by

> *e')* $T_\delta \in \mathcal{T}$ minimizes C_δ^T for $\delta \in [u, v) \bigcap (-\infty, \infty)$, where $-\infty \leq u \leq \delta^* < v \leq \infty$.

Then it is not difficult to see that if the δ analysis is restricted to $\delta \in [u, v) \bigcap (-\infty, \infty)$, then the *Results G.1 - G.6* (*G.4 ii)* if $u < 0$ and *G.6* if $\delta_2 = B(\delta_1) < v$) are still valid.

ii) Suppose that *a) - d)* hold, $S^T \leq r < \infty$ for all $T \in \mathcal{T}$, $\delta^* > -\infty$ and $T_{\delta^*} \in \mathcal{T}$ minimizes $C_{\delta^*}^T$ (the assumption *e)* is dropped). Then we can prove that $C(\delta^*) = 0$, i.e. T_{δ^*} is optimal. Let $\epsilon > 0$ be given and choose $T \in \mathcal{T}$ such that $B^T < \delta^* + \epsilon$. Then we have $0 \leq C(\delta^*) \leq C_{\delta^*}^T < \epsilon S^T \leq \epsilon r$. It follows that $C(\delta^*) = 0$.

iii) Standard numerical iterative methods, for example the bisection method or modified *regula falsi* [105] can in addition to the algorithm presented above be used to locate δ^*. We must then start with $\delta_a \leq \delta_b$ such that $\delta_a \leq B(\delta_a)$ and $\delta_b \geq B(\delta_b)$ (then $\delta_a \leq \delta^* \leq \delta_b$).

Bibliography

[1] Abraham, J. An improved algorithm for network reliability. *IEEE Trans. Reliability.* **20**, 58-61, 1979.

[2] American Institute of Chemical Engineers, *Guidelines for Hazard Evaluation Procedures.* 1985.

[3] Andersen, P.K. and Ø. Borgan, Counting process models for life history data: A review. *Scand. J. Statist.* **12**, 97-158, 1985.

[4] Andersen, R.A., MORT - User manual. (In Norwegian). Report University of Trondheim, Norwegian Institute of Technology, 1983.

[5] Apostolakis, G.E. (ed.), *Reliability Engineering and Systems Safety - Special Issue on Human Reliability Analysis.* **29**, 1990.

[6] Ascher, H. and H. Feingold, *Repairable Systems Reliability.* Marcel Dekker, 1984.

[7] Aven, T. Safety acceptance criteria for oil and gas production systems. Paper presented at Health, Safety and Environment Conference, The Hague, 11-14 November, 1991.

[8] Aven, T. *Reliability and Risk Analysis.* (In Norwegian.) The University Press, Oslo, 1991.

[9] Aven, T., On performance measures of multistate monotone systems. Submitted to *Reliability Engineering and System Safety,* 1992.

[10] Aven, T., Availability formulae for standby systems of similar units that are preventively maintained. *IEEE Trans. Reliability.* **39**, 603-606, 1990.

[11] Aven, T., Availability evaluation of flow networks with varying throughput-demand and deferred repairs. *IEEE Trans. Reliability.* **38**, 499-505, 1989.

[12] Aven, T., Some considerations on reliability theory and its applications. *Reliability Engineering and System Safety.* **21**, 214-223, 1988.

[13] Aven, T., Availability evaluation of oil/gas production and transportation systems. *Reliability Engineering.* **18**, 35-44, 1987.

[14] Aven, T., A counting process approach to replacement models. *Optimization.* **18**, 285-296, 1987.

[15] Aven, T., Reliability/availability evaluation of coherent systems based on minimal cut sets. *Reliability Engineering.* **13**, 93-104, 1986.

[16] Aven, T., Formulae for the average unavailability (MFDT) of a coherent system with periodically tested components. *Microelectronics and Reliability.* **26**, 283-288, 1986.

[17] Aven, T., Determination/estimation of an optimal replacement interval under minimal repair. *Optimization.* **16**, 743-754, 1985.

[18] Aven, T., Reliability evaluation of multistate systems of multistate components. *IEEE Trans. Reliability.* **34**, 473-479, 1985.

[19] Aven, T., Optimal replacement under a minimal repair strategy - a general failure model. *J. Appl. Prob.* **15**, 198-211, 1983.

[20] Aven, T. and B. Bergman, Optimal replacement times - a general set-up. *J. Appl. Prob.* **23**, 432-442, 1986.

[21] Aven, T. and S. Gaarder, Optimal replacement in a shock model; discrete time. *J. Appl. Prob.* **24**, 281-287, 1987.

[22] Aven, T. and S.P. Kirkeby, Reliability targets for oil/gas production systems. In *Reliability Achievement: The Commercial Incentive*, ed. T. Aven, Elsevier, 1989.

[23] Aven, T. and R. Østebø, Two new component importance measures for a flow network system. *Reliability Engineering.* **14**, 75-80, 1986.

[24] Barlow, R.E. og F. Proschan, *Statistical Theory of Reliability and Life Testing.* Holt, Rinehart and Winston, N.Y., 1975.

[25] Barlow, R.E. og F. Proschan, *Mathematical Theory of Reliability.* Wiley, New York, 1965.

[26] Barlow, R.E. and L.C. Hunter, Optimum preventive maintenance policies. *Operations Research.* **8**, 90-100, 1960.

[27] Bergman, B., On reliability theory and its applications, with discussions. *Scand. J. Statist..* **12**, 1-41, 1985.

[28] Bergman, B., Optimal replacement under a general failure model. *Adv. Appl. Prob.* **10**, 431-451, 1985.

[29] Bergman, B. and B. Klefsjö, *Quality.* (In Swedish). Linköping University, Institute of Technology, 1990.

[30] Bergman, B. and B. Klefsjö, *Reliability.* (In Swedish.) University of Luleå, 1989.

[31] Billingsley, P., *Probability and Measure.* Wiley, 1979.

[32] Billinton, R. and R.N. Allan, *Reliability Evaluation of Engineering Systems.* Pitman, 1983.

[33] Birnbaum, Z.W., On the importance of different components in a multicomponent system. In *Multivariate Analysis -II*, ed. P.R. Krishnaiah, Academic Press, 581-592, 1969.

[34] Block, H.W., W.S. Borges and T.H. Savits, A general age replacement model with minimal repair. *Naval Research Logistics.* **35**, 365-372, 1988.

[35] Brémaud, P., *Point Processes and Queues.* Springer-Verlag, 1981.

[36] British Standards Institution, BS 5760: Reliability of systems, equipment and components, 1982.

[37] Brok, J.F., Availability assessment of oil and gas production systems. *International Journal of Quality and Reliability Management.* **4**, 21-36, 1987.

[38] Brouwers, J.J., Probabilistic descriptions of irregular system downtime. *Reliability Engineering.* **15**, 263-281, 1986.

[39] Brouwers, J.J., P.H. Verbeek and W.R. Thomson, Analytical system availability techniques. *Reliability Engineering.* **17**, 9-22, 1987.

[40] Butler, D.A. , A complete importance ranking for components of binary coherent systems, with extensions to multi-state systems. *Naval Res. Log. Quart.* , **26**, 565-578, 1979.

[41] Center for Chemical Process Safety, *Chemical Process Quantitative Risk Analysis.* Guidelines, American Institute of Chemical Engineers, 1989.

[42] Chemical Industries Association, UK, *A Guide to Hazard and Operability Studies.* 1977.

[43] Comer, P.J., J.S. Fitt and R. Østebø. A drillers HAZOP method. *Society of Petroleum Engineers*, 1986.

[44] Doulliez, P. and J. Jamoulle, Transportation networks with random arc capacities. *RAIRO.* **3**, 45-60, 1972.

[45] Edwards, G.T. and D.W. Watson, A study of common-mode failures. United Kingdom Atomic Energy Authority, SRD R 146, 1979.

[46] Elliot, D.M. and J.M. Owen, Critical examination in process design. *Chem. Eng.*, Lond., 223, CE377, 1968.

[47] Elvik, R., Is travelling more dangerous than other activities? (In Norwegian.) Report Institute of Transportation and Economics, Oslo, 1989.

[48] Embrey, D.E. et. al., SLIM-MAUD: An approach to assessing human error probabilities using structural expert judgement. NUREG/CR-3518. U.S. Nuclear Regulatory Commision, Washington, D.C., 1984.

[49] Ford, L.R. and D.R. Fulkerson, *Flow in Networks.* Princeton University Press, Princeton New Jersey, 1962.

[50] Funnemark, E. and B. Natvig, Bounds for the availabilities in a fixed time interval for multistate monotone systems. *Adv. Appl. Prob.*, **17**, 638-655, 1985.

[51] Fussell, J.B., How to handcalculate system reliability and safety characteristics. *IEEE Trans. Reliability.* **24**, 1975.

[52] Gertsbakh, I.B., *Statistical Reliability Theory.* Marcel Dekker, Inc, 1989.

[53] Gjerstad, T., Safety and emergency response. Part of the documentation for the course Management of Petroleum Operations, PETRAD, The Norwegian Petroleum Directorate, 1991.

[54] Gjerstad, T., Developing a practical approach to offshore safety assessment. Paper presented at the IBM Seminar on Formal Safety Assessment, London, 5-6 December, 1990.

[55] Griffith, W. S., Multistate reliability models. *J. Appl. Prob.*. **15**, 735-744, 1980.

[56] Grundt, H.J., Reliability management in the planning period of development projects. General guidelines. (In Norwegian.) Technical standard, Statoil, 1990.

[57] Grundt, H.J., MIRIAM-software for system effectiveness evaluations. Presented at Ann. Mtg., Associate Members of Systems Reliability Service, Daresbury, UK, 1987.

[58] Grundt, H.J., Guidelines for reliability programmes. General Engineering Guide, Statoil, 1986.

[59] Green, A.J. and A.J. Bourne, *Reliability Technology.* Wiley, N.Y., 1972.

[60] Gut, A. and S. Janson, The limiting behaviour of certain stopped sums and some applications. *Scand. J. Statist.* **10**, 281-292, 1983.

[61] Hammer, W., *Handbook of system and product safety.* Prentice Hall, 1972.

[62] Hannaman, G.W. and D.H. Worledge, Some developments in human reliability analysis, approaches and tools. *Reliability Engineering and Systems Safety.* **22**, 235-256, 1988.

[63] Heikkila, M., A model for common mode failures. Second National Reliability Conference, 1979.

[64] Henley, E.J. and H. Kumamoto, *Reliability Engineering and Risk Assessment.* Prentice Hall, 1981.

[65] Holen, A.T., A. Høyland and M. Rausand, *Reliability analysis.* (In Norwegian.) Tapir, Trondheim, 1983.

[66] Hope, B. and P.A. Johannessen, *Safety Management in Offshore Development Projects.* Tanum-Norli, Oslo, 1983.

[67] Hovden, J. et. al., Evaluation of accidental risk. (In Norwegian.) Tapir, Trondheim, 1979.

[68] Hovden, J. et. al., Safe behaviour. (In Norwegian.) SINTEF-report, 1986.

[69] Hughes, R.P., Fault tree truncation error bounds. *Reliability Engineering.* **1**, 37-46, 1987.

[70] IEEE Transactions on Reliability, Reliability Discipline and Technology in Historical Perspective - 1930-1984. **33**, 1984.

[71] IEEE Service Center, Electronic Reliability Data. INSPEC, 1981.

[72] IEEE Standard 500.

[73] Ireson, W.G. and C.F. Coombs, Jr. (eds.), *Handbook of Reliability Engineering and Management.* McGraw-Hill, 1988.

[74] International Organization for Standardization (ISO), Quality Systems - 9001-9003, 1987.

[75] Jasmon, G.B. and S. Kai, A new technique in minimal path and cut set evaluation. *IEEE Trans. Reliability.* **34**, 136-143, 1985.

[76] Johnson, W.G., The Management Oversight and Risk Tree. MORT ERDA, SAN 821-2,1973.

[77] Juran, J.M. og F.M. Gryna, *Quality Planning and Analysis.* McGraw-Hill, 1980.

[78] Kapur, K.C. and L.R. Lamberson, *Reliability in Engineering Design.* Wiley, 1977.

[79] Karlin, S. and H.M. Taylor, *A First Course in Stochastic Processes.* Academic Press, 2. edition, 1975.

[80] Keller, A.Z. and M.T. Giblin, Optimal maintenance and replacement management. *International Journal of Quality and Reliability Management.* **44**, 42-50, 1985.

[81] Kirkeby, S.P. and R.S. McGlashan, Reliability engineering and life cycle costing in the Troll phase I gas project. In *Reliability Achievement: The Commercial Incentive*, ed. T. Aven, Elsevier, 1989.

[82] Kjellén, U. et. al., SMORT - Safety analyses of industrial organizations - offshore version. (In Norwegian). Yrkeslitteratur, Oslo, 1987.

[83] Lakner, A.A. and R.T. Anderson, *Reliability Engineering for Nuclear and Other High Technology Systems.* Elsevier, 1985.

[84] Lawless, J.F., *Statistical Models and Methods for Lifetime Data.* Wiley, 1982.

[85] Lawley, H.G., Size up plants hazards this way. *Hydrocarbon Processing*, **55**, 1976.

[86] Lawley, H.G., Operability studies and hazard analysis. *Chem. Eng. Progr.* **70**, 1974.

[87] Lee, K.W. et. al., A literature survey on the human reliability component in a man-machine system. *IEEE Trans. Reliability.* **37**, 24-34, 1988.

[88] Lee, K.W. et. al., Fault tree analysis, methods, and applications - a review. *IEEE Trans. Reliability.* **34**, 194-203, 1985.

[89] Lees, F.P., *Loss Prevention in the Process Industries.* Butterworths, 1980.

[90] Lindquist, B., Method Description ERAC, CARA Manual, SINTEF, 1989.

[91] Lindquist, B., Use of Markov analysis. Report Norwegian Institute of Technology, Trondheim, 1988.

[92] Lund, J., Accidents in Norway. (In Norwegian.) Presented at "Sikkerhetsdagene 88". Norwegian Institute of Technology, Trondheim, 1988.

[93] Lydersen, S., M. Sandtorv and M. Rausand, Processing and application of reliability data. SINTEF report, Trondheim, 1987.

[94] Mann, N.R., R.E. Schafer and Z.D. Singpurwalla, *Methods for Statistical Analysis of Reliability and Life Data.* Wiley, 1974.

[95] Martz, H.F. and R.A. Waller, *Bayesian Reliability Analysis.* Wiley, 1982.

[96] MIL-STD-1629A, Military standard - procedure for performing a failure mode, effects and criticality analysis. U.S. Dept. of Defense, Washington, D.C., 1984.

[97] Military Standardization Handbook, MIL-HNBK 217 B, Reliability Prediction and Electronic Equipment.

[98] Natvig, B. Strict and exact bounds for the availabilities in a fixed time interval for multistate monotone systems. Submitted to *Scand. J. Statist.*, 1991.

[99] Natvig, B., New light on measures of importance of system components. *Scand. J. Statist.* **12**, 43-54, 1985.

[100] Natvig, B., Multistate coherent systems. *Encyclopedia of Statistical Science.* **5**, eds. Johnson and Kotz, Wiley, 1984.

[101] Natvig, B. Recent developments in multistate reliability theory. Proceedings of the IUTAM symposium to the memory of Waloddi Weibull "Probabilistic methods in the mechanics of solids and structures", Stockholm, 19-21 June 1984. Springer-Verlag, Berlin, 1985.

[102] Natvig, B. and A. Streller, The steady-state behaviour of multistate monotone systems, *J. Appl. Prob.*, **21**, 826-835, 1984.

[103] Nelson, W., *Applied Life Data Analysis.* Wiley, 1982.

[104] Nordbø, B., Personal communication. Statoil, 1990.

[105] O'Connor, P.D.T., *Practical Reliability Engineering.* Wiley, N.Y., 1986.

[106] Norwegian Petroleum Directorate, Guidelines for safety evaluation of platform concepts, 1981.

[107] Norwegian Petroleum Directorate, Annual report, 1988.

[108] Norwegian Petroleum Directorate. Regulations concerning implementation and use of risk analyses in the petroleum activities, with supplementary guidelines, 1990. This Regulation replaces the guidelines from 1981.

[109] Olshausen, K.D., Cost-effective reliability - an example from offshore oil production. In *Reliability Achievement: The Commercial Incentive*, ed. T. Aven, Elsevier, 1989.

[110] OREDA (Offshore Reliability Data). OREDA-handbook, Box 370, 1322 N-Høvik, 1984.

[111] Park, K.S., *Human Reliability. Analysis, Prediction, and Prevention of Human Errors.* Elsevier, Amsterdam, 1987.

[112] Pierskalla, W. P. and J.A. Voelker, A survey of maintenance models: the control and surveillance of deteriorating systems. *Naval Res. Logist. Quart.* **23**, 353-388, 1976.

[113] RAC-NPRD-2 IIT-Research Institute, USA, Nonelectronic Parts Reliability Data. 1981.

[114] Rasmussen, J., Notes on human error analysis. In *Synthesis and Analysis Methods for Safety and Reliability Studies*, eds. Apostolakis, G. et. al. Plenum Press, N.Y., 1980.

[115] Rome Air Development Center, Nonelectronic Reliability Notebook. RADC-TR-75-22.

[116] Ross, S.M., *Applied Probability Models with Optimization Applications.* Holden-Day, 1970.

[117] Savits, T.H., A cost relationship between age and block replacement policies. Series in Reliability and Statistics, University of Pittsburgh, 1987.

[118] Shooman, M.L., *Probabilistic Reliability - An Engineering Approach.* McGraw Hill, N.Y., 1968.

[119] Silkoset, O., Policies on quality in petroleum operations. Part of the documentation for the course Management of Petroleum Operations, PETRAD, The Norwegian Petroleum Directorate, 1991.

[120] Smith, D.J., *Reliability and Maintainability in Perspective.* MacMillan Education, 1988.

[121] Smith, W.L. and M.R. Leadbetter, On the renewal function for the Weibull distribution. *Technometrics.* **5**, 393-396, 1963.

[122] Singh, C. and R. Billinton, *System Reliability Modelling and Evaluation.* Hutchinson, London, 1977.

[123] SINTEF, CARA - Computer Aided Reliability Analysis. 1990.

[124] SINTEF, *Handbook - accidental prevention*, 1989.

[125] SINTEF, Risk Analysis Requirements. Norwegian Standard (draft), 1990.

[126] SINTEF, Risk Analysis Requirements. Guidelines (draft), 1988.

[127] Sten, T., Accident statistics: Needs and tools. (In Norwegian.) SINTEF, Trondheim, 1987.

[128] Streller, A., A generalization of cumulative processes. *Elektr. Informationsverarb. Kybern.* **16**, 449-460, 1980.

[129] Swain, A.D., Human reliability analysis: need, status, trends and limitations. *Reliability Engineering and Systems Safety.* **29**, 301-313, 1990.

[130] Swain, A.D. and H.E. Guttmann, *Handbook of Human Reliability Data.* NUREG/CR-1278. Sandia Laboratories, 1983.

[131] Takacs, L., On certain sojourn time problems in the theory of stochastic processes. *Acta Math. Acad. Sci. Hungar.* **8**, 169-191, 1957.

[132] Taylor, H.M., Optimal replacement policy under additive damage and other failure models. *Naval Res. Logist. Quart.* **22**, 1-18, 1975.

[133] Tveit, O. Risk analysis, its use, benefits and limitations. In *Automation for safety in shipping and offshore petroleum operations*, eds. A.B. Aune and J. Vlietstra, North Holland Publishing Comp., 1980.

[134] UKAEA, Systems Reliability Services - SRS Data Bank. Wigshaw Lane, Culcheth, Warrington WA37JR, England.

[135] U.S.Atomic Energy Commision, Reactor Safety Study Report, WASH-1400, 1974.

[136] Valdez-Flores, C. and R.M. Feldman, A survey of preventive maintenance models for stochastically deteriorating single-unit systems. *Naval Res. Logist. Quart.* **36**, 419-446, 1989.

[137] Verwerd, M., Reliability data for use in offshore formal safety assessment studies. In *Offshore Safety and Reliability*, eds. Cox and Walter, Elsevier, 1991.

[138] Vinnem, J.E., Risk analysis - preliminary hazard analysis in occupational safety. (In Norwegian.) Vern og velferd, Oslo, 1981.

[139] Yearout, R.D., R. Prabhakar and D.L. Grosh, Standby redundancy in reliability - review. *IEEE Trans. Reliability.* **27**, 82-100, 1985.

Index